Palgrave Companions

Carolin Antos • Neil Barton • Giorgio Venturi
Editors

The Palgrave Companion to the Philosophy of Set Theory

palgrave
macmillan

Editors
Carolin Antos
Philosophy
University of Konstanz
Konstanz, Baden-Württemberg, Germany

Giorgio Venturi
Department of Civilizations and Forms of Knowledge
University of Pisa
Pisa, Italy

Neil Barton
National University of Singapore
Singapore, Singapore

IFIKK
University of Oslo
Oslo, Norway

ISSN 3004-9172 ISSN 3004-9180 (electronic)
Palgrave Companions
ISBN 978-3-031-62389-9 ISBN 978-3-031-62387-5 (eBook)
https://doi.org/10.1007/978-3-031-62387-5

Cover illustration: Lan Zhang/Getty Images

This Palgrave Macmillan imprint is published by the registered company Springer Nature Switzerland AG
The registered company address is: Gewerbestrasse 11, 6330 Cham, Switzerland

Contents

Notes on Contributors

Carolin Antos (editor) is an assistant professor in the Department of Philosophy at the University of Konstanz (Germany) with a focus on the philosophy of mathematics. She earned her PhD in mathematics at the Kurt-Gödel Research Center for mathematical logic at the University of Vienna. She is leading the research group *Forcing: Conceptual Change in the Foundations of Mathematics*.

Avner Ash is Professor of Mathematics at Boston College working in algebra and number theory. He has authored several books (with Robert Gross) about mathematics for a general audience, such as *Elliptic Tales* (Princeton University Press, 2012) and *Summing it up* (Princeton University Press, 2016).

Neil Barton (editor) is a lecturer at the National University of Singapore and Part-Time Associate Professor at IFIKK, University of Oslo. He obtained his PhD from Birkbeck, University of London, in 2017 and has previously worked in the Kurt Gödel Research Center at the University of Vienna and University of Konstanz. He works on the foundations of mathematics, specializing in pluralism, and the genesis and development of mathematical concepts. He has written a book *Iterative Conceptions of Set* which was published by Cambridge University Press in 2024.

Tim Button is Professor of Philosophy & Logic at UCL. He has written two books: *Philosophy and Model Theory* (OUP, 2018), co-authored with Sean Walsh, concerns the interactions between philosophy and model theory; *The Limits of Realism* (OUP, 2013) is about meta(meta)physics and the realism debate.

Ethan Brauer earned his PhD from Ohio State University in 2020 and is a Postdoctoral Research Fellow at IFIKK, University of Oslo. He was previously a Research Assistant Professor at Lingnan University. He has written on a variety of topics in logic and philosophy of mathematics, including the nature of infinity and the role of relevance in formal logic.

Sharon Berry received a PhD in Philosophy from Harvard. She is an assistant professor at Indiana University, Bloomington, having previously been Associate Professor of Philosophy at Ashoka University. Much of her research focuses on philosophy of mathematics and epistemology. Her book *A Logical Foundation for Potentialist Set* Theory was published by Cambridge University Press in 2022.

Justin Clarke-Doane is Associate Professor of Philosophy at Columbia University. He is also an honorary research fellow at the University of Birmingham, UK, an adjunct research associate at Monash University, Australia, and has been a periodic visitor at the Institute for Advanced Study, Princeton. He is the author of *Morality and Mathematics* (Oxford University Press 2020/paperback 2024), *Mathematics and Metaphilosophy* (Cambridge University Press 2022), and various articles. In addition to his primary areas of research, he has written on the mind-body problem, the philosophy of physics, and (with Kathryn Tabb) free will and psychopathology. He earned his BA in Philosophy/Mathematics from New College of Florida, and his PhD in Philosophy from NYU.

Luca Incurvati is Associate Professor of Philosophy at the University of Amsterdam. A recipient of an ERC Starting Grant, he is the author of Conceptions of Set and the *Foundation of Mathematics* (Cambridge University Press, 2020) and, together with Julian Schloeder, of *Reasoning with Attitude: Foundations and Applications of Inferential Expressivism* (Oxford University Press, 2023).

Deborah Kant is a postdoctoral researcher at the University of Hamburg and completed her PhD at the University of Konstanz. She is writing the monograph *Independence and Naturalness in Set-theoretic Practice* (Vittorio Klostermann, 2024) and is co-editor of Reflections on the Foundations of Mathematics: Univalent Foundations, Set Theory and General Thoughts (Synthese Library, 2019).

Toby Meadows is an associate professor at the University of California, Irvine. He is particularly interested in mathematical tools for comparing alternative extensions of ZFC. In the past, he has held positions at the Universities of St Andrews, Bristol, Aberdeen, and Queensland.

Chris Scambler holds a Postdoctoral Research Fellowship in Philosophy at All Souls College, Oxford He completed his PhD at New York University. His main areas of research interest are philosophy of mathematics and philosophical logic, focusing especially on the set-theoretic and semantic paradoxes.

Jeffrey Schatz received his PhD from the University of California, Irvine. He was previously a postdoctoral researcher at the University of Bergen, working on the project Mathematics with a Human Face: Set Theory within a Naturalized Wittgensteinean Framework. He is a JD student at Yale Law School.

Zeynep Soysal received her PhD in Philosophy from Harvard University in 2017. She is an associate professor in the Department of Philosophy at the University of Rochester. Her research primarily concerns issues at the intersection of the philosophy of language and the philosophy of mathematics.

Sourav Tarafder is Assistant Professor of Business Mathematics and Statistics at St. Xavier's College, Kolkata, India. His area of interest is the foundations of mathematics together with the classical and non-classical logics.

Claudio Ternullo is a Marie Skłodowska-Curie Seal of Excellence postdoctoral fellow at the University of Cluj-Napoca (Romania), having previously held postdoctoral positions at the Kurt Gödel Research Center for Mathematical Logic, University of Vienna, at the University of Tartu, and, until 2023, at the University of Barcelona. He specializes in the foundations and philosophy of mathematics. His work mainly focuses on the set-theoretic multiverse, new axioms and their justification, mathematical Platonism, abstraction principles, and the history of logic.

Giorgio Venturi (editor) is a tenure-track associate professor at the University of Pisa. Between 2017 and 2022 he was an assistant professor at the University of Campinas. He is an associated member of the University of Brasilia and of the Konstanz Center for Philosophy of Mathematics. He is Secretary of the European Society for the Philosophy of Mathematics (ESPM). He has a background in both mathematics and philosophy and his main area of research is logic and the foundations of mathematics, with particular emphasis in set theory and the philosophy of mathematics.

List of Table

1

Introduction

Carolin Antos, Neil Barton, and Giorgio Venturi

Philosophy of set theory is not only of interest to those working in set theory, but also for people focused on more general philosophical questions. In the last few years, the field has experienced rapid development due to the emergence of many new set-theoretic results. It has therefore become quite difficult for the general philosophical audience to keep pace with these numerous developments and to integrate these new discoveries in broader philosophical investigations. This collection of chapters aims to tackle this situation by exploring how contemporary research into the philosophy of set theory interacts with philosophy more widely.

There has been something of an explosion in the literature since the early 2000s. This has been driven by both (i) the desire of set theorists to provide a foundation that takes into account recent set-theoretic results and developments, and (ii) a renewed interest of philosophers willing to engage

C. Antos
Fachbereich Philosophie/Zukunftskolleg, University of Konstanz, Konstanz, Germany
e-mail: carolin.antos-kuby@uni-konstanz.de

N. Barton
National University of Singapore, Singapore, Singapore

IFIKK, University of Oslo, Oslo, Norway
e-mail: n.a.barton@ifikk.uio.no

G. Venturi (✉)
Department of Civilizations and Forms of Knowledge, University of Pisa, Pisa, Italy

C. Antos et al. (eds.), *The Palgrave Companion to the Philosophy of Set Theory*, Palgrave Companions, https://doi.org/10.1007/978-3-031-62387-5_1

with set theory. In this latter respect, the philosophy of set theory not only contributes to the foundations of mathematics but also allows us to analyse and inform more general questions in philosophy. There are several volumes and collections at least partly dedicated to the former line of research.[1] This collection of chapters contributes to the second (possibly somewhat neglected) track, examining the impact of contemporary set theory and its philosophy on wider philosophical questions.

The overall goals of the collection are twofold. On the one hand we would like to present the philosophy of set theory to a general philosophical audience as a multifaceted field of research with ramifications beyond the theoretical study of the sets themselves. For this reason we would like to include perspectives on this topic that have not formed the focus of collections thus far, such as how the study of set theory influences more general considerations about existence and our beliefs and knowledge concerning mathematical objects. On the other hand, we would like to present the new voices in this field that have the potential to influence the directions of philosophy of set theory in the coming years, showcasing that this is an active and still expanding field of philosophy.

We hope that the chapters collected in this volume can be of interest for both researchers in the philosophy of set theory and philosophers interested in how their area of research can interact with it. The collection is largely approachable by those with little experience in the philosophy of set theory as well as those with more who want to explore new lines of inquiry. Of course, there are places where details must be given and matters turn mathematical, but we hope that each chapter has something to offer philosophers who aren't necessarily engaged with the technical specifics of the discipline.

1 Overview of the Volume

In this collection of chapters we explore philosophy of set theory in interaction with three main areas: **Epistemology**, **Formal Theories**, and **Ontology**. While Ontology and Epistemology are standard areas of philosophy, the investigation of the relationship between set theory and other formal theories

[1]Some nice examples here: (1.) *Reflections on the Foundations of Mathematics: Univalent Foundations, Set Theory and General Thoughts*, Centrone, Kant, and Sarikaya (eds.) Synthese Library (2019), (2.) *Interpreting Gödel*, Juliette Kennedy (ed.) CUP (2014), (3.) *Infinity: New Research Frontiers*, Heller, Woodin (ed.) CUP (2011), (4.) *Kurt Gödel and the Foundations of Mathematics, Horizons of Truth*, Matthias Baaz (ed.) CUP (2011), (5.) *One Hundred Years of Russell's Paradox: Mathematics, Logic, Philosophy*, Godehard Link (ed.) Berlin, New York: De Gruyter, (2008).

attempts to incorporate the study of sets into a broader context within the Philosophy of Science.

Ontology, epistemology, and formal theories are areas that have profited from close ties to (philosophy of) set theory in the past. In each we have identified a few core themes of special importance in contemporary research. In Epistemology, we delve into the research on the nature of set-theoretic concepts and we find discussions connecting set theory with the classical problem of justification; in Formal Theories we deal with the relationship of set theory with other formal theories or logics (e.g. how they relate and compare); finally in Ontology, many chapters focus on the question of modality and the related discussion about potentialism, along with more classical questions on existence of sets and/or other mathematical objects.

Each of the chapters in the collection advances new research in the philosophy of set theory, connecting formal insights with classical and new philosophical perspectives and problems. Since there are often links drawn between these different areas, we find a good deal of interrelation between the chapters. We therefore hope to present both broad and in-depth discussions that showcase the fruitful interaction of philosophy of set theory and general philosophy.

The authors for this collection were chosen so as to represent the young and promising voices that will contribute to the development of the field in the future. They represent different academic profiles, from PhD student to tenured professor, and work with various methodological approaches from mathematical logic to empirical investigation into scientific practice.

2 Overview of the Chapters

Part I: Epistemology This part deals with classic questions in epistemology regarding our understanding, justification, and knowledge of mathematical objects. It does so, however, with some new twists. In particular, the authors consider ideas from contemporary philosophy of language and science, and examine how they play out in the set-theoretic case. We'll see this regarding metasemantics (Soysal), the idea of systematising data and intuitions (Clarke-Doane and Ash), the possible link between justification and metaphysics (Schatz), as well as the use of interview studies in examining justification (Kant).

Zeynep Soysal's chapter "The Problem of Existence for Descriptivism About the Reference of Set-Theoretic Expressions" discusses one of the most important problems in the epistemology of set theory: How is it possible to gain knowledge about sets? A possible solution, imported from the empirical sciences, consists in using language to describe reality and then to derive the semantic properties of our expression by their connection with reality. This is what normally goes under the name of descriptivism. Soysal investigates a challenge for descriptivism called "the problem of existence". Descriptivism, on the face of it, doesn't seem capable of explaining our set-theoretic knowledge, since it only explains the truth of conditionals (e.g. "If there are sets, then they are thus and so"), but fails to explain our knowledge of the existence of sets simpliciter. Soysal proposes to solve this problem via a form of descriptivism about logical expressions and the consistency of set theory.

Justin Clarke-Doane and Avner Ash's chapter "Intuition and Observation" deals with the issue of justification of set-theoretical statements: How are we to believe a theorem of set theory? The chapter analyses the classical answer provided by Russell and Gödel that there is a form of intuition that grants us some degree of access, and that axioms play the role of laws of nature meant to systematise mathematical data. Clarke-Doane and Ash suggest, however, that the analogy between mathematics and natural science is limited. Rather, they argue that it breaks down when we consider the consensus on the observable data that are to be systematised. The chapter ends by suggesting a more pragmatic answer to their starting question.

Jeffrey Schatz argues in "Arealism, Thin Realism, and the Problem of Extrinsic Evidence" for the reliability of extrinsic arguments in set theory. Intrinsic evidence is often taken to concern what *follows from a conception*, whereas extrinsic evidence is standardly concerned with *good consequences* of a statement or theory. Often, extrinsic evidence is pointed to as convincing for the adoption of new axioms. However, as Schatz points out, the problem arises if such indirect evidence, relying on secondary criteria such as fruitfulness, can be regarded as satisfactory. He argues that this problem about justification can be resolved by metaphysical means, in particular approaching set theory from an arealist point of view. If one supposes that there is no meaningful notion of mathematical truth, then arguments based on utility via means-ends relations remain as the only convincing evidence we have. Arealism is therefore strengthened as a viable candidate amongst promising positions in the debate about mathematical ontology.

Deborah Kant's chapter "The Hidden Use of New Axioms" investigates the set-theoretic practice of using extensions of the standard axiom system **ZFC** in their reasoning but then removing the use of additional axioms in the final version of their proofs. This highlights a seldom considered part of foundational practices, namely the use of axioms for heuristic purposes and how they contribute to knowledge production. Kant approaches this from a social-epistemological perspective substantiated by a large-scale, qualitative interview study. She argues that the hidden use of axioms provides extrinsic evidence and verification for them, therefore making a strong argument for the epistemic relevance of heuristic practices in set theory.

Part II: Formal Theories Philosophy of set theory stands right at the crossroads of mathematics, meta-mathematics, and philosophy. Studying set theory is therefore a perfect template for studying formal theories. Here we can investigate different interpretations, how they embed in one another, how to talk about their content, or their limitations and possible extensions. In particular, one can analyse what it means for a theory to provide a foundation and if and how that is connected to its mathematical content as in Button's and Meadows' chapters. Or one can compare different theories as to how they approach the question of mathematical truth, as Berry does in her study of geometry and set theory. Finally, one can also study how theories can agree on the mathematics they provide while being based on very different logics, as Tarafder does for paraconsistent logics.

Tim Button's "The Iterative Notion of Function and the Iterative Notion of Set" examines the idea that apparently different languages with different primitives can be regarded as providing the same foundation for metaphysics and mathematics. He does so by considering two approaches to foundations: conceiving of objects as *sets* and conceiving of them as *functions*. He provides a stage theory for functions, shows how the stages are eliminable (much as in the set-theoretic case), gives a quasi-categoricity result, and shows that a version of his function theory is synonymous with ZF. He then draws some morals from this discussion for mathematical and metaphysical foundations, before concluding that there is a sense in which both provide the *same* foundation.

Toby Meadows' "What Set Theory Could Not Be About" argues that we cannot specify the content for set theory in its capacity as a foundation. He formally supports this claim via relative interpretability of mathematical theories. He argues that there are four different theories each of which can fulfil the job of a foundation and which are equivalent to each other in a certain sense. However, he also argues that each appears to provide very different content. Meadows concludes that there is no serious connection between formal theories of sets and the concepts of collection and membership. However he also argues that this doesn't vitiate the use of set theory as a foundation—ideas of collection and membership can still be viewed as useful heuristics, even when it is not possible to pin down the content of set theory.

Sharon Berry's "Hamkins' Analogy Between Set Theory and Geometry: Pluralism by Leveling Up?" articulates a taxonomy for pluralism of different kinds. Pluralism by "levelling down" occurs when we deny the special status of a previously privileged theory, such as when we realise that Euclidean geometry does not have a special status as the true geometry of physical space. Pluralism by "levelling up" requires viewing different theories as telling us about something beyond mere "toy models" (possibly in addition to rejecting that one theory is privileged). She discusses the merits of this kind of pluralism in the context of set theory and geometry and argues that in the end such a pluralistic position cannot be the radical pluralism of Hamkins. She puts forward a counterpossible counterfactual variant that satisfies a levelling-up pluralism and therefore extends the recent debate about monism/pluralism with an interesting new candidate.

Sourav Tarafder's "Basic Mathematics Beyond Classicality" considers the use of models of paraconsistent logics in accounting for "basic mathematics": the kind of mathematics we use for everyday arithmetic. He shows that there are models of paraconsistent set theory (built via the technique of algebra-valued models) that agree with classical set theories on this basic mathematics. Using this, he discusses implications for the Quine-Putnam indispensability argument, and the idea that these paraconsistent set theories may in the end turn out to be indispensable. After giving some further examination of the role of independence proofs in this context, he leaves open the question of whether we may end up according these non-classical set theories the status of classical ones.

Part III: Ontology The chapters gathered in this section deal with the age-old question: What is there? The chapters in this part examine this issue along three different lines. The first such deals with the nature of sets. Luca Incrurvati's chapter investigates the iterative conception by proposing a deflationary, minimal account of it, against the backdrop of metaphysical dependence. The second examines the modal nature of sets, and in particular the idea that there could always be more sets (call this view *potentialism*). There are different forms of potentialism corresponding to different ways of obtaining new sets. The chapters by Ethan Brauer and Chris Scambler investigate the compatibility between two such forms, height and width potentialism, arriving at opposed conclusions. The third line considers how we should conceive of the nature of the existence of *models* of set theory as well as the sets themselves. Claudio Ternullo's chapter is devoted to the search for a theory of ontology that allows us to equally deal with sets and models, while Neil Barton and Giorgio Venturi's contribution deals with the ontological nature of models and whether we can separate them from sets, based on their nature. In more detail:

Luca Incurvati's "Iteration and Dependence Again" contributes to the elucidation of the iterative conception. The chapter has two goals. The first is to extricate the debate about understanding the iterative conception in terms of metaphysical dependence or a minimalist account from the actualism vs. potentialism debate. The distinction between these two discussions, Incurvati argues, is beneficial for a better understanding of the iterative conception. The second goal of the chapter consists in discussing and clarifying a criticism advanced by Mark Gasser against *ante rem* structuralism. Incurvati shows how part of this criticism can be addressed by a minimalist conception, while other aspects of it can be seen to suggest a more deflationary view of structuralism.

Ethan Brauer's "What is Forcing Potentialism?" examines the relationship between *height* and *width* potentialism. Height potentialism is the idea that we could always turn proper classes into sets, and in particular that there could be more ordinals than there are. This view is often mobilised in service of giving a response to the set-theoretic paradoxes. *Width* potentialism is the idea that there could always be more *subsets* than there are. Brauer examines in particular the *forcing* potentialist idea that holds that new subsets can always be added using the technique of forcing. He notes some problems in combining the two, in particular as it pertains to the iterative conception of set. He then responds by providing a height and forcing potentialist conception of sets, based on iterative ideas but employing a version of the logical conception of set with a notion of free-choice sequence added.

Chris Scambler's "On the Consistency of Height and Width Potentialism" also addresses whether these two forms of potentialism in set theory are compatible with each other. The main thesis of the chapter is that yes, it is possible to hold that the universe of sets can be extended along both axes of height and width. In order to do so the author addresses important objections against the compatibility and mutual consistency of these two forms of potentialism. This has both a philosophical and mathematical arm. On the philosophical side, he argues that a Stalnakerian conception of properties is needed. Mathematically, he introduces two formal theories that combine height and width potentialism and then presents some relative consistency results.

Claudio Ternullo's chapter "Higher-Order Platonism and Multiversism" addresses the question of which ontology can account for a multiverse perspective in set theory. Indeed, if set theory deals with sets and every object in set theory is a set, what is the background ontology for the study of the different universes of set theory? Ternullo's chapter analyses the specific case of Hamkins' multiverse and proposes to understand its ontological commitments in terms of Zalta and Linsky's Object Theory. The outcome of this proposal is not only that Object Theory can nicely accommodate the higher-order form of platonism implicit in Hamkins' multiverse conception, but also that Object Theory offers solutions to the challenges faced by Hamkins' view.

Neil Barton and Giorgio Venturi's "Language, Models, and Reality: Weak Existence and a Threefold Correspondence" explores an analogy between modelling in scientific practice and the use of models in set theory. They suggest a picture of set theory on which we consider whether models provide better or worse pictures of the way the universe might be, much as how in physics we examine how a particular mathematical model might be a better or worse picture of the way the physical universe might be. On this view, there's a threefold relationship between (i) the language we use, (ii) the models we employ, and (iii) the way the world is. They identify a problem of circularity in this context: since set-theoretic models are often thought to be sets themselves, it seems as though we need an understanding of the set-theoretic universe in order to understand the models. A response to this problem can be obtained, they argue, by examining notions of "weak existence", cases where an object can be said to exist merely in virtue of the way we use language. Contrasting this to forms of weak existence in the wider philosophical literature, they argue that weak existence delivers the models required to support the threefold correspondence.

Acknowledgments The idea of such a collection grew out of a series of online seminars organised by the editors between the University of Campinas and the University of Konstanz. We are grateful to all the participants and to the authors for their active and friendly participation. We also would like to thank Brendan George for handling our editorial work with the publisher. Chapters were given double-blind peer review, and we'd like to thank the many colleagues and friends that helped in the initial referee process, as well as an anonymous reviewer for Palgrave Macmillan.

Part I

Epistemology

2

The Problem of Existence for Descriptivism About the Reference of Set-Theoretic Expressions

Zeynep Soysal

1 Introduction

In this chapter, I argue that descriptivism as a metasemantic theory for mathematical and logical expressions can help explain our mathematical knowledge. A metasemantic theory for a class of expressions is an account of what explains or determines their semantic properties, such as having a particular meaning, reference, or truth-value. Descriptivism is a metasemantic theory according to which we associate some expression with a description, and this in part determines that this expression has the same meaning or reference as that description. For instance, on a descriptivist theory of the reference of 'electron,' we associate 'electron' with some description of the form 'the entity that plays such and such a role in physics and chemistry.' This association then in part determines that 'electron' refers to entities that play such and such a role in physics and chemistry. It only determines this "in part" because the world also needs to cooperate: 'electron' refers to entities that play such and such a role in physics and chemistry because we associate it with this description and because the world actually contains entities that satisfy this description. The "speaker associations" in the descriptivist thesis are intended to capture the speaker's use of the expression: to associate an expression E with a description

Z. Soysal (✉)
Department of Philosophy, University of Rochester, Rochester, NY, USA
e-mail: zeynep.soysal@rochester.edu

C. Antos et al. (eds.), *The Palgrave Companion to the Philosophy of Set Theory*, Palgrave Companions, https://doi.org/10.1007/978-3-031-62387-5_2

'the D' is, roughly, to be disposed to accept sentences of the form 'If E exists, then E is the D' come what may.

On the descriptivist view, families of expressions can similarly acquire their meanings and referents jointly by being associated with a description that involves them all. For instance, we can associate 'belief' with a description of the form 'the entity which, together with desire, plays such and such a role in folk psychology' and 'desire' with 'the entity which, together with belief, plays such and such a role in folk psychology.' Put another way, we can associate 'belief' and 'desire' together with a certain folk theory, T('belief,' 'desire'), that involves both expressions. As before, 'belief' and 'desire' then refer to entities that play such and such a role in folk psychology, if the world actually contains such entities; or, in other words, 'belief' and 'desire' have referents that make T('set,' 'membership') true, if the world actually contains such referents.[1] In this form, descriptivism is also known as 'the Ramsey–Carnap–Lewis method' or 'the method of implicit definitions,' and a theory is sometimes said to 'implicitly define' the expressions it is associated with.[2]

Descriptivism has been thought to be a plausible metasemantic theory for mathematical expressions. The so-called Hilbertian Strategy according to which axiomatic theories such as Peano Arithmetic (PA) or Zermelo–Fraenkel Set Theory with Choice (ZFC) implicitly define their mathematical vocabulary is a version of descriptivism.[3] And descriptivism about mathematical expressions has been advanced as explaining our knowledge of mathematics: the project of explaining our knowledge of mathematics in terms of our knowledge of meaning that originated with the logical empiricists is based on a descriptivist metasemantics for mathematical expressions.[4] What I will call 'the problem of existence for descriptivism about mathematical expressions' takes issue with descriptivism's ability to explain our knowledge of mathematics. The problem arises because descriptivism leaves part of the determination of meaning to the world: if we need to know that the world contains mathematical entities in order to know that our associated mathematical theories are true, then descriptivism doesn't seem capable of explaining our mathematical knowledge.[5] My aim in this chapter is to develop a solution to the problem of

[1]See, e.g., Lewis (1974) or Stalnaker (1984) for defenses of such a view.

[2]See Ramsey (1978), Carnap (1975), and Lewis (1970). Lewis (1984) calls his view 'descriptivism.'

[3]This is a strategy arguably advanced by Hilbert (1899). See Waxman (2022) for a recent explanation and defense of the Hilbertian Strategy.

[4]Those who worked on this project include Ayer (1936), Carnap (1939), Poincaré (1952), Hahn (1959), or more recently Peacocke (1992) and Boghossian (1996).

[5]Those who raise this problem include Field (1984), Horwich (1997), Hale and Wright (2000), Berry (2015), Linnebo (2018, 105), Warren (2020b, 204), and Waxman (2022).

existence. I will focus on the problem as it arises for descriptivism about the reference of set-theoretic expressions, but my solution applies more broadly to any mathematical expression.[6] I will argue that descriptivism about set-theoretic expressions, together with descriptivism about logical expressions, can explain our set-theoretic knowledge provided that we have reason to think that the theory we associate with our set-theoretic expressions is consistent. As I will explain, this is because the theory we associate with quantifiers plausibly entails the Hilbertian claim that the consistency of a theory associated with mathematical expressions suffices for the existence of the relevant mathematical entities. Along the way, I will compare my descriptivist approach and solution to the problem of existence to the approach and solution of the neo-logicists Hale and Wright (2001), the conventionalist Warren (2020b), and the pluralists Balaguer (1998) and Clarke-Doane (2022).[7]

2 Descriptivism

My focus in this chapter will be on descriptivism as a theory of the determination of the extensions of sub-sentential expressions. As usual, the extension of a singular term (i.e., its referent) is an individual (e.g., the extension of 'Morocco' is a particular country), the extension of a general term is a class or a set (e.g., the extension of 'cat' is the set of all cats), the extension of a predicate is a property or perhaps a class or a set (e.g., the extension of 'hot' is hotness or the set of all hot things), and so on. Throughout I will say that expressions "refer to their extensions," and I will use 'entity' as a broad ontological category that includes all extensions, whether they are sets, or classes, or individuals, or

[6]I focus on the case of set-theoretic language, among other reasons, due to the foundational status of set theory, and because my solution here complements my solution to the problem of indeterminacy (see footnote 7), which is most concerning as it arises for descriptivism about the reference of set-theoretic expressions.

[7]The other most important problem standardly raised against descriptivism about mathematical expressions is the problem of indeterminacy (see, e.g., Putnam 1975 or more recently Button & Walsh 2018): It seems that no matter what theory we associate with our mathematical expressions, we can't rule out that they refer to entities that are clearly unintended as their referents. Thus, either our mathematical language is radically indeterminate, or descriptivism isn't the correct metasemantic theory of it. In (Soysal 2020), I develop a solution to the problem of indeterminacy that involves the idea that the theories we associate with our mathematical expressions, together with other definitional truths about reference, entail that entities that are clearly unintended as the referents of our mathematical expressions aren't their referents. My solutions to the problem of indeterminacy and the problem existence have some important similarities; in particular, they both stem from an understanding of what it is to associate a description with some expressions that allows that the descriptions we associate with our mathematical and logical expressions are much richer than they are standardly construed, and that they are "anchored" in ordinary, informal language.

properties, and so on. It is standard to take descriptivism as a theory of the determination of *idiolect* meaning or reference; here I will assume the same.[8]

Mathematical expressions that don't have explicit definitions usually come in families within mathematical theories, such as 'line' and 'point' in geometry, or 'number,' 'zero,' and 'successor' in number theory. Descriptivists take the extensions of such "primitive" mathematical expressions to be fixed jointly by a description that involves them all. In set theory, the relevant primitive expressions are 'membership' (or '$\in$') and 'set.'[9] On the descriptivist view, then, set theorists associate 'set' and 'membership' with a certain theory, T('set,' 'membership'), and this in part determines that in these set theorists' idiolects, 'set' and 'membership' have extensions that make T('set,' 'membership') true.[10] The view leaves open that not all set theorists associate the same theory with set-theoretic expressions. Distinct associated theories T_1('set,' 'membership') $\neq$ T_2('set,' 'membership') could yield the same extensions for 'set' and 'membership,' but they needn't. The view thus also leaves open that different set theorists use 'set' and 'membership' with different extensions, and thus different meanings.

Descriptivism is part of a family of views according to which the use of an expression determines its meaning. Accordingly, to *associate* an expression E with a description 'the D'—or expressions $E_1, \ldots, E_n$ with a theory $T(E_1, \ldots, E_n)$—is to be disposed to use the expression(s) in a certain way. Here, I will understand it more specifically as being disposed to accept 'If E exists, then E is the D'—or the sentences in $T(E_1, \ldots, E_n)$ conditionalized on the existence of the $E_1, \ldots, E_n$—come what may. Note that on the assumption that E and the E_i are implicitly defined by their associated theories, these conditionals should be understood as logically equivalent to 'If there is something that is

[8]See, e.g., Burge (1979, 90), Kripke (1980, 8, 71), or Putnam (1980, 704).

[9]The predicate 'set' is one of the relevant primitives, for even when the formal set-theoretic language is defined as having the one mathematical primitive '$\in$,' the range of the quantifiers is standardly intended to give the extension of the informal (English) language word 'set.' Moreover, in the context of implicitly defining set-theoretic expressions, the formal set-theoretic language itself is usually taken to have a 'set' predicate; see, e.g., Button and Walsh (2018), Warren (2020b), and Waxman (2022).

[10]Descriptivism could be understood as a "bottom-up" metasemantic view in the sense that "the semantic facts concerning sub-sentential expressions explain or generate the semantic facts concerning sentences" (Warren 2020b, 215). For we can take the extensions of the sub-sentential expressions to be fixed "first" (in an explanatory sense) as extensions that would make the associated theory true, and then, together with the structure of the sentences, to "make true" (in an explanatory sense) the associated theory. But like all use-based metasemantic theories, descriptivism is clearly a "top-down" metasemantic theory in the sense that our dispositions towards *sentences* involving the sub-sentential expressions determines the semantic properties of both these expressions and the sentences (where the explanatory "order" of this determination can be left open). This is a reason to prefer this second sense of "top-down" in characterizing use-based metasemantic theories.

the D, then E is the D' or to the sentences in $T(E_1, \ldots, E_n)$ conditionalized on the existence of entities $e_1, \ldots, e_n$ that satisfy $T(x_1, \ldots, x_n)$, respectively.[11] For ease of expression, I will usually discuss the former types of conditionals and existence clauses, but they should be understood to be interchangeable with the latter in my discussion throughout. Then, here is how I will unpack what it is to "be disposed to accept a sentence come what may."

First, to *accept* a sentence ϕ means to believe that it expresses a true proposition in one's idiolect. Such a belief can manifest itself in various ways. For instance, one can be disposed to utter ϕ, to say 'Sentence ϕ is true,' to base one's reasoning on the truth of ϕ, and so on. Rational agents who have this belief will also typically have dispositions to adjust their relevant beliefs to new information. For instance, if I am rational and believe that 'If E exists, then E is the D' is true, then if I learn that the referent of 'the D' is F, I will also believe that the referent of E is F.[12]

Second, someone's being *disposed* to accept a sentence *come what may* here means that one would accept the sentence no matter what first-order evidence one were to have, if one were to consider the question of whether the sentence is true. By "first-order" evidence, I mean evidence that isn't about one's own belief-forming process. So, for instance, a set theorist who is disposed to accept 'If there are sets and a membership relation, then every set has a union' come what may would judge that this sentence expresses a true proposition even if they were to learn that, say, category theory has physical applications that set theory doesn't, or that some famous set theorist denies that this sentence is true, or that if this sentence is true, then so is some other sentence (perhaps surprisingly).[13] But they might come to believe that this sentence is false if they were to learn that they have ingested a drug that makes them mathematically inept.[14]

As I have explained it, the notion of "being disposed to accept a sentence come what may" closely resembles Chalmers' notion of a sentence being

[11] For Lewis (1970, 431), the "analytic postulates" of an associated theory are defined as the latter types of conditionals.

[12] Kipper and Soysal (2022) unpack what it is to associate a property with an expression in terms of such dispositions.

[13] As I explain below, I adopt a sense of "being disposed to accept a sentence come what may" on which it is possible for a speaker to be disposed to accept a sentence ϕ come what may without being disposed to accept come what may some sentence ψ that is logically entailed by ϕ. But any logical entailment known with certainty should also be accepted come what may, and thus taken to be part of the relevant associated theory. See also footnote 35 for references on how to construe the relevantly non-ideal epistemic scenarios.

[14] Strictly speaking, we should be concerned with what people would judge upon *considering* that they have such and such evidence, not what they would actually judge if they were to receive this evidence. But it is easier to use the latter formulation, and, as Chalmers (2012a, 211–214) argues, these should be equivalent in rational agents on pain of change of meaning.

"epistemically necessary."[15] According to Chalmers, we have the ability to judge whether a sentence is true if we are given sufficient information about the world. The same applies if the information given is hypothetical, describing merely possible states of the world. If we judge a sentence to be true no matter which (actual or hypothetical) information we are given, then this sentence is epistemically necessary. In Chalmers' account, the judgments in question are highly idealized: they are judgments the subjects would make if they were given a huge amount of information about the world (a "canonical description" of a possible world) and when they are allowed to have "arbitrary concepts, thoughts of infinite complexity, arbitrary amounts of calculation, no missteps in reasoning, and ideal judgment throughout" (Chalmers 2012a, 65).[16] I won't adopt this full-blown idealization here among other reasons because, as we will see in Sect. 3, I want it to be possible for a speaker to be ignorant of whether the theory she associates with some expression is consistent. But there must be some idealization in the notion of a disposition to accept a sentence come what may, because, for instance, we need to rule out errors in judgment.[17] I will leave it open in this chapter how to best analyze the required idealization.[18] I will only assume that on any way of ultimately analyzing it, the judgments one actually makes when one is in normal conditions are at least evidence for what one is disposed to judge come what may. I will also leave it open how exactly to understand "normal" conditions, but they should include conditions in which, for instance, one isn't inattentive, incapacitated, inebriated, overly emotionally attached to the truth or falsity of the sentence, and so on.[19] So, for instance, if a reasonably competent speaker in normal conditions judges that 'If there are sets and a membership relation, then every set has a union' is true no matter what scenarios she is asked to consider, this is evidence that 'Every set has a union' is part of the theory she associates with set-theoretic expressions. Assumptions of this type are needed on any view on which facts about the actual use of expressions play a role in determining meaning and reference.

---

[15] See, e.g., Chalmers (2002, 2012a). Chalmers (2012b) himself proposes to understand "being disposed to accept come what may" in the descriptivist thesis in terms of epistemic necessity.

[16] Chalmers (2012a) also restricts the account to judgments of "insulated" idealized agents, i.e., "cognizers whose rational processes are practically insulated from higher-order belief" (Chalmers 2012a, 104) to rule out that these agents judge some sentence to be false because of higher-order evidence.

[17] This is famously raised as a problem for "dispositionalist" views such as descriptivism by Kripke (1982).

[18] For instance, one might follow Warren (2020a)'s recent analysis on which the dispositions relevant to determining meaning are ones that are stable in the vast majority of normal conditions.

[19] Warren's (2020a, 271) definition of normal conditions as conditions under which no internal or external factors are interfering with the person's cognitive functioning could be used here.

3 The Problem of Existence

On the descriptivist view as I have construed it in Sect. 2, part of the determination of extensions and truth-values is done by speaker associations, and part of it is left to the world: speakers associate expressions with a theory, and, if the world contains entities that satisfy the theory, then the expressions refer to these entities and the theory is true.[20] Associated theories usually entail the existence of the entities defined.[21] So, for instance, T('set,' 'membership') will entail 'there are sets,' and thus if it turns out that there are no sets, T('set,' 'membership') will be false. Now, say we know that descriptivism is true and that T('set,' 'membership') is associated with 'set' and 'membership.' We then know that if there are sets and a membership relation, then 'set' and 'membership' refer to these and T('set,' 'membership') is true. But how do we know that there are sets and a membership relation? Descriptivism by itself doesn't seem to answer this question. A standard answer is that the existence of sets and the membership relation follows from our standard set theory, which we know to be true. But if that is all we can say about how we know that set-theoretic entities exist, then descriptivism about set-theoretic language is clearly idle in explaining our knowledge of set-theoretic truths: it can only explain our knowledge of set-theoretic truths if we assume that we already know set-theoretic truths. This is an instance of the problem of existence for descriptivism.[22]

One response to this problem is to say that descriptivism about mathematical language at least explains our knowledge of mathematical truths conditional on the existence of mathematical entities. For if descriptivism is true, then by standard disquotational principles, for any $\phi \in$ T('set,' 'membership'), 'if there are sets and a membership relation, then'$^\frown\phi$ is true. So, if we know that descriptivism is true, we also know that these conditionals are true. From here, we could also argue that the truth of descriptivism alone suffices for these conditionals to have some positive epistemic status for whoever associates 'set' and 'membership' with T('set,' 'membership'). For instance, we could argue that if descriptivism is true, then a speaker's beliefs that these conditionals are true are epistemically blameless, or the method of associating theories with expressions yields reliably formed beliefs in these

[20] Once again, I take the order of determination of extensions and truth-values to be left open by descriptivism—see footnote 10.

[21] See Lewis (1970, 429).

[22] See footnote 5 for references to some who raise this problem.

conditionals.[23] Horwich (1997, 426ff.), who raises the problem of existence for descriptivism, considers this response on behalf of descriptivists. This is also arguably Lewis and Carnap's response to the problem of existence, for on their view the "analytic" truths that descriptivism yields are the sentences in $T(E_1, \ldots, E_n)$ conditionalized on the existence of entities $e_1, \ldots, e_n$ that satisfy $T(x_1, \ldots, x_n)$ (what Lewis 1970, 431 calls 'the Carnap sentences'). On a natural interpretation, neo-logicists Hale and Wright advocate for a closely related type of response to the problem of existence by putting an "anti-arrogance" constraint on implicit definitions, which says that associated theories can't require for their truth "the obtaining of conditions of which it's reasonable to demand an independent assurance" (Hale & Wright 2009, 465), i.e., although implicit definitions don't need to have a conditional form, they can't be such that there is a condition that is necessary for their truth for which we need independent justification.[24]

As Horwich (1997, 426) notes, this response is quite concessive, given that the original motivation behind the view that mathematical theories implicitly define their primitive expressions was to explain our knowledge of (non-conditional) mathematical truths. As I noted in Sect. 1, for logical empiricists, for instance, the idea that all mathematical truths are analytic—i.e., in our framework, that they are all part of the theory we associate with our mathematical expressions—was supposed to provide an empiricist explanation of our knowledge of mathematics.[25] Waxman (2022, 118f.) has recently made this point against Hale and Wright's proposal to implicitly define number-theoretic expressions with Hume's Principle: if, as Hale and Wright (2009) argue, Hume's Principle is anti-arrogant because it doesn't entail the existence of infinitely many entities without assuming orthodox second-order logic, then our associating number-theoretic expressions with Hume's Principle can't by itself explain our knowledge of arithmetic.

A far less concessive response to the problem of existence can be found in recent work by Warren (2020b). Warren defends an "unrestricted" inferentialist metasemantics for logical and mathematical expressions. Although there are differences between inferentialism and descriptivism, Warren's unrestricted inferentialism about mathematical language suggests a natural counterpart

---

[23] See, e.g., Peacocke (1992), Boghossian (2003), or Warren (2020b).

[24] See, e.g., Waxman (2022) for this type of interpretation of the neo-logicists' anti-arrogance constraint. Neo-logicists clearly don't advocate for implicit definitions to be in conditional form since they advance Hume's Principle as an implicit definition of number-theoretic expressions. But they do argue that PA or ZFC are objectionably arrogant in entailing the existence of the relevant mathematical entities (Hale & Wright 2009).

[25] See footnote 4 for references to logical empiricists who held this view.

"unrestricted descriptivism" about mathematical language.[26] I will attribute unrestricted descriptivism to Warren and defend my attribution in footnotes 27, 28, 30, 31, 32, and 37, but the view is promising in its own right and so should be considered by descriptivists even if my attribution turns out to be incorrect.

On unrestricted descriptivism, if one is disposed to accept a sentence come what may, then that sentence is automatically true. This is an "unrestricted" type of descriptivism, because the thesis is understood to hold for *any* sentence.[27] So, on one way of putting the difference with the descriptivist view I outlined in Sect. 2, unrestricted descriptivism says that our associating 'set' and 'membership' with some theory T('set,' 'membership') *fully* determines that 'set' and 'membership' have referents that make T('set,' 'membership') true, where to *associate* 'set' and 'membership' with T('set,' 'membership') in the unrestricted descriptivist's sense is to be disposed to accept T('set,' 'membership') *itself* (i.e., not conditionalized) come what may.[28] Moreover, this holds even if T('set,' 'membership') isn't consistent, consistent with other (empirical) truths, or conservative, i.e., if it entails sentences of the original language without 'set' and 'membership' that weren't already provable in the original language.[29]

The second component of Warren's response to the problem of existence is the claim that we associate (now in the unrestricted descriptivist's sense) mathematical expressions with standard axiomatic theories; in particular, we associate 'set' and 'membership' with ZF formulated in a language with a

[26] Warren (2022) himself notes that inferentialism and descriptivism about mathematics are formally equivalent. One difference between inferentialism and descriptivism as I explicated it here as general metasemantic theories is that inference rules could be linked causally to the world (as Warren 2020b, 130f. allows), whereas descriptivism associates bits of language with other bits of language. This is why a global descriptivist thesis should instead be formulated in terms of associations between expressions and *properties* (see, e.g., Jackson 1998 or Kipper and Soysal 2022).

[27] Warren's inferentialism is "unrestricted" in that "any collection of inference rules used for and involving an expression can be meaning constituting for that expression" (Warren 2015, 4), and he allows rules without premises (Warren 2020b, 40). So, the rules 'one may infer ϕ from the empty set of premises' for $\phi \in T(E)$ would be meaning constituting rules for E in Warren's sense. Finally, being disposed to follow such a rule is plausibly equivalent to accepting all sentences in T(E) come what may, because for Warren following a rule requires having the disposition to infer according to the rule "when given the chance," i.e., "when someone or something brings the conclusion to your direct attention, perhaps by querying you on the matter" (Warren 2020b, 46).

[28] Warren (2020b, 214f.) would understand truth to be determined prior, in an explanatory sense, to reference, which I allow in my descriptivism (see footnote 10).

[29] It could also fail to be *Field-conservative*, i.e., entail sentences of the original language about the original ontology that weren't already provable in the original language, which is the constraint neo-logicists consider on implicit definitions (Hale & Wright 2001, 297).

'set' predicate.[30] On this view, our associated theories in mathematics are thus "ontologically committed" to the existence of mathematical entities in the sense that they entail the relevant existence sentences; in particular, T('set,' 'membership') entails '$\exists x \mathrm{Set}(x)$.' But then, unrestricted descriptivism entails that standard set-theoretic axioms and their consequences—including existence sentences such as '$\exists x \mathrm{Set}(x)$'—are automatically true fully because of our associations. From here, we can again draw epistemic conclusions, but this time about non-conditional mathematical truths. So, for instance, if we know that unrestricted descriptivism is true and that we associate 'set' and 'membership' with T('set,' 'membership') $\supseteq$ ZF, then we know that ZF is true and that '$\exists x \mathrm{Set}(x)$' is true. We can then also argue that the truth of unrestricted descriptivism suffices for speakers to be "automatically entitled" to believe that the sentences in T('set,' 'membership') are true, and so on.[31] Warren's unrestricted descriptivism thus purports to fully explain the truth and our knowledge of (non-conditional) set-theoretic truths. Call this 'the bold solution to the problem of existence.'

Key to the bold solution is the (bold) claim that descriptivism is unrestricted. Warren explains how it could be that blatantly false or even inconsistent associated theories could be made true by our associations in terms of what he calls 'the Translation Mistake.' Here is an illustration. Say some speaker of standard English begins to associate (in the unrestricted descriptivist's sense) 'phlogiston' with a description of the form 'the substance that plays such and such a role in combustion and that exists.' According to unrestricted descriptivism, 'Phlogiston exists' is thereby true in that speaker's new idiolect, while it is false in English. For Warren, this is because some expressions of the speaker's new idiolect are no longer synonymous with expressions of his old idiolect, which is why it would be a mistake to translate the new idiolect homophonically into the old. For instance, 'phlogiston' in the speaker's new idiolect doesn't mean the same as 'phlogiston' in English: in English, we clearly don't associate 'phlogiston' with 'Phlogiston exists.'[32] Similarly, some "old" terms in the theory the speaker associates with 'phlogiston' also have new extensions. For instance, 'combusting objects release a substance' is false in English but true in the speaker's idiolect, so some of 'combusting object,' 'release,' and 'substance' must have different extensions in the speaker's new

[30]Warren claims we follow the relevant inference rules from empty premises in the case of both PA (Warren 2020b, 200f.) and ZF (Warren 2020b, 272).

[31]See, e.g., Warren (2020b, 295).

[32]Warren makes this point about a speaker who is disposed to follow the rule "you may infer 'God exists' from the empty set of premises" (Warren 2020b, 223).

idiolect. Warren's general idea is that on any view on which use determines meaning, speakers should be interpreted charitably; so, if a speaker accepts a sentence come what may, they should be interpreted as expressing a true proposition with that sentence (Hirsch & Warren 2019; Warren 2020b, 91f.). But sometimes, this means that we also have to interpret some of their old vocabularies in new ways. In effect, this is to say that if one is disposed to accept $T(E_1, \ldots, E_n)$ come what may, then $E_1, \ldots, E_n$ *and* all the old terms in $T(E_1, \ldots, E_n)$ have extensions such that $T(E_1, \ldots, E_n)$ is true, where an *old* term can be defined as a term in $T(E_1, \ldots, E_n)$ other than $E_1, \ldots, E_n$ that already had a meaning in the idiolect without $E_1, \ldots, E_n$. On Warren's view, then, the old terms in an associated description shouldn't always be taken to keep their old meanings after that description is associated with the new terms.[33]

Applying this idea to set-theoretic language, Warren's view is that since we accept ZF come what may, we have to be interpreted as expressing a true proposition with each ZF axiom. But this means that some of the old logical vocabulary—and, in particular, the quantifiers—has to be interpreted so as to make the ZF axioms true: upon adopting the disposition to accept ZF come what may, we also adopt "new ways of using sentences containing 'there are' and 'exists'" (Hirsch & Warren 2019), and thus also a (perhaps new) meaning of 'exists' on which these axioms are true. Our dispositions to accept ZF come what may thus *guarantee* that 'exists' has a meaning such that the axioms of ZF and their consequences, including '$\exists x \mathrm{Set}(x)$,' are true. Presumably, it could be that before we adopted the disposition to accept ZF come what may, 'exist' already had a meaning on which 'there exist entities that satisfy ZF' came out true. But in any case, after adopting the disposition, 'exist' is guaranteed to have such a meaning. As Warren (2020b, 222) explains, his view presupposes a kind of pluralism about quantifiers: there can be multiple different meanings for 'exist' that all satisfy the standard syntactic rules for the unrestricted existential quantifiers.

I agree with the following assumption of unrestricted descriptivism: If a reasonable speaker really is disposed to accept a sentence come what may—if they consistently act as though they believe that the sentence is true, if they aren't disposed to adjust their belief to contrary evidence, and so on—then they express a true proposition with that sentence, even if this means that many of their expressions have non-standard meanings.[34] But I disagree with the bold solution to the problem of existence, specifically with the claim

[33]Note that the terms $E_1, \ldots, E_n$ don't strictly need to be new, but just have their meanings determined by the association.

[34]Kipper and Soysal (2022) use these types of considerations to argue for descriptivism.

that we do or would want to "associate" our expressions with ontologically committing theories in the unrestricted descriptivist's sense. In particular, I claim that we aren't actually disposed to accept a theory such as ZF come what may: we are only disposed to accept it conditionalized on the existence of the relevant mathematical entities, and for good reason. If I am right about this, then the bold solution to the problem of existence doesn't go through: unrestricted descriptivism about set-theoretic language, too, only explains the truth of conditional set-theoretic claims. So, we still need an account of the truth and our knowledge of the antecedents of these conditionals.

I claim we aren't disposed to accept non-conditional (existentially committing) theories come what may because there is at least one scenario on which we would judge them to be false, viz., the scenario in which the theories are (syntactically) inconsistent. If we learned or considered that the theory we associate (now in my sense from Sect. 2) with some expressions is inconsistent, we would judge that the associated expressions fail to refer and that the theory is false (or, at least, that the existence-entailing, non-conditional part of the theory is false).[35] I think this claim holds generally, but here I will only discuss it for the case of mathematics.

Say that some mathematician decides, at some time *t*, to implicitly define a new expression E with three existentially committing axioms A, B, and C (let us call this theory 'ABC'), i.e., E is implicitly defined as the entity such that ABC. But say that this mathematician is thereby disposed, from time *t* onward, to accept ABC itself, non-conditionalized, come what may. Now assume that the mathematician learns, at some later time t', that ABC is inconsistent. If she really is disposed to accept ABC come what may, then at t', she would still judge that ABC and 'There are Es' are true. But assuming classical logic—which most mathematicians do—if ABC is inconsistent, then everything is derivable from it.[36] We can assume that our mathematician also knows this. So, if our mathematician actually assumes classical logic and is disposed to follow the classical rules of inference, she would, it seems, thereby also be disposed to conclude any given sentence from ABC. But this is highly implausible: it seems obvious that ordinary mathematicians wouldn't start accepting every sentence

[35] It is a non-trivial question how to model such scenarios as epistemic possibilities, since on standard ways of construing these, they are consistent (see, e.g., Chalmers 2011). One option might be to model them as impossible worlds (see, e.g., Berto & Jago 2019 or Soysal 2023a). In (Soysal 2023b), I consider an alternative way of modeling failures of logical omniscience in a possible-worlds framework.

[36] More specifically, assuming that the mathematicians' logic with ABC is truth-functionally complete and contains modus ponens, which most standard mathematical axiom systems do.

come what may if they discovered an inconsistency in ABC, and that it would be highly practically irrational for them to do so.

Warren (2015) in effect provides an argument for this last claim concerning a related type of case involving Prior's (1960) connective 'Tonk.' Remember, Tonk's introduction rule allows inferring 'ϕ Tonk ψ' from ϕ, and its elimination rule allows inferring ψ from 'ϕ Tonk ψ,' for any sentences ϕ and ψ. Warren's argument (transposed as a descriptivist's) starts with the claim that someone who actually follows the Tonk-rules (called 'the Tonker') would in effect be disposed to accept every sentence come what may, and should thus, by unrestricted descriptivism, be interpreted as using a *trivial* language, i.e., a language in which every sentence expresses a necessary truth. But this also means, by the Translation Mistake, that the Tonker's language should be interpreted in a highly non-standard way. For instance, 'It is not the case that' doesn't mean negation in the Tonker's idiolect, since 'It is raining' and 'It is not the case that it is raining' both express necessary truths in the Tonker's idiolect. As Warren notes, trivial languages are highly impractical, for they can't be used to convey any contingent information (Warren 2015, 9). Moreover, it is practically highly undesirable for one's dispositions to use a new expression to alter the meanings of one's old expressions:

> [I]f we cannot translate homophonically into our new language, then we will not be able to take for granted that the sentences we previously accepted should still be accepted after the change. We will, so to speak, have to start over epistemically. And starting over is hard work, so we tend to embrace expansions of our language only if the change preserves our old reasons for accepting our old sentences. (Warren 2015, 22)

For these reasons, Warren concludes that it would be practically irrational to adopt the Tonker's language. We can make the same point about our mathematician: it would be practically irrational for our mathematician to adopt the dispositions to accept ABC come what may at t, because this adoption would render her entire language trivial in case ABC is inconsistent. The mathematician would clearly be better off having a language that isn't trivial and that is translatable homophonically into her language as it was before t in case ABC is inconsistent. But in order to have such a language, she shouldn't accept ABC "come inconsistency," and thus come what may.

Warren (2015) discusses another option for interpreting people who accept inconsistent theories come what may despite accepting classical logic, which he illustrates with the "Sophisticated Tonkers." The Sophisticated Tonkers use 'Tonk' only in certain restricted cases, for instance, "in religious ceremonies or on other scattered occasions" (Warren 2015, 15). Otherwise, they don't accept 'tonk'-involving proofs in mathematics, physics, or everyday life. According to Warren, we should translate the Sophisticated Tonkers' language to English homophonically except for the 'tonk'-involving part of this language; we should treat the 'tonk'-involving fragment of the Sophisticated Tonker's language "as a self-contained fragment that doesn't bleed out into the larger language" (Warren 2015, 15) by, for instance, translating it as expressing logical truths of English. Our mathematician could be "sophisticated" like the Sophisticated Tonkers: in case she learns that ABC is inconsistent, she could continue to accept ABC without thereby being disposed to accept everything come what may. In that case, as Warren argues, we should treat the E-involving fragment of the mathematician's language as a "self-contained fragment" that expresses necessary truths. For instance, we could say that our sophisticated mathematician expresses logical truths of English with every sentence of ABC. The rest of her language could then be interpreted as keeping its meaning from before t.[37]

I think it is just as obvious that ordinary mathematicians wouldn't act like Sophisticated Tonkers, and that it would be practically irrational for them to do so. Let us assume that ABC turns out to be inconsistent. Our sophisticated mathematician uses the language of ABC in combination with some other vocabulary, for she implicitly defines E in terms of some old vocabulary. This old vocabulary at least includes some logical expressions; for instance, the sophisticated mathematician accepts that ABC proves a contradiction (i.e., that it proves both P and $\sim P$ for some P), and so on. But then, the old vocabulary used in combination with the language of ABC must change its meaning, since ABC is interpreted as true even though inconsistent. For instance, since our mathematician accepts ABC and accepts that ABC is inconsistent and thus proves everything in the language

[37]Warren considers the case of Frege and early set theorists who developed inconsistent axiom systems, and claims that either their set-theoretic language was trivial or they didn't actually "have a disposition to apply the rules naïvely," and claims it is a "descriptive/historical" question which disjunct is correct (Warren 2020b, 285). Unlike what I have done here, he characterizes the first disjunct as matching the Naïve Tonkers and the second disjunct as matching the Sophisticated Tonkers. I think, however, that the scenario in which the mathematician's *entire* language isn't trivial but only the set-theoretic fragment is trivial matches the Sophisticated Tonkers better in that respect. Moreover, I interpret the second disjunct as the one that I am arguing for here. This is because, plausibly, a mathematician who doesn't accept ABC "naïvely" is one who doesn't accept ABC in the scenario in which ABC is inconsistent. This interpretation is supported by a similar example Warren uses involving the truth predicate: Warren argues that we arguably don't follow the disquotational truth rules naïvely, in the sense that we follow them for every sentence of the language "*except for liar-like sentences*" (in a way that would need to be made precise) (Warren 2015, 17). My claim here is that there was practical reason for these set theorists to not have followed unconditional rules, and thus that it would be more charitable to interpret them as having followed conditional rules.

of ABC, she also accepts ∼A, ∼B, and so on. So, '∼' can no longer mean negation in her new idiolect. Moreover, it isn't clear how easy it would be to "cordon off" the problematic fragment of the sophisticated mathematician's language. Unlike Sophisticated Tonkers, who presumably only use 'tonk' in sentences involving a small class of (e.g., religious) expressions, mathematicians commonly use their mathematical expressions across mathematical domains and even outside of mathematics. The expression 'set,' in particular, is used in all areas of mathematics outside of ZFC, and even outside of mathematics. So, if ABC is inconsistent, a large part of our mathematician's vocabulary will need to be interpreted in a non-standard way. Moreover, since the old vocabulary—and, in particular, the old logical vocabulary—can in principle be used in combination with any other part of our mathematician's idiolect, the inconsistency of ABC threatens to trivialize the entirety of our mathematician's language, and to collapse the sophisticated mathematicians' case into the Tonker's case. To avoid this, one needs to postulate that those expressions that are used in combination with ABC become polysemous: they are interpreted in a non-standard way whenever they are used in combination with ABC, and in a standard way otherwise. Many of our sophisticated mathematician's expressions, including her logical expressions, thus either have non-standard meanings or are polysemous. For the same types of reasons as above, either option yields a highly undesirable kind of language. There is a much simpler and practically rational set of dispositions to adopt: our mathematician could judge that ABC and 'There are Es' are false, if it turns out that ABC is inconsistent. None of the practical problems discussed arise for this set of dispositions: if the mathematician judges that ABC is false in the scenario in which ABC is inconsistent, then in case ABC really is inconsistent, the mathematician's language isn't trivial, or highly non-standard, or polysemous. This makes it clear that mathematicians don't—and practically shouldn't—accept existentially committing sentences come what may.

Running this illustration with ZF or ZFC instead of ABC is slightly more complicated, because there do seem to be scenarios where ZF is inconsistent about which mathematicians would still judge that sets exist. Consider, for instance, a scenario in which the standard Powerset Axiom is the main culprit in deriving a contradiction from ZF, and where a slightly restricted Powerset Axiom, together with the rest of the ZF axioms, yields a consistent theory. Set theorists might be inclined to judge that, in that scenario, 'There are sets' and 'Sets are extensional, well-founded entities' are still true, but that sets just turn out to be somewhat different from what set theorists thought. However, this only reveals that the complete set of ZF axioms is plausibly not part of the theory set theorists associate (in either sense) with set-theoretic expressions:

after all, our disposition to give up some axioms in some scenarios just means, on the descriptivist view, that those axioms aren't part of our associated theory T('set,' 'membership'). It might be that the theory we associate with ZF only contains a part of ZF, for instance, a part that includes the axioms of Extensionality and Foundation, but perhaps not much more; and indeed, it does seem plausible to think that a discovery of an inconsistency from Extensionality and Foundation would lead set theorists to judge that there are no sets after all. It might also be that the theory we associate with set-theoretic expressions is just more complex. As I have argued elsewhere (Soysal 2024), we plausibly associate set-theoretic expressions with sentences that describe certain theoretical roles that sets are supposed to play in mathematics, such as the role to "maximize interpretive power" as formulated by Steel (2000). In that case, if it turned out that the theoretical role we associate with sets is itself inconsistent, we would likely judge that there are no sets. Alternatively, it might be that our associated theory contains conditionals of the form 'If ZF is consistent, then sets are as ZF describes, but if ZF is inconsistent, then, if ZF′ is consistent, sets are as ZF′ describes,' and so on. In any case, there will always be *some* inconsistency scenarios about which we would judge that there are no sets. In particular, we would judge that if the existentially committed theory T('set,' 'membership') we associate with and that thus implicitly defines our set-theoretic expressions—whatever exactly that theory is—is inconsistent, then there are no sets and the theory is false.

I argued that we don't—and that, for practical reasons, we shouldn't—accept existentially committing mathematical theories come what may, just like we don't and shouldn't accept existentially committing theories about physical entities come what may. The descriptivist view that I accept thus works uniformly across mathematical and non-mathematical expressions. But this still leaves open the problem of existence. A large part of the appeal of Warren's bold solution to the problem of existence was that it explained mathematical existence and truth fully in terms of facts about language; as Warren notes, "offloading the mathematical existence facts to the world is not in the spirit of conventionalism" (Warren 2020b, 204). I take descriptivism to be a type of conventionalism, and I agree with Warren that mathematical truth and existence should be explained purely in terms of facts internal to language and language use. In the following, I will outline a solution that attempts to do this.

The general idea of this solution is that we accept sentences of the form 'If T('set,' 'membership') is consistent, then sets and the membership relation exist' come what may, as part of or as entailed by the theory we associate with 'exist.' According to the descriptivist view I outlined in Sect. 2, this is evidence that such sentences are (analytically) true. But then, this means that knowledge of the consistency of T('set,' 'membership') suffices to know that there are sets and a membership relation. If we also know that descriptivism is true, then we know that T('set,' 'membership') is true. From here we could then also argue, as before, that the truth of descriptivism and (our reasons for) the consistency of T('set,' 'membership') suffice for us to be justified in believing that T('set,' 'membership') is true, and so on. This solution only "offloads" the mathematical existence facts to facts "internal" to language: it explains mathematical existence fully in terms of facts about what is part of the theory we associate with the quantifiers and facts about the consistency of our associated theories.[38]

This solution is more modest than Warren's, in that according to it, descriptivism only explains knowledge or justification of mathematics on the assumption that we have knowledge or justification of the consistency of our associated theories. But this isn't a disadvantage of the solution. If inconsistent mathematical theories are always false (contra Warren), then every account of our knowledge of mathematics must presuppose that the consistency of our mathematical theories has some sort of positive epistemic status.[39] Moreover, knowledge of consistency plays a crucial role even on Warren's solution, for on his view we can't know that our mathematical language isn't trivial without knowing that our associated theory is consistent; and, as Warren agrees, it is clearly practically important for mathematicians to know that their mathematical languages aren't trivial because, as discussed above, trivial languages in a classical background prove everything thus is "mathematically uninteresting" (Warren 2020b, 289).[40]

Before defending my solution to the problem of existence, let me explain its main advantages. First, this solution doesn't require us to say that adopting the disposition to accept a mathematical theory come what may changes the meaning of 'exist' or guarantees that the word 'exist' has a meaning on which the associated theory is true. Usually, when we implicitly define some expressions, we intend the old expressions in the associated description to retain their old meanings. Let an *anchor* of $T(E_1, \ldots, E_n)$ be an old term of $T(E_1, \ldots, E_n)$ that retains its meaning in the idiolect that includes $E_1, \ldots,$

[38] Warren himself notes that facts about consistency are "internal to the [linguistic] conventions themselves" (Warren 2020b, 105) and thus that adding a consistency constraint on the determination of mathematical truths would be in the spirit of conventionalism.

[39] At least if we accept a standard closure-under-known-entailment principle on knowledge.

[40] See also Warren (2020b, 289).

E_n with their meanings given by $T(E_1, \ldots, E_n)$.[41] My claim is that we often intend our associated descriptions to be anchored in this sense. In particular, we usually intend 'exist' to be an anchor: When we accept 'If there are sets and a membership relation, then the empty set exists' in considering a given scenario, we usually do so on the assumption that 'there are' and 'exists' keep their meanings from our previous idiolect. This is manifested in our dispositions to use these expressions. For instance, if asked, we would say that we intend 'exists' to have its standard meaning, we would be disposed to treat 'exists' as synonymous with 'exists' in the old idiolect, and so on. In line with Warren, I think it is certainly possible to be disposed to accept existential statements themselves come what may, with the understanding that 'exists' will then be guaranteed to have a (perhaps new) meaning on which the existential sentence is true—just as it is certainly possible to decide to mean something altogether new and different with the word 'exists' in one's idiolect. But my claim is that we don't usually do this in the context of acquiring dispositions to use new mathematical expressions. The solution that I am defending respects this.[42]

Second, the view that the consistency of a theory of pure mathematics suffices for the existence of the relevant mathematical entities is popular in metaphysics, philosophy of mathematics, and even among mathematicians themselves. Under the slogan 'consistency implies existence,' this idea is usually traced back to Hilbert's views in (Hilbert 1899) and in the subsequent exchanges with Frege, for instance, to the famous passage:

> You write: 'I call axioms propositions … From the truth of the axioms it follows that they do not contradict one another.' I found it very interesting to read this very sentence in your letter, for as long as I have been thinking, writing and lecturing on these things, I have been saying the exact reverse: if the arbitrarily given axioms do not contradict one another with all their consequences, then they are true and the things defined by the axioms exist. This is for me the criterion of truth and existence. (Gabriel, Hermes, Kambartel, Thiel, Veraart, McGuinness, and Kaal 1980, 39)

---

[41] I also use this notion in (Soysal 2020).

[42] This criticism also applies to Berry (2015)'s explanation of mathematicians' "freedom" to introduce new kinds of mathematical objects in terms of quantifier variance: although Berry, unlike Warren, claims that coherence is a necessary condition on the truth of (non-conditional) implicit definitions, on her view the act of stipulative definition "not only gives meaning to newly coined predicate symbols and names but can change the meaning of expressions like 'there is,' in such a way as to ensure the truth of the relevant hypothesis" (Berry 2015, 8).

The Completeness Theorem later proved by Gödel (1986) is sometimes taken to provide a precise mathematical formulation and confirmation of Hilbert's idea, but it doesn't fully embody the idea of "consistency implies existence" that I claim is a conceptual truth here, because it doesn't guarantee that associated theories have *intended* models.[43] For instance, if T('set,' 'membership') is even first-order and thus in the purview of the Completeness Theorem, the latter only guarantees that the consistency of T('set,' 'membership') suffices for it to have a set-sized model, not a class model. One might add that the models given by the Completeness Theorem aren't guaranteed to contain the intended referents of our mathematical terms (i.e., numbers, sets, etc.), but only proxies for them (e.g., linguistic entities).[44] Nonetheless, it is highly plausible that Hilbert and others sympathetic to the "consistency implies existence" slogan would also think that the consistency of T('set,' 'membership') suffices for the existence of sets and the membership relation in the intended sense, in particular, where the extension of 'set' is a class-sized entity.[45]

In the philosophy of mathematics, the view that the consistency of a theory of pure mathematics is sufficient for the existence of the relevant mathematical entities is entailed by many so-called 'pluralist' views (Balaguer 1998; Clarke-Doane 2022), and arguably also by 'multiverse' views (Hamkins 2012).[46] For instance, Balaguer's plenitudinous platonism is, in outline, the view that every logically possible mathematical object exists (Balaguer 1998, 5), which Balaguer argues entails that

> all (consistent) mathematical descriptions and singular terms will refer, and any (consistent) representation of a mathematical object that someone could construct will be an accurate representation of an actually existing mathematical object. (Balaguer 1998, 43)

On Hamkins (2012)'s multiverse views, in turn, every consistent theory that includes ZF (or perhaps only part of ZF) is instantiated. Although there is wide

---

[43]For historical discussion of the relation between Hilbert's principle and the Completeness Theorem, see, e.g., Doherty (2017) or Dean (2021).

[44]If one is a mathematical structuralist, it doesn't matter which objects instantiate the associated theories, but if not, one would want the referents of number-theoretic terms, say, to be *numbers*, and not sets or linguistic expressions, and so on. See also Soysal (2020) for further discussion of these points as they arise with respect to model-theoretic arguments for the indeterminacy of reference.

[45]Linnebo (2018, 5–7) discusses the 'consistency implies existence" view and attributes it also to Cantor and to structuralists such as Parsons, Resnik, and Shapiro.

[46]Although Clarke-Doane (2022, 34) requires arithmetic soundness instead of consistency; see also footnote 49.

disagreement about whether multiverse views are correct, these disagreements aren't usually about whether consistent mathematical theories are instantiated. Opponents of multiverse views don't usually question that there are "universe of sets"-like entities that one could have described using some consistent mathematical theory; rather, they question that all of these different "universe of sets"-like entities are eligible referents for *our* expressions 'set' and 'membership,' and thus that the reference of our set-theoretic expressions is radically indeterminate.[47] Generally speaking, then, the view that the consistency of a theory of pure mathematics is sufficient for the existence of the relevant mathematical entities—and, more specifically, the view that the consistency of T('set,' 'membership') is sufficient for the existence of sets—enjoys broad appeal. But such views aren't usually defended from more basic principles. For instance, Balaguer (1998, 152) himself argues that there is no fact of the matter whether plenitudinous platonism or fictionalism is true. For other defenders of pluralist or multiverse views, the view that consistency suffices for existence in mathematics is only supported indirectly, on the basis that it helps us explain our knowledge of mathematical truths (Clarke-Doane 2022) or the practice of set theory (Hamkins 2012). On my account, consistency implies existence in mathematics because this is something that we (or, at least, many of us) hold true come what may, and thus something that is analytically true (for many of us). This respects the intuition that there is nothing much more basic to say about why consistency implies existence, and that it might even seem that there is no fact of the matter about it. Let me now turn to my reasons for holding this account.

On the descriptivist picture that I outlined in Sect. 2, there is no principled reason to think that the theories we associate with mathematical or logical expressions are couched in a formal language, or that they are exhausted by standard formal axioms or introduction and elimination rules. As I mentioned above, there is good reason to think that T('set,' 'membership') contains more than axioms of ZFC, and that it is anchored in ordinary informal language. In particular, as I argued in (Soysal 2020, 2024), whatever reason we have to think that some axioms of ZFC are accepted come what may apply to sentences of informal language such as 'Sets are abstract objects' and to sentences of

---

[47] See, e.g., Martin (2001); Koellner (2013). Field (2001, 335) argues that the motivation for being a plenitudinous platonist is to claim that our set-theoretic language is radically indeterminate, or that there isn't a unique "universe of sets"-like entity that our set-theoretic expressions refer to. But this isn't so clear; Balaguer (1998, 62f., 66), for instance, thinks that it is plausible that our "full conception" of sets captures at least a unique class of intended "universe of sets"-like entities—perhaps even one on which independent sentences such as CH have determinate truth-values—and that this is perfectly compatible with the motivations of plenitudinous platonism.

the form 'Sets are fruitful for such and such in mathematics': set theorists accept these sentences; they seem unwilling to give them up no matter what scenario they consider; they base their set-theoretic practice on the truth of such sentences, for instance, their practice of accepting or rejecting new axioms of set theory, such as $V = L$, or the Axiom of Inaccessible Cardinals, and so on. The same points can be made about logical expressions, such as the quantifiers. The standard rules of introduction and elimination are plausibly part of the meaning-constituting rules for 'exist,' but they needn't exhaust them, and 'exist' can be associated with a theory, just like any other mathematical expressions or expressions of ordinary language. This means that it is at least possible to associate 'exist' with a theory that includes or entails 'If T('set,' 'membership') is consistent, then sets and the membership relation exist.' The fact that this claim is held by many people as a basic fact is then evidence that it is, at least among these people, accepted come what may, and thus that it is part of or entailed by the theory associated with 'exist.'

The sentence 'If T('set,' 'membership') is consistent, then sets and the membership relation exist' could be a "basic" analytic truth about existence, i.e., be held true come what may not on the basis of being inferred from some other sentence.[48] But it could also be entailed by some more basic analytic truths about existence. A natural such candidate analytic truth would be plenitudinous platonism itself. Alternatively, the analytic truth in question could be a more general maximalism about existence of the kind outlined by Eklund (2006, 2009), viz., the view that if the existence of Fs is consistent with certain basic truths, then Fs exist. Even if platitudinous platonism and maximalism aren't ultimately viable, this doesn't undermine the idea that certain specific sentences—such as our original 'If T('set,' 'membership') is consistent, then sets and the membership relation exist'—are part of the theory we associate with 'exist.'[49] This point applies more generally: It is in general

[48] See Warren's related notion of "basic" rule of inference as a rule that the "speaker accept[s] without any epistemic reason for doing so, [...] *because they do* [...]" (Warren 2020b, 66).

[49] A standard objection to plenitudinous platonism and maximalism is that there are consistent mathematical theories that posit seemingly incompatible objects, such as ZFC + CH and ZFC + ~CH (Balaguer 1998, 59; Restall 2003), or PA and a theory of "anti-numbers" that entails 'there are no numbers' (Eklund 2006, 112; Schiffer 2003, 55). Many responses have been given to this type of problem. For the first example, one strategy is to use something like Warren's Translation Mistake argument: Although both ZFC + CH and ZFC + ~CH fix the referents of 'set,' strictly speaking ambiguity is introduced, viz., ZFC + CH describes a universe of sets$_{CH}$, while ZFC + ~CH describes a universe of sets$_{\sim CH}$ (Balaguer 1998, 59, Rabin 2007). Warren (2020b, 303–306) defends this type of response for pairs of theories of the form T and T + ~Con(T), which Clarke-Doane (2022, 33–35) discusses as a problem for pluralists, leading him to use arithmetic soundness instead of consistency as a sufficient condition for existence. For the second example, Eklund (2006) and Schiffer (2003) propose the idea that consistency is only sufficient for existence in conservative theories, and the theory of anti-numbers isn't conservative.

difficult to find necessary and sufficient conditions for the applications of our terms, or to find simple theories that systematize our use of an expression. But this doesn't undermine the idea that some sentences are clearly part of the theory we associate with an expression. For instance, it is notoriously difficult to find a simple theory we associate with 'knowledge,' but we should nonetheless be highly confident in the claim that 'If something is known, then it is true' is part of that theory.[50] My proposal here doesn't require finding a formulation of maximalism or plenitudinous platonism that avoids all possible objections; rather, my claim is that there are clear cases of theories that are about pure mathematical abstracta that don't conflict with other empirical or abstract entities—such as T('set,' 'membership')—where by our (or at least many people's) notion of existence, the consistency of the theory introducing these mathematical entities suffices for their existence.[51] This is my proposed modest, but not too concessive, solution to the problem of existence: descriptivism can yield knowledge of non-conditional mathematical truths, provided we have reason to think that our associated theories are consistent.

Note, finally, that the solution I proposed relies on there being *some* concept of existence on which it is analytically true that the consistency of an associated mathematical theory is sufficient for the existence of the relevant mathematical entities. I argued above that many of us express this concept with our existential quantifiers. But my solution also gives a recipe for solving the problem of existence on other concepts of existence. For instance, if one expresses a more demanding concept of existence with one's existential quantifiers, then the question of whether the relevant mathematical entities exist will have to be evaluated according to the demands of their concept of existence, i.e., according to the theory they associate with their existential quantifier. On some "demanding" theories associated with the quantifiers, it might turn out that there are no mathematical objects, and, thus, that our mathematical theories are strictly speaking false. I would argue that there are practical reasons to not adopt such associated theories; it is, after all, much simpler to have a

One could also say that the theory of anti-numbers entails a negative existence claim, which should be evaluated according to the theory associated with 'exist.' So, in entailing that numbers don't exist, the theory of anti-numbers entails that the theory introducing numbers is inconsistent, which is (very likely) false. Others, such as Thomasson (2015, Ch. 8) have developed solutions to cases of incompatible objects outside of mathematics.

[50]This point is made assuming a descriptivist theory of the reference of 'knowledge,' but the same point could be made more generally about finding necessary and sufficient conditions for knowledge.

[51]See Waxman (2022) for proof and discussion of a result that shows that the consistency of ZFC suffices for Field-conservativeness. If T('set,' 'membership') isn't first-order, Waxman (2022) shows that a slightly weaker result still applies.

theory on which mathematicians do utter truths, do talk about mathematical objects, and so on. But I leave the questions of what is part of other theories one might associate with existential quantifiers, and of what are the practical benefits of different associated theories, outside the scope of this chapter.

4 Conclusion

Descriptivism gives an account of how the reference of set-theoretic expressions is determined, if there are set-theoretic entities to begin with. The problem of existence is that because descriptivism leaves part of the determination of reference and meaning to the world, it appears to be epistemologically idle. I critically examined two solutions to the problem of existence—one concessive and another one bold—and found each of them wanting. I then argued for a more middle-ground solution according to which it is part of the theory that we—or, at least, many of us—associate with our existential quantifier that the consistency of an associated mathematical theory is sufficient for the existence of the relevant mathematical entities. Descriptivism can thus explain the truth and our knowledge of mathematical theories, provided that we have reason to think that these theories are consistent. Although this solution is more modest than the bold one, it is just as explanatorily strong: For the only unconditional knowledge one can acquire on the bold solution is knowledge that certain sentences are true. But without knowledge of consistency, we can't know what these sentences mean, or even what a large part of our language means. As we saw, this is because this language becomes trivial if there is an inconsistency. The added value of this knowledge is thus marginal at best. From here, the next step is to investigate the question of whether we have reason to think that our mathematical theories are consistent. This is a widely discussed question, and one that whoever gives an account of our knowledge of mathematics must face.[52]

References

Ayer, A. J. (1936). *Language, truth and logic*. Victor Gollancz.

Balaguer, M. (1998). *Platonism and Anti-platonism in mathematics*. Oxford University Press.

[52]For very helpful feedback, I am grateful to Matti Eklund, Jens Kipper, Tobias Alexius Ohlsson, James R. Shaw, two anonymous reviewers, and audiences at the University of Pittsburgh and Uppsala University.

Berry, S. (2015). Chalmers, quantifier variance and mathematicians' freedom. In A. Torza (Ed.), *Quantifiers, quantifiers, and quantifiers: Themes in logic, metaphysics, and language* (pp. 191–219). Springer.

Berto, F., & Jago, M. (2019). *Impossible worlds.* Oxford University Press.

Boghossian, P. (1996). Analyticity reconsidered. *Noûs, 30*(3), 360–391.

Boghossian, P. (2003). Blind reasoning. *Aristotelian Society, Supplementary, 77,* 225–248.

Burge, T. (1979). Individualism and the mental. In P. French, T. E. Uehling, Jr, & H. K. Wettstein (Eds.), *Midwest studies in philosophy IV: Studies in metaphysics* (pp. 73–121). University of Minnesota Press.

Button, T., & Walsh, S. (2018). *Philosophy and model theory.* Oxford University Press.

Carnap, R. (1939). Foundations of logic and mathematics. *International Encyclopedia of Unified Science, 1*(3), 1939.

Carnap, R. (1975). Observational language and theoretical language. In J. Hintikka (Ed.), *Rudolf carnap: Logical empiricist* (pp. 75–85). D. Reidel Publishing Company.

Chalmers, D. J. (2002). On sense and intension. *Philosophical Perspectives, 16,* 135–182.

Chalmers, D. J. (2011). The nature of epistemic space. In A. Egan & B. Weatherson (Eds.), *Epistemic modality* (Chap. 2). Oxford University Press.

Chalmers, D. J. (2012a). *Constructing the world.* Oxford University Press.

Chalmers, D. J. (2012b). Reference magnets and the grounds of intentionality. In *Constructing the world, extended edition.* http://consc.net/books/ctw/extended.html

Clarke-Doane, J. (2022). *Mathematics and metaphilosophy.* Cambridge University Press.

Dean, W. (2021). On consistency and existence in mathematics. *Proceedings of the Aristotelian Society, 120*(3), 349–393. https://doi.org/10.1093/arisoc/aoaa017

Doherty, F. T. (2017). Hilbert on consistency as a guide to mathematical reality. *Logique Et Analyse, 237,* 107–128.

Eklund, M. (2006). Neo-fregean ontology. *Philosophical Perspectives, 20,* 95–121.

Eklund, M. (2009). Carnap and ontological pluralism. In D. J. Chalmers, D. Manley, & R. Wasserman (Eds.), *Metametaphysics* (pp. 130–156). Oxford University Press.

Field, H. (1984). Critical review of Frege's conception of numbers as object, by crispin wright. *Canadian Journal of Philosophy, 14*(4), 637–662.

Field, H. (2001). Which undecidable mathematical sentences have determinate truth values? In *Truth and the absence of fact.* Oxford University Press. https://doi.org/10.1093/0199242895.003.0012

Gabriel, G., Hermes, H., Kambartel, F., Thiel, C., Veraart, A., McGuinness, B., & Kaal, H. (1980). *Gottlob frege: Philosophical and mathematical correspondence.* Blackwell.

Gödel, K. (1986). On the completeness of the calculus of logic. In S. Feferman, J. W. Dawson, S. C. Kleene, G. H. Moore, R. M. Solovay, & J. van Heijenoort (Eds.), *Kurt godel: Collected works. Vol. 1: Publications 1929–1936*. Oxford University Press.

Hahn, H. (1959). Logic, mathematics and knowledge of nature. In A. J. Ayer (Ed.), *Logical positivism* (pp. 147–161). The Free Press.

Hale, B., & Wright, C. (2000). Implicit definition and the a priori. In P. Boghossian & C. Peacocke (Eds.), *New essays on the a priori* (pp. 286–319). Oxford University Press.

Hale, B., & Wright, C. (2001). *The reason's proper study*. Oxford University Press.

Hale, B., & Wright, C. (2009). Focus restored: Comments on John Macfarlane. *Synthese, 170*(3), 457–482.

Hamkins, J. D. (2012). The set-theoretic multiverse. *The Review of Symbolis Logic, 5*(4), 419–449.

Hilbert, D. (1899). *The foundations of geometry*. Open Court, La Salle, Illinois (10th ed.). Translated by Leo Unger.

Hirsch, E., & Warren, J. (2019). Quantifier variance. In *The routledge handbook of philosophy of relativism* (pp. 349–357). Routledge.

Horwich, P. (1997). Implicit definitions, analytic truth and a priori knowlege. *Noûs, 31,* 423–440.

Jackson, F. (1998). Reference and description revisited. *Noûs, 32,* 201–218.

Kipper, J., & Soysal, Z. (2022). A Kripkean argument for descriptivism. *Noûs, 56*(3), 654–669.

Koellner, P. (2013). *Hamkins on the multiverse*. Manuscript, Ms.

Kripke, S. (1980). *Naming and necessity*. Harvard University Press.

Kripke, S. (1982). *Wittgenstein on rules and private language*. Harvard University Press.

Lewis, D. (1970). How to define theoretical terms. *Journal of Philosophy, 67,* 427–446.

Lewis, D. (1974). Radical interpretation. *Synthese, 27*(3/4), 331–344. http://www.jstor.org/stable/20114928

Lewis, D. (1984). Putnam's paradox. *Australasian Journal of Philosophy, 62,* 221–36.

Linnebo, Ø. (2018). *Thin objects: An abstractionist account*. Oxford University Press.

Martin, D. A. (2001). Multiple universes of sets and indeterminate truth values. *Topoi, 20*(1), 5–16. https://doi.org/10.1023/a:1010600724850

Peacocke, C. (1992). *A study of concepts*. The MIT Press.

Poincaré, H. (1952). *Science and hypothesis*. Dover. Republication of the first English translation, published by Walter Scott Publishing (1905).

Prior, A. N. (1960). The runabout inference ticket. *Analysis, 21,* 38–39.

Putnam, H. (1975). The meaning of 'meaning'. In K. Gunderson (Ed.), *Language, mind and knowledge*. Minnesota Studies in the Philosophy of Science (Vol. 7, pp. 131–193). University of Minnesota Press.

Putnam, H. (1980). Models and reality. *Journal of Symbolic Logic, 45,* 464–482.

Rabin, G. (2007). Full-blooded reference. *Philosophia Mathematica, 15*(3), 357–365. https://doi.org/10.1093/philmat/nkm027

Ramsey, F. P. (1978). Theories. In D. H. Mellor (Ed.), *Foundations: Essays in philosophy, logic, mathematics and economics* (pp. 101–125). Routledge & Kegan Paul.

Restall, G. (2003). Just what is full-blooded platonism? *Philosophia Mathematica, 11*(1), 82–91.

Schiffer, S. (2003). *The things we mean*. Clarendon Press.

Soysal, Z. (2020). Descriptivism about the reference of set-theoretic expressions: Revisiting putnam's model-theoretic arguments. *The Monist, 103*(4), 442–454. https://doi.org/10.1093/monist/onaa016

Soysal, Z. (2023a). An algorithmic impossible-worlds model of belief and knowledge. *The Review of Symbolic Logic, 17*, 1–25. https://doi.org/10.1017/S1755020323000059

Soysal, Z. (2023b). A metalinguistic and computational approach to the problem of mathematical omniscience. *Philosophy and Phenomenological Research, 106*(2), 455–474. https://doi.org/10.1111/phpr.12864. https://onlinelibrary.wiley.com/doi/abs/10.1111/phpr.12864

Soysal, Z. (2024). A Carnapian solution to the puzzle of extrinsic justifications. In S. Arbeiter & J. Kennedy (eds.), *The Philosophy of Penelope Maddy, Outstanding Contributions to Logic* (pp. 319–336). Springer.

Stalnaker, R. (1984). *Inquiry*. MIT Press.

Steel, J. (2000). Mathematics needs new axioms. *Bulletin of Symbolic Logic, 6*(4), 422–433.

Thomasson, A. L. (2015). *Ontology made easy*. Oxford University Press.

Warren, J. (2015). Talking with tonkers. *Philosophers' Imprint, 15*(24), 1–24.

Warren, J. (2020a). Killing Kripkenstein's monster. *Noûs, 54*, 257–289.

Warren, J. (2020b). *Shadows of syntax: Revitalizing logical and mathematical conventionalism*. Oxford University Press.

Warren, J. (2022). Inferentialism, conventionalism, and a posteriori necessity. *Journal of Philosophy, 119*(10), 517–541.

Waxman, D. (2022). Rescuing implicit definition from abstractionism. In G. Oliveri, C. Ternullo, & S. Boscolo (Eds.), *Objects, structures, and logics: Filmat studies in the philosophy of mathematics* (pp. 97–129). Springer International Publishing.

3

Intuition and Observation

Avner Ash and Justin Clarke-Doane

The motivating question of this chapter is: 'How are our beliefs in the theorems of mathematics justified?' This is distinguished from the question 'How are our mathematical beliefs reliably true?' We examine an influential answer, outlined by Russell, championed by Gödel, and developed by those searching for new axioms to settle undecidables, that our mathematical beliefs are justified by 'intuitions', as our scientific beliefs are justified by observations. In this view, axioms are analogous to laws of nature. They are postulated to best systematize the data to be explained. We argue that there is a decisive difference between the cases. There is agreement on the data to be systematized in the scientific case that has no analog in the mathematical one. There is virtual consensus over observations, but conspicuous dispute over intuitions. In this respect, mathematics more closely resembles stereotypical philosophy. We conclude by distinguishing two ideas that have long been associated—realism (the idea that

Thanks to Barry Mazur, Will Cavendish, Michael Harris, Juliette Kennedy, and two anonymous referees for helpful comments.

A. Ash
Department of Mathematics, Boston College, Chestnut Hill, MA, USA
e-mail: ashav@bc.edu

J. Clarke-Doane (✉)
Columbia University, IAS, Princeton, NJ, USA
e-mail: jc4345@columbia.edu

C. Antos et al. (eds.), *The Palgrave Companion to the Philosophy of Set Theory*, Palgrave Companions, https://doi.org/10.1007/978-3-031-62387-5_3

there is an independent reality) and objectivity (the idea that in a disagreement, only one of us can be right). We argue that, while realism is true of mathematics and philosophy, typical claims from these domains fail to be objective. One upshot of the discussion is that even many questions of fundamental physics fail to be objective in roughly the sense that the question, 'Is the Parallel Postulate true?', understood as one of pure mathematics, fails to be. Another is a kind of pragmatism. Factual questions in mathematics, modality, logic, and evaluative areas go proxy for non-factual practical ones.

1 Informal and Formal Proof

How do we know mathematical truths? The obvious answer is: on the basis of proofs (Rachels, 1998, 3)! For example, we know Fermat's Last Theorem on the basis of the proofs of Wiles and Taylor-Wiles and know the 4-dimensional Poincaré Conjecture on the basis of Freedman's proof. But what is a mathematical proof? To most philosophers, it is a *formal* proof: a finite sequence of formulas, each of which either is an axiom or follows from a previous formula by a rule of formal inference, the last line of which is the formula proved. Mathematical proofs are not remotely so finicky.[1] A formal proof of practically any mathematical conjecture of interest would be incomprehensible to human beings. Mathematical proofs are closer to (informal) arguments that convince experts that *there exists* a formal proof. Such arguments are sometimes sketches, or outlines, of formal proofs. But they are not so in general.

Why think that formal proof is the final arbiter of mathematical validity? Because, on one hand, a formal proof of (a formal statement of) a conjecture is widely agreed to establish that conjecture, even if the proof is deemed otherwise non-ideal. For example, the proofs of the Four Color Theorem and the Kepler Conjecture make apparently ineliminable use of computers, given the limitations of human computing ability with respect to time, patience, and accuracy. These proofs are unsatisfying insofar as they cannot be 'surveyed'. But practically no one denies that the theorem follows (or, at least, follows in classical logic—see Sect. 11).[2] On the other hand, the independence results concern *formal* proofs, but are widely agreed to settle whether the conjecture

[1]Indeed, formal proofs are at best rational reconstructions of actual mathematical reasoning. See Ash and Gross (2008), Harris (2012), and Mazur (2012).

[2]For a contemporary illustration, consider the purportedly almost complete formal verification in *Lean* of Holmes' proof of Con(*NF*). If this is indeed completed, then it will establish the (relative) consistency of Quine's system, even if no one besides Holmes can follow the proof.

in question is (informally) provable. For example, the proof that Con(*ZFC*) → Con(*ZFC* + *CH*) and Con(*ZFC*) → Con(*ZFC* + ∼*CH*) is an argument that there exists no *formal* proof of (a formal statement of) the *Continuum Hypothesis* (*CH*) or its negation from (a formal statement of) *Zermelo-Fraenkel* set theory with the *Axiom of Choice* (*ZFC*), if that theory is (formally!) consistent. This establishes that anyone seeking to (informally) prove the *CH* or ∼*CH* (from accepted principles) is doomed to fail. At best, they might identify new axioms that mathematicians come to accept, along with the others.[3]

It is a good question of how mathematicians know that there is a formal proof on the basis of the (informal) arguments that convince them. Most mathematical theorems have no (concrete) formal proofs to back them up (though some in the interactive theorem-proving community have begun to rectify this situation). But even if this question were answered, two puzzles would remain. First, how do mathematicians know what formally follows from what? Second, how do they know the axioms? We focus on the second question, but return to the first in Sect. 11.

A preliminary point: logicians partition axioms into *foundational* and *structural* ones (Shapiro, 1997, 41 & 50). Outside of set theory, arithmetic, and real analysis in some treatments, mathematicians tend to use axioms to describe anything that satisfies certain conditions, not to characterize an antecedently given collection of things. The question of correctness (over and above consistency) only arises in connection with foundational axioms. So, we focus on them.

2 Knowing the Axioms

There is a familiar narrative according to which the axioms of mathematics are *self-evident* (Greene, 2013, 184). Historically, this story is associated with Euclid's *Elements* and Descartes' clear and distinct ideas. In the twentieth century, self-evident axioms were often deemed *analytic*—truths that one can know merely by virtue of their linguistic competence (Coffa, 1982; Singer, 1994, 8). Perhaps it is just part of what we mean by 'set' that every set occurs

[3]One might think that proofs by diagram show that not all informally valid proofs can be formally verified. But, again, informal proofs are not, in general, outlines of formal ones. So, in order for this to be the case, there would need to be a diagrammatically provable conjecture that either lacks a formal analog or whose proof does. Whether there is such a thing depends on what 'formal analog' should mean. This is outside the scope of the present chapter.

at some level of the cumulative hierarchy, for example—i.e., that the *Axiom of Foundation* is true (Boolos, 1971, 498; Shoenfeld, 1977, 327).[4] One who lacks an even defeasible disposition to believe Foundation must lack competence with the predicate '$\in$' (and concept of membership).

However, even if it were analytic that all sets are well-founded, it is not analytic that there are no non-well-founded set-like things. If we were worried that some sets fail to be well-founded, then, under the assumption that Foundation is analytic, we should just worry that there are shmets—where it is no part of the concept of shmet that all shmets are well-founded. The claim that there are no shmets is like the claim that there are no superstrings. This is not analytic on anyone's view. But the claim that there are no shmets is just the 'transposition' of the claim that there are no non-well-founded sets, under the assumption that Foundation is analytic. Of course, there might be sets *as well as* shmets.[5] But then the question of whether all sets are well-founded would be like the question of whether two straight lines intersecting another so as to make less than a 180° angle on one side must intersect on that side. It depends on what you mean by 'line'! If 'line' is interpreted to mean Euclidean line, then yes. If it is interpreted to mean, say, hyperbolic line, then no. There is no metaphysical, as opposed to semantic, question to dispute.[6]

Perhaps the axioms of mathematics are self-evident, but not analytic. Many reject the analytic/synthetic distinction on independent grounds (Quine, 1951). But they still allow that some claims are self-evident—or, anyway, initially plausible or intuitive. Consider the proposition that *here is a piece*

[4]We use 'true' throughout in the ordinary Tarskian sense (as per Benacerraf, 1973). So, e.g., 'There is an inductive set' is true just in case there is an inductive set. Mathematicians, of course, frequently use the term differently. One often calls the conclusion of a proof 'true' in ordinary mathematical practice when one thinks that the proof is correct by the ordinary standards that are regulative there, independent of one's belief in the literal truth of its conclusion. Ordinary truth, not 'mathematical correctness', is what is relevant to the comparison with empirical science.

[5]In the development of Warren (2020), it is apparently analytic that there are shmets as well as sets.

[6]Shoenfeld himself writes, 'It is... possible that there is a completely different analysis of the notion of a set, and this might lead to quite a different set of axioms' (1985, 324). But he does not seem to appreciate the deflationary methodological ramifications of this possibility. Boolos is bolder. But Incurvati responds: 'we cannot believe in... Foundation on the... grounds that the iterative conception is, as Boolos suggested... 'the only natural and (apparently) consistent conception of set we have'' (2014, 206). He continues, '[Aczel's] graph conception [for example] ... emerges as a candidate to be the conception of set embodied by a set theory centered around the idea of a set being depicted by a graph... [I]f sets are what is depicted by an arbitrary graph, then most of the axioms of *ZFA* are justified' (2014, 205). Bar-Hillel, Fraenkel, and Levy note, 'To an extreme Platonist the question of the consistency of a system of set theory is not really a central foundational problem. He will discard a theory at the point when it turns out not to fit the 'true facts' about the universe, even if it is consistent. To put it more strongly, an extreme Platonist, like a physicist, will prefer an inconsistent theory some parts of which give a faithful description of the real situation to a demonstrably consistent theory which gives 'wrong information' about the universe' (1973, 326).

of paper. This is not analytic. It is conceivable that we are brains in vats. But we are presented to—perceptually—as being in a world, with external objects, including pieces of paper (Pryor, 2000). Do we not enjoy analogous experiences in connection with pure mathematics (Bealer, 1999; Bengson, 2015; Chudnoff, 2013; Huemer, 2005)?[7] Set theory presents to us—albeit intuitively, and not perceptually—as being about a cumulative hierarchy, where every set has a Choice function, sets under definable functions also form sets, and so on. Gödel writes,

> [D]espite their remoteness from sense experience, we do have a perception also of the objects of set theory, as is seen from the fact that the axioms force themselves upon us as being true. (1947, 483–4)

One problem with Gödel's proposal is that even if *some* mathematical propositions 'force themselves on us as being true', these do not seem to include most of the axioms of standard mathematics (Boolos, 1999, 130–131; Maddy, 1988a).[8] For example, notwithstanding Foundation, sets that contain themselves, or those with infinitely descending chains of membership, are important in some parts of mathematics, linguistics, and computer science (Azcel, 1988, Introduction, Moss, 2018). They also let us preserve something of the naive concept of set, according to which there is a universal set, complements of arbitrary sets, and sets of all sets with n members (Quine, 1937, 1969). Meanwhile, it is not beyond dispute that Foundation is even coherent (Ferrier, 2019; Potter, 2004, Sec. 3.3; Rieger, 2011, 17–18). What, after all, does it *mean* to say that an object, like V, is 'built in stages' via repeated applications of the *Powerset* and *Union* axioms? What kind of priority is involved? Not temporal priority.[9] But it cannot be 'metaphysical' possibility

[7]The conception of intuition common to these authors is that an intuition that P is an experience as of P (not just dispositions to judge that P) which defeasibly justifies one in believing that P. The key difference between intuition and observation, on this conception, is the content that is experienced. Ironically, given that these authors all accept an a priori/a posteriori distinction, one of us has argued from this conception to a new basis for skepticism about that distinction (Clarke-Doane, 2020, Section 3.6 & 3.9).

[8]*Cf.* Boolos: 'I am by no means convinced that any of the axioms of infinity, union, or power [set] . . . force themselves upon us or that all the axioms of replacement that we can comprehend do . . . .[T]hat there are doubts about the power set axiom is of course well known . . . .[T]here is nothing *unclear* about the power set axiom . . . .But it does not seem to me unreasonable to think that . . . it is not the case that for every set, there is a set of all its subsets' (1999, 130–1, italics in original).

[9]An anonymous referee requested an argument for this. Two immediate problems are that it is unclear whose construction would be canonical and how anyone could complete a transfinite construction in a finite time. Note also that, in light of the Relativity of Simultaneity, there may be no Lorentz invariant fact as to what sets existed if they were constructed in time. But at least what sets of *events* exist is something all observers should agree to!

either. There is no possible world in which a set exists but its powerset fails to.[10] Perhaps there is a hyperintensional respect in which earlier stages 'ground' or 'constitute' later ones? Perhaps. But whether such hyperintensional relations are well-founded is itself doubtful (Cameron, 2008, 4–5; Dixon, 2016; Rabin & Rabern, 2016).

The more promising intuition-based epistemology of mathematics draws an analogy between the axioms of mathematics and the laws of empirical science. Russell hoped to 'prove' all of mathematics in logical as well as epistemic senses of that term (Russell, 1903/2022; Russell & Whitehead, 1910/1963). He hoped to begin with indubitable 'laws of logic' (somewhat like Frege) and deduce.[11] Unfortunately, he soon discovered the paradox that bears his name, according to which, even the seeming banality that every predicate has an extension (perhaps empty) must be false. He further realized that his project required axioms that practically no one claimed were self-evident, like the axioms of *Infinity*, *Reducibility*, and *Choice*. He concluded,

> We . . . believe the premises [axioms] because we can see that their consequences are true, instead of believing the consequences because we know the premises . . . .But the inferring of premises from consequences is the essence of induction; thus the method in investigating the principles of mathematics is really an inductive method, and is substantially the same as the method of discovering general laws in any other science. (1907/1973, 273–274)[12]

[10] On the orthodox interpretation, anyway. We return to different interpretations of 'possible' in Sect. 11.

[11] These laws included, of course, what we would today regard as substantial mathematics, since the (higher-order) logic that Russell used is incomplete.

[12] Russell's changing view on the epistemological status of mathematics seems to be related to his abandonment of logic in favor of metaphysics, epistemology, and social philosophy at the height of his logical powers. He recalls,

> I wanted certainty in the kind of way in which people want religious faith. I thought that certainty is more likely to be found in mathematics than elsewhere. But I discovered that many mathematical demonstrations, which my teachers expected me to accept, were full of fallacies, and that, if certainty were indeed discoverable in mathematics, it would be in a new field of mathematics, with more solid foundations than those that had hitherto been thought secure. But as the work proceeded, I was continually reminded of the fable about the elephant and the tortoise. Having constructed an elephant upon which the mathematical world could rest, I found the elephant tottering, and proceeded to construct a tortoise to keep the elephant from falling. But the tortoise was no more secure than the elephant, and after some twenty years of very arduous toil, I came to the conclusion that there was nothing more that I could do in the way of making mathematical knowledge indubitable. (1952/2023)

Despite his remark about axioms 'forcing themselves upon us', Gödel agrees. He writes,

> [Russell] compares the axioms of... mathematics with the laws of nature and logical evidence with sense perception, so that the axioms need not... be evident in themselves, but rather their justification lies (exactly as in physics) in the fact that they make it possible for these 'sense perceptions' to be deduced... .I think that... this view has been largely justified by subsequent developments, and it is to be expected that it will be still more so in the future. (1944/1990, 121)

Russell's mature proposal is that the *epistemological* priority of mathematical principles typically precedes their *logical* priority.[13] We deduce theorems from axioms. But we are justified in believing the axioms because we are justified in believing the theorems that they imply, rather than the other way around (contra the aspirations of Frege's logicism). Assuming that our mathematical 'sense perceptions' are revisable, Russell's method is, accordingly, just Rawls' method of *reflective equilibrium*—the method of 'testing theories against judgments about particular cases, but also testing judgments about particular cases against theories, until equilibrium is achieved' (Blackburn, 2008, 312).[14] We begin with 'intuitions', or plausibility judgments. We then abduct to the best axiomatization of those judgments. Explanatory considerations may lead us to amend some of the judgments with which we began. Likewise, compelling intuitions may 'disconfirm' an otherwise attractive theory. We vacillate between considerations of plausibility and systematicity until (ideally) we achieve harmony between the two.[15]

---

[13]'Typically' because this need not always be so. For instance, the axioms of Euclidean geometry (which Frege [1900/1997], unlike Russell, denied were provable from 'logic' alone) seem to be more like a compendium of our intuitions about space than a post facto axiomatization of still more concrete intuitions. But Russell (1907/1973) explicitly takes the axioms of arithmetic to be 'inductively' supported. Similarly, Weyl reminds us, 'A set-theoretic treatment of the natural numbers such as that offered by Dedekind may indeed contribute to the systematization of mathematics; but it must not be allowed to obscure the fact that our grasp of the basic concepts of set theory depends on a priori intuition of iteration and of the sequence of natural numbers' (1918/1994, 109).

[14]See Rawls (1974, 8) for one characterization of the method.

[15]*N.B.*: Ordinary mathematicians sometimes use the word 'intuition' for a kind of mental state that differs from the one appealed to by Gödel and those searching for new axioms. Intuitions in this sense are reflective of mathematicians' knowledge of the rules of the practice, not of their conviction as to the literal truth of a mathematical claim in dispute. Conspicuously, an expert may have an 'intuition' as to how to solve a problem. By contrast, intuitions in the sense of impressions as to the literal truth of (non-logical) mathematical statements (e.g., *PCC*, mentioned in the next paragraph) are the ones that are relevant to the comparison with empirical science. (By 'non-logical', we mean not intuitions about what follows from what. We return to logical intuitions in Sect. 11.)

We can illustrate the method with reference to *AC*. *AC* says that every set has a Choice function. It implies a range of plausible propositions. For example, *AC* implies that the union of countably many countable sets is countable and that either any two sets have the same cardinality or one has greater cardinality—that is, the *Principle of Cardinal Comparability* (*PCC*). But it also implies what are widely agreed to be counterintuitive results, like the *Banach–Tarski Paradox*, and that the set of real numbers is well-orderable (it is relatively consistent even with *ZFC* that there is no *definable* well-order on the reals). Perhaps these results show that we ought to reject *AC*. In that case, we might adopt a weakening of it, like *Dependent* or *Countable*, Choice, or (given the consistency of sufficiently large cardinals) the *Axiom of Determinacy* (which happens to imply Countable Choice). But maybe *AC*'s implications instead show that our intuitions about, say, mathematical space, need to be revised. What exactly they do show depends on which alternative would facilitate equilibrium among our beliefs.[16]

3 Justification and Reliability

Russell's epistemology has considerable appeal. It has been widely adopted by those searching for new axioms to settle undecidables, like the Continuum Hypothesis (Woodin, 2010, 1; Kennedy & van Atten, 2004, 2008; Koellner, 2006; Maddy, 1988b; Martin, 1998; Steel, 2014). How do we know the axioms of mathematics? In the way that we know the laws of nature! The difference is that in science, laws are postulated to systematize observations, while in mathematics, laws (axioms) are postulated to systematize intuitions. But observation and intuition play the same epistemic role. They both present the world to us as being a certain way. They are the data that we try to 'save'.[17]

Unfortunately, there is a complication. At most, the *justification* of our mathematical beliefs is analogous to the justification of our empirical scientific

[16]What precisely constitutes equilibrium is a notorious problem (Quine & Ullian, 1978; Daniels, 2020). In particular, we do not assume that harmony is ever achieved. Maybe it is a regulative ideal. This will not be important.

[17]Compare Gödel: 'I don't see why we should have less confidence in this kind of perception, i.e., in mathematical intuition, than in sense perception, which induces us to build up physical theories and to expect that future sense perceptions will agree with them . . . ' (1973/1907, 273–274). (We of course use 'data' in its Latin sense, not in a statistical one. We restrict ourselves to fundamental, or basic, givens, as is standard in these discussions. In keeping with the analogy to observation, the data in the mathematical case concern the likes of numbers and sets, not *our experiences as of* numbers and sets. The data in the scientific case (with the exception of the psychological sciences) concern pointers, planets, and particles, not our *experiences as of* pointers, planets, and particles.)

beliefs. It does not follow that the *reliability* of our mathematical beliefs parallels that of our scientific ones. Benacerraf notes,

> without an account of *how* [mathematical propositions] 'force themselves upon us as being true,' the analogy with sense perception and physical science is without much content....[T]here is a *superficial* analogy....[W]e 'verify' axioms by deducing consequences from them concerning areas in which we seem to have more direct 'perception' (clearer intuitions). But we are never told how we know even these, clearer, propositions. (1973, 674, italics in original)

To see the problem, consider again how we know that *here is a piece of paper*. Knowledge is justified and reliable belief (where specifying the relevant sense of 'reliable' is the Gettier Problem). We can explain the justification of our belief by citing its perceptual evidentness. It seems to us that *here is a piece of paper*, and we have no reason to doubt appearances. But what about its reliability (Field, 1989, Introduction)? Although we are far from having a complete theory, an evolutionary account of how we came to have a reliable mechanism for perceptual belief, and a neurophysical account of how that mechanism works such that it is reliable, suggests itself (Schechter, 2010). This contrasts with the mathematical case. We can explain the justification of our belief that, say, $1 + 1 = 2$ by citing its intuitive evidentness. But what explains the *reliability* of this belief? '$1 + 1 = 2$' is about *numbers*. It says that the plus function maps the number *1* and itself onto *2*. How could that have any bearing on our ancestors' reproductive fitness? More relevant are *(first-order) logical truths* about things in our ancestors' environments—like that if there is 'exactly one' lion to the left, and 'exactly one' lion to the right, and no lion to the left is to the right, then there are 'exactly two' lions to the left or right (where the phrases 'exactly one' and 'exactly two' are abbreviations for constructions out of ordinary quantifiers and the identity sign and do not refer to numbers) (Clarke-Doane, 2012).

Indeed, even if there were a theory of how we came to have a reliable mechanism for mathematical belief, it would remain to explain how that mechanism works such that it is reliable. We have no established psychophysics of number perception—not to be confused with the perception of cardinalities of (very small) collections, as with the Approximate Number Sense (*ANS*). Numbers are not like hands, deflecting photons that stimulate our optic nerves![18]

[18] Presumably we know this much, despite not even knowing what numbers would be, or whether the question of what they would be is somehow ill-formed (Benacerraf, 1965).

The upshot is that while the *justification* of our mathematical beliefs may be explicable in analogy with the justification of our scientific beliefs, the reliability of those beliefs is not.

4 Science and Mathematics

What makes science work? Part of the story is that it relies on observation. Observation is shared. We disagree about evolution, but we all agree on the observations that are supposed to support it. Well, not exactly. For one thing, 'all' is too strong. Observations can be impaired. Not *everyone* perceives what we do. Also, observation is theory-laden. In claiming to observe a fossil, we betray commitment to a theory according to which the world is old enough for the object in question to have fossilized. This is not common ground in all contexts. Nor could we avoid every such difficulty by appealing to more primitive kinds of judgment. Even claims about what we *seem* to see presuppose a theory to which some object (Churchland, 2014, Sec. 2.5).

Nevertheless, there is characteristic conformity about the data that our scientific theories aim to systematize. It is remarkable that a believer in dark matter can agree in the ways that matter for experiment with supporters of *MOND* (Milgrom, 2002) even though the former believes that a quarter of the universe is made of a substance whose existence the latter denies. The dispute between advocates and detractors of dark matter is about what best accounts for our observations, not over our observations themselves. While observation is theory-laden, it is, on the whole, shared.

If mathematics is like science, at least *at the level of justification* (even if not at the level of reliability), then intuition should play an analogous role. It should be the common ground, for all practical purposes, against which theorists with very different conceptions of mathematical reality can test those conceptions.

Prima facie, it does just that. Mathematics constantly surprises, sometimes undercutting hopes or expectations in the process. It does not submit to our wishes like a lump of clay. But the picture is complicated on inspection. It is clear that there are rules to which the overwhelming majority of mathematicians implicitly agree, whose violation is a matter on which they also generally concur—independent of any position that they may harbor on the nature of mathematical reality.[19] Ultimately, those rules answer to a formal system like

[19]The current case of Mochizuki's claimed proof of the *ABC* conjecture may be an exception. Kiran Kedlaya writes that 'it is potentially *unknowable* whether Mochizuki's proof is correct . . . .[I]n principle one could try to translate the proof into a formalized system (e.g. *Lean*) and then verify mechanically the

(first-order) *ZFC*, as per Sect. 1. We say 'ultimately' because there is a division of cognitive labor. Algebraic geometers can usually rely on others in their field to check the validity of a proof, disregarding questions of logic and set theory. But if questions are raised about background set-theoretic (or category-theoretic) assumptions, then outside experts may be consulted. In principle, the matter may face the tribunal of logicians who determine whether the theorem can be proved in *ZFC*.[20]

The agreement in practice among mathematicians is almost unique in the academy. But agreement in practice does not amount to agreement in theory unless there is also agreement on axioms (and inference rules). There is agreement in practice in chess (or baseball, or bridge, etc.) too. But there is no corresponding theory about the world to which all parties would subscribe.[21] Generally speaking, mathematicians follow the rules that they have been taught without seriously inquiring as to their correctness (or into what it might mean to say that an axiom or inference rule is correct) (Martin, 1998, 219). This is no indictment of mathematicians, contra Frege (1980/1884, 11). Arguably, mathematicians bracket foundational questions *in order to* prove theorems. If they had to first settle the question of what axioms (and inference rules) are correct, then they might never get started (Easwaran, 2008).

However, for purposes of arguing that there is an epistemic parity between intuition and observation, it matters what theorists believe, not what they do. Since attempting to determine what axioms are true (or what logical rules are necessarily truth-preserving) is not a normal mathematical activity, we must turn to the (small!) minority of theorists who concern themselves with foundational matters. Do intuitions play the role in the foundations of mathematics that observations play in fundamental physics? Are they common ground against which disparate theories are tested?

correctness of the proof. However, if one does end up confirming the existence of a correct proof, this still leaves the question of authorial intent: does the formal proof actually correspond to what Mochizuki had in mind? . . . .I wouldn't be surprised if we ended up with a proof that meets community standards but requires substantial work beyond what is in his papers . . . In that case, whose proof is it anyway? . . . And is the original attempt correct, incomplete, or something else?' (personal correspondence, quoted with permission).

[20]We do not suggest that this is a common occurrence!

[21]One might think that all parties would agree on how you are *permitted* to move chess pieces, and what moves are *possible*, in various circumstances. But any such theory (construed literally) would involve commitment to evaluative and modal facts, which we will see in Sects. 10 and 11, are at least as controversial as mathematical ones.

5 The Problem of Conflicting Intuitions

In a word: no. Among those (very few) theorists who concern themselves with foundational questions, there is infamous disagreement. Fraenkel, Bar-Hillel, and Levy write that there is 'far-going and surprising divergence of opinions and conceptions of the most fundamental mathematical notions, such as set and number' (1973, 14). And Bell and Hellman remark,

> Contrary to the popular (mis)conception of mathematics as a cut-and-dried body of universally agreed upon truths . . . as soon as one examines the foundations of mathematics one encounters divergences of viewpoint . . . that can easily remind one of religious, schismatic controversy. (2006, 64)

It is conceivable that all such disagreement concerns the best systematization of our shared mathematical intuitions, not the contents of those intuitions themselves. But a cursory glance at the literature tends to undercut this hypothesis. The following quotations are representative:

> There is almost a continuum of beliefs about the . . . world of set theory. (Cohen, 2005)
>
> Let me try to be as accurate, explicit, and forthright about my belief about the existence of κ [= the least ordinal greater than all $f(i)$, where $f(0) = \aleph_0$ and $f(i+1) = \aleph_{f(i)}$] as I can . . . I . . . think it probably doesn't exist . . . .I am also doubtful that anything could be provided that should be called a *reason* and that would settle the question. (Boolos, 1999, 121, italics in original)
>
> Without any doubt the most problematic axiom of set theory is the axiom of choice . . . .The current situation with *AC* is that the contestants have agreed to differ. (Forster, forthcoming, 72)[22]
>
> The set-theoretical axioms that sustain modern mathematics are self-evident in differing degrees. One of them—indeed, the most important of them, namely . . . the so-called axiom of infinity—has scarcely any claim to self-evidence at all. (Mayberry, 2000, 10)
>
> *ZF* does not embody a philosophically coherent notion of set. There is a coherent constructivist position . . . .There is also a coherent anti-constructivist position . . . .But *ZF* is an uneasy compromise between these two: it pays lip-

[22]Although Forster is a logician, it should not be thought that his assessment only reflects that of logicians with interest in foundations. For instance, Villani writes, 'The axiom of choice . . . is absolutely not constructive; youngsters like me want to get rid of it . . . ' (Cartier et al., 2016, 67), to which Cartier responds, 'In my generation too, the axiom of choice was far from being uncontroversial, van der Waerden was reluctant to use it in his book . . . and . . . [a]s for Deligne, he clearly admitted that he uses the axiom of choice out of convenience, but that he does not believe in it' (*ibid.*, 67).

> service to constructivism without . . . meaning it . . . . Only a non-well-founded [set] theory can . . . be shown to modify the naive conception as much as, but no more than, is required . . . . (Rieger, 2011, 17–18)[23]
>
> Set theorists say that $V = L$ has implausible consequences . . . . [They] claim to have a direct intuition which allows them to view these as so implausible that this provides 'evidence' against $V = L$. However, mathematicians [like me] disclaim such direct intuition about complicated sets of reals. Many say they have no direct intuition about all multivariate functions from N into N! (Friedman FOM, 5.25.00)[24]
>
> I have had numerous opportunities to discuss the alternatives in set theory with mathematicians from other disciplines. I am always surprised at how rapidly and with what certainty they express their opinion. Their opinions are divided and reflect, in my view, a pre-existing orientation. (Jensen, 1995, 401–2, n. 18)

Of course, the mere existence of disputes like those above does not *imply* that theorists have conflicting intuitions (Stratton-Lake, 2020). But the face-value interpretation of the debates is that opposing parties do often have different intuitions. To borrow Jensen's language, the disagreements are reflections of 'deeply rooted differences in mathematical taste' (1995, 401).[25]

6 Theory Ladenness

Perhaps, however, while disagreement in mathematics does turn on conflicting intuitions, controversial mathematical intuitions are analogous to controversial observations, not to *here is a piece of paper*. They are theoretical. Disagreement over abstruse matters of set theory, like whether there are more subsets than L allows, is like disagreement over the extent to which a borderline biopsy sample harbors dysplasia. The latter is an *observational* matter over which there is controversy.

---

[23]Compare (Potter, 2004, p. 36): '[I]n an attempt to make the history of the subject read more like an inevitable convergence on the one true religion, some authors have tried to find evidence of the iterative conception quite far back in the history of the subject'.

[24]Similarly, Devlin thinks that $V = L$ 'is . . . a natural axiom, closely bound up with what we mean by 'set' . . . . [and] tends to decide problems in the 'correct' direction' (1977, 4). On the virtues of $V = L$, see especially Jensen (1995).

[25]Compare Mogensen discussing moral disagreement: 'Diagnosing a clash of intuitions . . . will typically involve attempting a careful hermeneutic reconstruction of the underlying dialectic, designed to reveal that the dispute rests ultimately with certain . . . premises that one side finds intuitive and the other does not. Any such reconstruction is bound to be controversial . . . . [W]hereas many philosophers agree that some questions boil down to . . . differences in intuition, there is considerable disagreement as to exactly which questions those are' (2016b, 24).

The problem with this response is that, in the empirical case, we can contrast observational judgments of, e.g., the degree of dysplasia with 'practically bedrock' observational judgments about the overt appearance of the biopsy sample ('. . . not very circular'). The contrast is vague (Hanson, 1958, Ch. 1). There is no bright line between observations that serve as a common arbiter of scientific dispute and those that fail to. However, in the mathematical case, not even a vague distinction suggests itself. No claims appear to be off limits from serious disputes. Consider:

> It will be recognized . . . that in any wording [the Least Upper Bound Axiom] is false. (Weyl as quoted in Kilmister, 1980, 157)[26]
>
> The reason for mistrusting the induction principle [of arithmetic] is that it involves an impredicative concept of number . . . .A number is conceived to be an object satisfying every inductive formula. (Nelson, 1986, 1)
>
> I am a platonist . . . [but] I deny even the . . . Peano axiom that every integer has a successor . . . . (Zeilberger, 2004, 32–3)
>
> This standard concept of even the potential infinite is no less dubious than our standard concept of the actual infinite . . . . (Benardete, 1964)
>
> I have seen some . . . go so far as to challenge the existence of 2^{100} . . . . (Friedman, 2002, 4)
>
> There are no such things as mathematical objects. And so . . . .[even s]entences like '4 is even' are not true. (Balaguer, 2018)

Of course, remarks like the above often reflect intuitively philosophical commitments. Weyl and Nelson, who evidently lack an intuition in support of the *Least Upper Bound* and (unrestricted) *Induction*, axioms, respectively, subscribe to more or less radical versions of predicativism, the philosophical doctrine that it is not coherent to define an object in terms of a totality to which it belongs.[27] It might be objected that the unsullied among us share the intuition that every non-empty set of real numbers with an upper bound has a least upper bound and that for *any* predicate, *F*, in the language of arithmetic,

---

[26]Note that Weyl accepts classical logic here. So, his objection is different from that of the intuitionists (a group Weyl later joined). See Weyl (1918/1994). Indeed, none of the quotations to follow is from a theorist who explicitly rejects classical logic. But some do seem to be committed to non-classical logic. For instance, the *Successor Axiom*, rejected by Zeilberger (below) and sometimes, it seems, by Nelson (1986, 176, 2013), is a (classical) logical truth in a language with the successor sign. By the axioms of identity, $Sx = Sx$. By *Existential Introduction*, $\exists y(y = Sx)$. And by *Universal Generalization*, $\forall x \exists y(y = Sx)$. So, Zeilberger, at least, would seem to be committed to a free logic.

[27]Predicativism was introduced by Russell (1906) and Poincaré (1906), although not in the radical form advocated by Nelson.

if *0* satisfies F and $n+1$ satisfies it whenever n does, then all natural numbers satisfy F. Only those corrupted by philosophy lack this good sense.

There are three problems with this objection, however. First, who knows what intuitions we would have if all philosophical considerations were bracketed? How could one even form a view as to whether, for example, *AC* is true bracketing the philosophical question of whether sets depend on us for their construction? (If they do, how could one hold that there is a well-order on the reals—assuming that we cannot construct the *Ord* of *V*, and, hence, *L*?) Second, insofar as we can make sense of such questions, mathematical disputes often appear to turn on untutored intuition, not philosophy. Consider the intuitions that are traded in debates over the Axiom of Foundation, for example. The key question is whether there are sets that contain themselves or sets with infinitely descending chains of membership. The standard view is that such sets are 'pathological' (Boolos, 1971, 491; Maddy, 1988a). But, as Rieger (above) remarks, others have a very different reaction. Again, the dispute over Foundation appears to be more indicative of 'deeply rooted differences in mathematical taste' than philosophical priming (1995, 401).

Finally, it is irrelevant to the contrast whether intuition explains theory, or theory explains intuition. Either way, intuition varies in a way that observation does not. *If we could check philosophical theories independently of our intuitions*, then this difference might have epistemic import. But, as critics of reflective equilibrium have long stressed, we do not seem to be able to.

7 Heretics

Maybe an appeal to poll numbers would reestablish a parity between observation and intuition. There are always heretics, including geniuses. Perhaps mathematicians with apostate intuitions are in the overwhelming *minority*. And maybe *this sociological fact* affords (defeasible) evidence that the heretics are 'hallucinating'. For instance, Koellner writes of the proposed axiom, *Projective Determinacy* (*PD*), which is inconsistent with the *Axiom of Constructibility*, $V=L$, that '[*PD*] has gained wide acceptance by the set-theorists . . . who know the details of the constructions and theorems involved in the case that has been made for *PD*', apparently citing this as evidence for *PD* and against $V = L$ (2013, 21–22).[28]

[28]The sense in which $V = L$ says that all sets are 'constructible' is *not* the sense in which constructivist mathematicians (i.e., intuitionists) like Brouwer, say this (to say nothing of 'ultra-intuitionists' like Esenin-Volpin). L has the same ordinals as V and takes them as given prior to the 'construction' process.

However, this rejoinder does not stand up to scrutiny. First, it is not clear that there are *any* prominent scientists with heretical observations. One can be a fine mathematician—as, e.g., Nelson was—while having wildly heterodox 'tastes' (in Jensen's sense). But it is hard to see how one could make major contributions to any empirical science while doubting the observational analog of the totality of exponentiation (rejected by Nelson). Of course, there are scientists who are colorblind or otherwise impaired. But there is an *independent* account of what is inaccurate about their observations, as evinced by the fact that they are convinced of their observational limitations. A colorblind scientist believes that she is colorblind, not that the rest of the scientific community hallucinates wavelengths! By contrast, a prevalent and exasperating feature of clashes in intuition is that, as Forster writes, 'the contestants have agreed to differ'.

Second, the *relevant* group to poll would presumably consist of theorists who actually work on the disputed axioms and intimately related topics—not merely those 'who know the details of the constructions and theorems involved in the case' for the standard axioms.[29] However, as Russell underscores, specialist knowledge tends to turn 'something so simple as not to seem worth stating' into something that admits of reasonable debate (1918, 514). Forster quips,

> [F]or people who want to think of foundational issues as resolved . . . [standard axioms provide] an excuse for them not to think about [them] any longer. It's a bit like the role of the Church in Medieval Europe: it keeps a lid on things that really need lids! (forthcoming, 15)

In other words, *among the relevant sample group*, skeptics about Choice are unlikely to be in the overwhelming minority, although they may well be among mathematicians more generally—simply because most of the ones who understand *AC* and its implications uncritically adopt it.

The final problem with appeal to poll numbers is that it is unclear why they would matter. Arguably, poll numbers matter in the empirical case because we should expect ourselves to be reliable detectors of facts about our surroundings. Insofar as empirical science is an extrapolation from everyday environmental inquiry, we should predict that scientists would converge toward the truth

[29] Devlin draws a similar distinction when he writes: 'Currently I tend to favour $V = L$ . . . .At the moment I think I am in the majority of informed mathematicians, but the minority of set theorists . . . ' (1981, 205). Chow similarly remarks, 'You can stick to your guns, seek out like-minded people, develop theories, and evangelize. What I don't think you can do is to judge your efforts by how popular they seem to be' (FOM 7.25.2023).

(and insofar as science floats free of observational check, as with *GUT*s, or *TOE*s, we should be unsurprised by intractable disagreement). Of course, this prediction is premised on a scientific theory—evolutionary theory. It has no force in connection with disagreement over evolution itself. But *assuming* the truth of evolutionary theory, we have the following contrast. There is no reason to think that true mathematical intuitions—as opposed to true scientific observations—would be popular *independent of their contents*. (*Given* the truth of *AC*, say, and an argument that it was likely for us to *believe AC*, we of course have reason to think that the intuitions that we do have, which happen to be true, would be popular. But that is not reason to believe that *true intuitions per se* would be (Clarke-Doane, 2020, 5.5)). The mathematical intuitions that we have often appear to owe themselves to where we went to graduate school, who we studied with, and other factors that are irrelevant to the mathematical truth.[30] Martin remarks, 'For individual mathematicians, acceptance of an axiom is probably often the result of nothing more than knowing that it is a standard axiom' (1998, 218). And Cohen laments, '[T]he attitudes that people profess towards the foundations [the question of what axioms are true] seem to be greatly influenced by their training and their environment' (1971, 10).

8 A Misleading Contrast?

It might be thought that the apparent contrast between intuition and observation owes itself to a sleight of hand. In the mathematical case, we cited foundational theorists. But in the scientific case, we referred to the likes of (ordinary) physicists. Arguably, the latter take a similar attitude toward the Standard Model of particle physics, General Relativity, or inflationary cosmology, as ordinary mathematicians take toward *AC*. If what matters is belief, not practice, then we should compare foundational theorists in mathematics to *foundational theorists* in science. However, controversy in the foundations of science—where questions of literal truth and falsity are at issue—seems to be as prevalent as disputation in the foundations of mathematics. Carroll recalls,

> At a workshop attended by expert researchers in quantum mechanics....Max Tegmark took an ... unscientific poll of the participants' favored interpretation of quantum mechanics....The Copenhagen interpretation came in first with

[30] Indeed, Koellner, quoted above, studied with the most prominent advocate of *PD*, Hugh Woodin. (Would he have the same intuitions concerning 'simple' sets of reals had he studied with Jensen?)

thirteen votes, while the many-worlds interpretation came in second with eight. Another nine votes were scattered among other alternatives. Most interesting, eighteen votes were cast for 'none of the above/undecided.' And these are the experts. (2010, 402, n. 199)

Disputes about the interpretation of quantum mechanics are not about the meanings of words—notwithstanding the term 'interpretation'. They are about the reality (or lack thereof) behind the Schrodinger Equation (or its relativistic surrogate) and the Born Rule. Is there really a state vector? Does it collapse? Does consciousness have anything to do with this? Attempts to answer these questions can make different experimental predictions. For instance, the Ghirardi–Rimini–Weber (*GRW*) 'interpretation' predicts collapse in circumstances that the Copenhagen Interpretation does not. So, disagreement about the interpretation of quantum mechanics is disagreement about the non-semantic world, like disagreement in the foundations of mathematics.

But while the foundations of physics also exhibit 'divergences of viewpoint . . . that can easily remind one of religious, schismatic controversy', this does not conflict with the thesis that intuition varies in a way that observation does not.[31] Disputes over the interpretation of quantum mechanics are still over how best to systematize the experimental record, not the record itself. In this way, they are more like the dispute over dark matter and unlike debates over the axioms of mathematics. For example, all interpretations agree that the Bell Inequalities have been violated. But the Everett interpretation explains this by denying that experiments have unique outcomes, the Bohmian and *GRW* interpretation do this by postulating action-at-a-distance, and superdeterminists allow that the choice of measurement is correlated with the measured system.

Is there any other way to argue that the data that is systemized in mathematics is as uniform as the data that is systematized in science? Only, it seems, if one could argue that mathematics answers to observations to the exclusion of intuitions. For instance, if the data of mathematics were limited to the observed outputs of computers, or observations of collections of things (as per Mill [1882/2009]), then mathematics would be on a par with empirical science at the level of justification. It would *be* an empirical science at the

[31] Rickles writes, 'There exist very many ways to 'make sense' of quantum mechanics: Copenhagen, modal, relative-state, many-worlds, many-minds, Bayesian, Bohmian, Qbist, spontaneous collapse, etc. However, what one person considers to be a perfectly rational account, another might consider to be outright lunacy. There is an almost religious fervour concerning the holding of a particular stance on quantum mechanics: 'the church of Everett' versus 'the church of Bohm" (2016, 132).

level of justification. Actually, Quine points out a more promising way that mathematics could be an empirical science. He writes,

> Objects at the atomic level and beyond are posited to make the laws of macroscopic objects, and ultimately the laws of experience, simpler . . . .Moreover, the abstract entities which are the substance of mathematics . . . are another posit in the same spirit. Epistemologically these are myths on the same footing with physical objects . . . neither better nor worse except for differences in the degree to which they expedite our dealings with sense experiences. (1951, 42)

In one reading,[32] Quine is explaining how our belief in typical theorems of mathematics, like *Euclid's Theorem*, could be empirically justified, even though there is no useful sense in which we observe infinitely many prime numbers. Our belief in that theorem may be like our belief in gluon confinement—a theoretical postulate 'to be vindicated . . . by the indirect systematic contribution which [it] make[s] to the organizing of empirical data in the natural sciences' (1958, 4). But not even Quine suggests that *all* of mathematics can be so vindicated. He writes,

> I recognize indenumerable infinities only because they are forced on me by the simplest known systematizations of more welcome matters. Magnitudes in excess of such demands, e.g., $\beth_\omega$ or inaccessible numbers, I look upon only as mathematical recreation and without ontological rights. (1986, 400)[33]

The existence of '[m]agnitudes in excess of such demands, e.g., $\beth_\omega$' is implied by *ZFC*. So, Quine is rejecting standard axioms of mathematics (presumably, the *Axiom of Replacement*).

In sum, even if a fragment of mathematics is justified by observation, the rest of mathematics must be justified differently. If that justification derives

[32] Putnam writes that, as he prefers to formulate it, the 'argument was never intended to be an 'epistemology of mathematics.' If anything, it is a *constraint* on epistemologies of mathematics from a scientific realist standpoint' (2015, 63, emphasis in original). In other words, the indispensability of at least some mathematics to science shows that one cannot be a realist about science without being a realist about some mathematics. It does not explain the justification of our belief in mathematics. We will invoke a weakening of this constraint in Sect. 11.

[33] Many hold that Quine was too sanguine about how much mathematics is indispensable to empirical science. Feferman argues that Weyl's system, *W*, exhausts what is indispensable for empirical science and that '*W* [itself can be] treated in an instrumental way, its entities outside the natural numbers are regarded as 'theoretical' entities, and the justification for its use lies in whatever justification we give to the use of *PA*' (1992, 451). Field goes so far as to maintain that we can avoid 'all appeal to mathematical entities in explanations when the chips are down: it must be possible, for instance, to develop theoretical physics without any appeal to mathematical entities' (1989, 6).

from intuition, as we have assumed, then the data supporting scientific laws is uniform in a way that the data supporting mathematical axioms is not.

9 Mathematics as Philosophy

We have been discussing analogies between mathematical and scientific knowledge. We have supposed that the justification of mathematical axioms is similar to that of scientific laws. Following Russell, axioms are postulated to systematize mathematical intuitions, as laws are postulated to systematize observations. But the analogy misleads in a way that has not been sufficiently appreciated. Mathematical disagreement characteristically bottoms out in disputes over the data to be systematized, while scientific disagreement tends not to. The view that 'mathematics begins with a small number of shared, self-evident assumptions' (Greene, 2013, 401) is indeed a cartoon. But mathematics is not just like science even at the level of justification.

Where does that leave mathematics in the science-humanities continuum? *Insofar as mathematics is theoretical*—making claims on the world, like the sciences—it leaves mathematics on a par with philosophy. This includes modality (the theory of possibility and necessity), (meta)logic (the theory of what follows from what), and value (the theory of what is morally, intellectually, or otherwise good or obligatory). It is a familiar point that, with respect to *reliability*, our philosophical and mathematical beliefs may be alike. Bealer writes,

> The lack of... an explanation [of our reliability] in the case of [philosophical] intuitions makes a number of people worry about relying on intuitions. (This really is just Benacerraf's worry [quoted above] about mathematical knowledge.) (1999, 52, n. 22)

The problem of explaining the reliability of our philosophical intuitions closely resembles the problem of explaining the reliability of our mathematical intuitions. Neither an evolutionary account of how we came to have a reliable mechanism for philosophical belief, nor a neurophysical explanation of how that mechanism works such that it is reliable, is forthcoming. So, if our philosophical beliefs are reliable, this fact also requires a different kind of explanation.

What has been less widely recognized is that the *justification* of our mathematical beliefs seems to be the same as the justification of our philosophical beliefs. In both cases, we are left to apply the method of reflective equilibrium to 'data points' *which are not generally accepted.* For any two mathematicians with positions on foundational questions, the equilibria that they approach will commonly seem mutually inconsistent, as in philosophy.[34] Lewis writes,

> Our 'intuitions' are simply opinions: our philosophical theories are the same. Some are commonsensical, some are sophisticated; some are particular; some general; some are more firmly held, some less . . . .[A] reasonable goal for a philosopher is to bring them into equilibrium. Our common task is to find out what equilibria there are that can withstand examination, but it remains for each of us to come to rest at one or another of them . . . . (1983: x–xi)

It is easy to miss this similarity between mathematics and philosophy on account of their divergent aims. Mathematicians want standards by which their work can be judged. The reality (or lack thereof) behind those standards is secondary, even personal. By contrast, philosophers want to know the reality behind the standards that are regulative in an area. While an oversimplification, Kreisel is not far off when he suggests that philosophy seeks 'reasons *for* axioms' while mathematical 'practice is concerned with deductions *from* axioms' (1967, 191, italics in original).[35]

We submit that this similarity between mathematics and philosophy may be characteristic of apparently a priori, or what, following Williamson (2005), we will call 'armchair', areas.[36] *Armchair disagreement tends to bottom out in disputes about the data to be systematized, while empirical scientific disagreements tend not to*. This difference illuminates why disagreements over what axioms to accept, how the world could have been, what follows from what, and what we ought to do are intractable in a way that disagreements over whether there are gravitons seem not to be. We agree on the data that is pertinent to the latter question to an extent that we do not agree on the data that is pertinent to the former.[37]

---

[34]'Seem' for reasons that will emerge in Sect. 11.

[35]One reason that this is an oversimplification is that, ideally, at least, 'Self-criticism feeds back into mathematics and transforms it' (Bevan & Paseau, 2020, 446).

[36]The term 'armchair' is neutral on whether the relevant areas are a priori or whether there is even a useful a priori/a posteriori distinction. As Quine (1951) argued, mathematics, logic, and fundamental ontology may merely be especially abstract branches of the empirical scientific project whose results we are particularly reluctant to amend.

[37]Insofar disagreement over the existence of gravitons *does* appear intractable, this is arguably because the parties to it disagree over *armchair* questions like the extent to which we are justified in believing in whatever entities are referred to in the best explanation of our observations, i.e., 'inference to the best explanation' (Harman, 1965).

10 Illustration: Modal Metaphysics

Consider modal metaphysics, the theory of how the world could have been. 'Applied' modal claims certainly turn in part on observation, like applied mathematical ones. For instance, assuming the *Necessity of Identity* (*NI*), $\forall x \forall y[x = y \rightarrow \Box(x = y)]$, if the mind (consciousness) is identical to the body (a physical property), then it necessarily is. The actual identity of a physical property, like *being a synchronized 35–75 hertz neural oscillation in the sensory areas of the cortex*, is an empirical matter. But basic modal principles, like *NI* itself, do not seem to admit to observational check. Disputes over them appear to bottom out in clashing intuitions, like debates over *AC* or the Axiom of Foundation. The following quotations are illustrative.

> The necessity of identity follows if we have (i) in every world 'a' and 'b' refer to what they actually refer to *and* (ii) in every world 'a' and 'b' co-refer. But we have no right to the latter claim . . . .It follows from . . . (i) and the necessity of identity; but of course one can't rely on that without begging the question. (Cameron, 2006)[38]
>
> Lewis thinks this 'unrestricted composition' . . . is . . . necessary . . . .[But w]hy suppose that . . . it is impossible for the world to have different principles governing the part-whole relation? (Nolan, 2005, 36)
>
> [T]he metaphysically possible . . . is constrained by the laws of nature. (Edgington, 2004, 1)
>
> [C]ould this table have been made from a completely different block of wood? . . . . [T]hough we can imagine making a table out of another block of wood . . . identical in appearance with this one, and . . . in this very position in the room, it seems to me that this is not to imagine this table as made of wood . . . but rather it is to imagine another table, resembling this one . . . made of another block of wood . . . . (Kripke, 1980, 113–114)
>
> In order to provide a decisive argument for absolute motion using [Newton's] bucket (or globes) [thought experiment] we would have to empty the universe of all other matter. Though we might think we can do this in thought experiments . . . our intuitions might differ [and] how are we to decide between them in such cases? (Rickles, 2016, 46)
>
> [T]he criteria to which one appeals in . . . reflection [on the essence of things] are sufficiently obscure to leave me, at least, with an embarrassingly large number of undecided cases . . . .The existence of such cases does show the distinction

[38] For defenses of modal logics with contingent identity, see Wilson (1983), Girle (2017, 7.4, 8.5, and 8.6), Gibbard (1975), and Priest (2008, Ch. 17).

> [between essential and accidental attributes] is a good deal less clear than essentialists are wont to suppose. (Cartwright, 1968, 626)

Cartwright's lament resembles Friedman's remark that 'mathematicians [like me] disclaim . . . direct intuition about complicated sets of reals'. Different theorists have different intuitions, and some lack relevant intuitions altogether. But intuitions are the data to which modal theories, like set theories, respond. How are we to theorize absent agreement on the data to be 'saved'?

Modal metaphysicians could follow the lead of (ordinary) mathematicians and adopt some generalizations (and inference rules), thereafter bracketing arguments against them (leaving those to the 'shphilosophers').[39] This would generate the appearance of metaphysical progress, as more results were 'proved'. But metaphysicians could only do this in bad faith. The sense of proof in question would be logical, not epistemic. What the 'proofs' showed about, e.g. whether the mind is identical to the body, would remain controversial among those who concerned themselves with questions of literal truth. And, yet, as Williamson emphasizes, 'our task', in modal metaphysics, is precisely 'in a scientific spirit to build and test theories that codify putatively true generalizations of the sort at issue, to find out which are true' (2013, 423).

11 Disagreement, Objectivity, and Practice

Disagreement in intuition has long motivated skepticism about philosophical questions. Mogensen writes,

> Moral disagreement is widespread. Some disagreements are shallow, but others seem deep: they apparently cannot be traced to ignorance of relevant non-moral facts or failures of epistemic rationality, reflecting instead fundamental differences in people's moral intuitions . . . . [W]e ought to suspend judgment when we encounter disagreements of this kind. (2016a, 1)

In light of the above, one might be tempted to argue from disagreement to skepticism about mathematics, modality, logic, and 'armchair' questions more generally. Even in the case of logic, Williamson reminds us, 'If we restricted it to uncontroversial principles, nothing would be left' (2012). For example, paracomplete logicians reject the *Law of the Excluded Middle* (*LEM*), which says that, for any *P*, (*P* v ~*P*) is a logical truth. And paraconsistent logicians

[39] It is arguable that this is what certain segments of the metaphysical community have actually done.

allow that contradictions—claims of the form (P & $\sim P$)—can, as a matter of logical possibility, be true. Unfortunately, armchair skepticism cannot be contained. It engenders empirical scientific skepticism. There is no way to be agnostic about mathematics, possibility, logic, *and* every kind of value while maintaining any belief of interest about the empirical world (Nagel, 1997). Even fictionalists, like Field (1980), require knowledge of possibility and (meta)logic in order to make the case that mathematics is conservative over 'nominalistic' theories or that—at the appropriate energies—the world is just like the Standard Model says it would be if (ignoring gravity and dark matter and energy) there were mathematical entities (Rosen, 2001, 75).

There is an alternative to skepticism about armchair areas, however. Instead of amending our epistemology, we could amend our metaphysics. Mackie writes,

> Disagreement on questions of history, biology, or cosmology does not show that there are no objective issues for investigators in these fields to disagree about. But such scientific disagreement results from speculative inferences or explanatory hypotheses based on inadequate evidence, and it is hardly plausible to interpret moral disagreements in the same way. (1977, 37)[40]

By 'objective' Mackie means mind-and-language independent. Again, we cannot give up on the mind-and-language independent truth of mathematics, logic, and so forth without giving up on the mind-and-language independent truth of the scientific theories that presuppose them.[41] However, there is another sense of 'objective' in which we might give up on the objectivity of armchair facts, including (pure) mathematical ones, while retaining belief in their mind-and-language independence—i.e., while retaining commitment to armchair *realism*.

Consider the ordinary Platonist's conception of pure (rather than applied) geometry. Platonists do not believe that the truth of the Parallel Postulate depends on our minds or languages. That postulate would have been true (of Euclidean space) even if no one had existed. But it is *as if* its truth depended on our conventions. Where a relativist claims that the Parallel Postulate is true according to the framework of Euclidean geometry and false according to the framework of, say, hyperbolic geometry, a Platonist claims that the Parallel

[40]See Leiter (2009) for a related argument.

[41]Of course, some do give up on this. Dummett (1991) argues from intuitionistic logic to a kind of physical antirealism. But we assume realism—to be shortly distinguished from *objectivism*—about the physical world throughout this chapter.

Postulate (syntactically individuated) is true of the $lines_{Euclidean}$ and false of the $lines_{hyperbolic}$. Rather than switching frameworks on which to relativize, the Platonist switches kindred subjects. In both cases, the methodological consequences are the same: 'the conflict between the divergent points of view... disappears ....[B]efore us lies the boundless ocean of unlimited possibilities' (Carnap 1937/2001, XV). There is enough *mind-and-language independent reality* to go around.

This suggests the following proposal. Maybe *armchair questions quite generally are like the Parallel Postulate question, according to the mathematical Platonist.*[42] Does every set have a Choice function? All $sets_{ZFC}$ do, but some $sets_{ZF+AD}$ do not. Could Kripke's table have been made of a different piece of wood? This is $possible_{Logical}$, but not $possible_{Metaphysical}$. Does 'the moon is made of green cheese' follow from 'snow is white and snow is not white'? It $follows_{Classical}$ but does not $follow_{Paraconsistent}$. And so on. While each of these claims is true independent of us, its truth is like that of the Parallel Postulate, for the Platonist. A claim 'just like' its negation is also true.

How does this view help with the problem of mathematical and, more generally, armchair knowledge? If *armchair pluralism* (as we will call the resulting view) is correct, then intuitions about *AC* may vary in the banal sense that our intuitions about the Parallel Postulate do. Different intuitions may be about different subjects. Intuitions supporting *AC* may be about one class of set-like universes, while intuitions against it may be about another. All of these are real. So, there is no question of who is really right. The classicist is right of classical consequence and the paraconsistent logician is right of paraconsistent consequence!

Armchair pluralism, thus, invites a kind of *pragmatism.* The only non-semantic questions (i.e., questions that are not just about what we happen to mean by words) in armchair debates are *practical.* Instead of asking whether every set has a Choice function, we ask the practical question of *whether to use* a concept of set that satisfies the Axiom of Choice (for a purpose).[43] Instead

[42]Objectivist realists often note that there are important differences between the Parallel Postulate and set-theoretic undecidables like the Continuum Hypothesis (*CH*) (Kennedy & van Atten, 2004, Sec. 5.1). For example, the Parallel Postulate is undecidable in Hilbert's second-order axiomatization, but second-order *CH* is decidable. The significance of this difference is doubtful, though. The sense in which *CH* is 'decidable' is that, given a full semantics for second-order logic, $ZFC_2 \models CH$ or $ZFC_2 \models \neg CH$. *CH* is not decided in the usual, syntactic sense (second-order logic is incomplete). Also, second-order logic assumes a metatheory—some set theory or something just as epistemically (if not ontologically) obscure. Even if Quine was wrong that second-order logic is 'set theory in sheep's clothing', knowledge of second-order consequence is hardly more intelligible than knowledge of mathematics.

[43]We speak of concepts for readability. One who is skeptical of them, like Quine, is free to replace our talk of concepts with talk of (formal) theories (set theories, modal theories, evaluative theories, and so on).

of asking whether the mind could have been different from the body, we ask whether to use a concept of possibility that satisfies the Necessity of Identity. And so forth. Since divergent theories are all true—albeit of subtly different subjects—truth factors out. It is as if armchair claims did not admit of truth at all. Intractable armchair questions get traded for practical ones.[44]

It might be worried that armchair pluralism robs armchair questions of interest. But this assumes that their intellectual value depends on their answering to a unique and independent reality. Why should that be? Even if there were 'one true V', for example, this may be less interesting than universes of our imagination—as it presumably is, according to advocates of 'large' large cardinals, if $V = L$ is true. Metaphysics does not settle questions of intellectual value. This is just an application of Hume's point that one cannot derive an (in this case, epistemic) 'ought' from an 'is'.

It remains to say how armchair pluralism applies to questions of value themselves. It radicalizes Hume's dictum. If evaluative pluralism is correct, then there is a property answering to Mill's utilitarianism alongside a property answering to Kant's deontology—and a property answering to every other such theory (as in the pure geometric case). Thus, as in set theory, modality, or logic, there is no useful question of whether, e.g., we *really* ought to kill the one to save the five (in some concrete situation). We ought$_{\text{utilitarian}}$, and ought not$_{\text{deontological}}$! Instead, we need to ask *whether to use* a utilitarian or deontological concept of ought. But there is an intimate connection between *this* practical question and a question of fact from the target domain. Evaluative facts, unlike mathematical, modal, or logical ones, are supposed to tell us what to do. So, what happens when we replace the practical question of whether to use a given concept with the factual question of whether we *ought* to use it? The argument reapplies! We ought$_{\text{utilitarian}}$ to use ought$_{\text{utilitarian}}$, and ought$_{\text{deontological}}$ to use ought$_{\text{deontological}}$ (and so on for every concept of ought).[45] The question of *which concept to use* remains open. This cannot be the question of which concept we *ought* to use on pain of triviality. (We ought to use ought, ought*

[44]A distinct argument for armchair pluralism about mathematics, in particular, comes from the history of mathematics. Unlike physics, chemistry, or biology, mathematical theories of the past (even going back to antiquity) do not seem to be jettisoned in light of newer theories. They are absorbed into the body of accepted mathematics through a process of reinterpretation and widening of the relevant axiomatic system. See Crowe (1975). A liability of this argument is that it may trade on an equivocation with 'theories'. *Sentences* are certainly preserved. But what matters is whether the *propositions expressed by them* are. This is far less straightforward.

[45]Or, at least, for any self-sanctioning concept of ought. 'Self-hating' concepts of ought, ought**, such that we ought** not to use ought** are presumably not attractive guides to decision.

to use ought*, and so forth.) Hence, the facts, *even the evaluative ones* (if there are any), fail to settle what to do.[46]

12 Implications for Physics

Before concluding, let us illustrate some consequences of armchair pluralism for fundamental physics. Just as armchair anti-realism engenders scientific anti-realism, armchair pluralism leads to scientific pluralism. The physical world becomes *perspectival* to a surprising degree.

Why is this the case? After all, the ordinary Platonist's pluralism about pure geometry does not commit them to pluralism about physical space(time). It merely shows that, of the myriad pure geometries, at most one of them precisely models our universe. The reason is that the set theoretic, logical, and modal cases are different from pure geometry in a crucial respect. Different geometries can all be realized in a single metatheory, like some set theory. But every set theory, plus a logic and modal theory, *is* a metatheory, with implications for all reasoning.

Consider the intersection of a family of sets of points in spacetime, $\bigcap x$.[47] Does it exist? In order to prove that it does, we must form $\{y: y \in \bigcup x \;\&\; (\forall z)(z \in x \rightarrow y \in z)\}$. This requires the (restricted) *Comprehension Axiom* schema, which says that, for any set, z, and any formula, Φ, there is a set containing those members of z that satisfy Φ. That is, $(\forall z)(\exists y)(\forall x)(x \in y \longleftrightarrow (x \in z \;\&\; \Phi))$ (where y is not free in Φ). The existence of the union of the sets in x, $\bigcup x$, is given by the *Union Axiom*, that for any set, z, there is a set, $\bigcup z$, containing the members of members of z. In symbols: $(\forall z)(\exists y)(\forall x)(x \in y \longleftrightarrow (\exists w)(w \in z \;\&\; x \in w))$. Finally, the points that get collected in the first place are represented by ordered quadruples, understood as sets, constructed out of natural numbers—whose existence is given by the *Axiom of Infinity*—by way of repeated applications of Comprehension and the *Powerset Axiom* that, for any set, z, there is a set containing the subsets of z, $P(z)$, written $(\forall z)(\exists y)(\forall x)(x \in y \longleftrightarrow (\forall w)(w \in x \rightarrow w \in z))$. But not all set theories countenance Infinity and Powerset. For example, Kripke–Platek set theory lacks the Powerset, and sometimes Infinity, axioms. So, whether $\bigcap x$ exists—

[46] How do we decide what to do if knowledge of the facts fails to settle this? We decide by practical reasoning! For approaches to the logic and semantics of practical reasoning, non-factually construed, see Blackburn (1984), Gibbard (2003), Schroeder (2010), and Silk (2015).

[47] This example is a variation on that of Clarke-Doane (2022, 2.1).

what might have seemed to be an objective physical question—depends on the set theory that we use to check.[48]

The logical and modal cases are similar. Let T be a regimented physical theory, such as General Relativity or the Standard Model.[49] Does T imply O, where O is some observation statement that has been refuted (i.e., we have that $\sim O$)? This depends on whether T is (classically) consistent. If $\sim \mathrm{Con}(T)$, then T implies everything, including O. So, prima facie, one way that it could fail to be objective whether T has been refuted is if it failed to be objective whether $\mathrm{Con}(T)$—and, if $\mathrm{Con}(T)$, then T did not imply O. We say 'prima facie' because, if $\mathrm{Con}(T)$, then $\sim \mathrm{Con}(T)$ is only true in a nonstandard model where 'proof' and 'theory' are interpreted non-standardly so that, perhaps, $\sim \mathrm{Con}(T)$ does not say *of* T that it is *inconsistent*. (There is a similar prima facie argument that there would be no objective fact as to whether a physical computer computing a Turing uncomputable function would eventually halt.) But there is another way that it could fail to be objective whether T implies O. It may fail to be objective *what logic is correct*, as it is if armchair pluralism is true. In that case, if O is classically, but not intuitionistically, derivable from T, then the question of whether T has been refuted by O would be like the question of whether the Parallel Postulate is true, understood as one of pure mathematics.

Finally, consider a quantum spin system, S, with eigenstates $|\uparrow>$ and $|\downarrow>$. Could S have been in the indeterminate state of being *neither* $|\uparrow>$ *nor* $|\downarrow>$ *nor* $a|\uparrow> + b|\downarrow>$, for all complex numbers, a and b? What about the contradictory state $|\uparrow>$ *and* $|\downarrow>$? As a matter of metaphysical, a fortiori physical, possibility, the answer to both of these questions is, of course, 'no'. But if the kind of possibility at issue is paracomplete, allowing for the possibility of indeterminacies, then the state, *neither* $|\uparrow>$ *nor* $|\downarrow>$ *nor* $a|\uparrow> + b|\downarrow>$, will be in S's state space. If it is paraconsistent, then the state $|\uparrow>$ *and* $|\downarrow>$ will be in that space. And if *First-Degree Entailment* (*FDE*) governs the possible states of the system, then *neither* $|\uparrow>$ *nor* $|\downarrow>$ *nor* $a|\uparrow> + b|\downarrow>$ as well as $|\uparrow>$ *and* $|\downarrow>$ willall be possible

[48] A simpler argument with a more complicated conclusion appeals to the Banach–Tarski paradox. Is it physically possible to cut a (solid) ball into a finite number of pieces and, using only rigid motions, construct two balls of the same size? Yes, relative to *ZFC*, and no, relative to *ZF* + *AD*. Or, to take an example from the mathematical physics literature, 'the spacetime manifold M is partially ordered by the chronology relation, and any timelike curve g on M is simply ordered by the same relation. Therefore, *on the strength of the maximum principle* [which is equivalent to *AC*] there is a subset h of M, which is simply ordered by the chronology relation, contains g and is maximal with respect to these properties of M ....[T]he set h is a timelike curve' (Heller, 1986, 64, italics added).

[49] We indulge here in the fiction that there is a provably consistent axiomatization of quantum field theory (in four dimensions) with interactions.

states of the system, S.[50] In general, if there is no objective fact as to which kinds of possibility are real, then there is no objective fact as to the state space of a physical system.

To be clear: this does not mean that we should expect to observe indeterminate or contradictory states, any more than the existence of metaphysically possible but physically impossible states in which a projectile accelerates above c shows that we should expect to observe those (or, indeed, the existence of metaphysically *impossible* states shows that we should expect to observe them). One question is how the world could have been. This is not objective, on the current view. Another question is how it *will* be—which is objective, modulo pluralism about set theory.

Although these implications may seem hard to swallow, the alternative is arguably more so. It is widely conceded that not all aspects of our best theories reflect objective features of the world. Those theories include a representational apparatus that we bring to the table. For instance, the world itself may not decide the fine details of spacetime structure. But if that just means that, e.g., whether $\cap x$ above exists depends on our beliefs, then we seem to be committed to the view that spacetime would be amorphous if none of us believed otherwise.[51] Armchair pluralism allows that all of the facts of spacetime are equally independent of us. The sense in which some of them are not objective is just that reality is rich enough to witness any conception of them that we might have had. Which conception is 'right' is a practical question, not settled by the facts.[52]

13 Armchair Pluralism

We have argued that intuitions vary in a way that observations do not and that this favors an anti-objectivist, but realist, conception of armchair

[50] For details on how to combine non-classical and modal logics, see Priest (2008, Sec. 11).

[51] A tame variation on this idea is due to Grünbaum (1973), who argues that spacetime is metrically amorphous.

[52] It might be thought that one could appeal to the concept of *naturalness* in the sense of Lewis (1983) and Sider (2011) in order to give epistemic (or 'theoretical') content to the claim that one conception of set, possibility, consequence, or value right. But either the claim that some such conception is natural is itself evaluative (with implications for which is *good* or which *ought* to use) or not. If it is, then the problem of pluralism rearises vis-à-vis naturalness. Conception$_1$ may be natural$_1$ but not natural$_2$. What could the claim that naturalness$_1$ is *real* naturalness amount to if not just something about what we happen to mean by the word 'naturalness'? Alternatively, the claim that some conception is natural is not normative. But then it is neither here nor there from the methodological standpoint of which to use, for familiar is/ought reasons. See Clarke-Doane (2020, Sec. 6.6).

areas—armchair pluralism. Armchair reality is so rich that any set-theoretic, modal, logical, or evaluative theory that we might have adopted is true of a corresponding subject. While we can always ask what is packed into the concepts of set, possibility, implication, and obligation that we find ourselves with, this autobiographical exercise is of no *methodological* import. It just raises the question of whether to use our present concepts rather than alternatives.[53] The only non-semantic questions are *practical*: which concepts to use?

The foundations of armchair pluralism require in-depth treatment. Three questions are pressing.

First, with respect to what subjects exactly is pluralism tenable? Pluralism itself would seem to be an armchair theory. It is a product of philosophy, which we have arrived at by reasoning. Moreover, our views on pluralism seem to share the contingent profile of our mathematical, modal, logical, and moral ones. So, the above argument would appear to show that whether pluralism is correct *itself* lacks an objective (even if not independent) answer (because there are subtly different senses of 'objective' giving intuitively opposite verdicts—as with the Parallel Postulate). Is that so? Or is armchair pluralism itself objectively (and independently) correct?[54]

Second, how liberal should our pluralism about a subject be? It is natural to be as liberal as we can consistently be. But Gödel's Second Incompleteness

[53]In other words, if armchair pluralism is correct, then the '[m]any set theorists, including Gödel', who 'believe that conceptual analysis will eventually lead to an idea of a set so clear and distinct that the answer to the continuum question will become apparent' (Huber-Dyson, 1991, 9) are misguided *even if* their belief is true.

[54]It might be thought that there are subtly different senses of (counterfactually) 'independent' giving rise to different concepts of realism as well. But if these senses correspond to more or less inclusive classes of worlds with respect to which to evaluate the counterfactuals, then realism is not thus ambiguous assuming the following principle is true.

> *Counterfactual Absoluteness* (Clarke-Doane, 2019, Section 7): Suppose that $\forall P$ ($[N+]P \rightarrow [N]P$), but not conversely. Then, if (A [] $\rightarrow$ B) is non-vacuously true with respect to a model, $N = <D, S, V, @>$, where $w \in D$ just in case w is N-possible, then (A [] $\rightarrow$ B) remains true with respect to a model, $N^{+} = <D', S', V', @>$ where $w \in D'$ just in case w is N^{+}-possible, whenever N is a submodel of N^{+} (where N and N^{+} are models of propositional conditional logic, where S is a class of relations, one for each formula of the language, D is a domain, V is a valuation, and @ is the actual world).

Counterfactual Absoluteness says that once one finds a witness for the antecedent of a true counterfactual in some N-possible world, one cannot flip its truth-value by ascending to a stricter kind of necessity, $N+$. It is, therefore, a relativized version of Nolan's 'Strangeness of Impossibility' condition (Nolan, 1997). If counterfactuals like 'had our mathematical beliefs been different, the truths would have been the same' are non-vacuously true, then they remain true as we add more worlds to our model. So, the counterfactually independent truth of our mathematical beliefs is indifferent to this variation.

Theorem says that it is consistent to say false things about consistency (if weak theories of arithmetic are consistent).[55] Is a more stringent criterion needed? Even if not, logical pluralism generates different versions of the consistency requirement, depending on the logic.[56] So, in order to know how rich armchair reality is, we must know how rich logical reality is. In Clarke-Doane (2019), one of us argues that logical reality is *indefinitely extensible.* Given a logic, we can always intelligibly weaken it (but the 'trivial logic', according to which nothing follows from anything, is unintelligible as an all-purpose reasoning device). And, yet, as the locution 'can . . . weaken it' suggests, indefinite extensibility is normally taken to be a *modal* notion. We cannot take it that way if modal reality—the way the world *could have been*—is itself indefinitely extensible too.

Finally, what kind of framework exactly is armchair pluralism? It cannot be a theory in the formal sense. Any formalization of it would assume a metatheory of its own which would wrongly take itself to be maximal. Is it a truth that one can 'show but not say'? Or is it an attitude, not apt for representing the world, a la Carnap's Principle of Tolerance?[57] If our intuitions are not 'seriously inadequate and badly distorted' (Mackie, 1977, 75), each of these questions has an answer.

---

[55]This is rough because, again, one could argue that, e.g., $ZF + \sim\mathrm{Con}(ZF)$ asserts the 'shinconsistency' of the 'Sh*ZF* axioms', not the inconsistency of the *ZF* axioms. See, e.g., Azzouni (forthcoming).

[56]Note that armchair pluralism does not countenance inconsistent mathematics (in the sense of Mortensen, 2022) just because it encompasses paraconsistent logics. Logics are about what follows from what, not what (non-metalogical) claims are (actually) true (though of course the claim that *S* is valid implies that *S* is true). The claim that any contradiction is actually true is a further commitment on which paraconsistent logics are silent.

[57]In the set-theoretic case, pluralism (or something like it) is often treated as an unformalizable theory (Balaguer, 1995, 6). See also Hamkins (2012, 417). Shelah writes, 'My mental picture is that we have many possible set theories . . . .I do not feel 'a universe of *ZFC*' is like 'the Sun', it is rather like 'a human being'' (2003, 211). But Shelah does not go on to say what the force of calling something a 'mental picture' is. (Cohen's method of forcing was the impetus for pluralism among set theorists. Bell (1981, 358) writes, 'The techniques [Cohen] invented have led to an enormous proliferation of essentially different models of set theory and the rise of a 'relativistic' attitude toward the set-theoretical foundations of mathematics. This attitude involves abandoning . . . the idea that mathematical constructions should be viewed as taking place within an 'absolute' universe of sets with fixed and predetermined properties. Instead, one works in suitably chosen *models* of set theory having the properties required to carry out the construction in question' (italics in original). However, for reasons discussed here, we do not see an *epistemological* difference between axioms whose independence is proved via forcing and the others, like Foundation or Infinity. We are also skeptical of a metaphysical/semantic difference. See Clarke-Doane, 2012.)

Bibliography

Ash, A., & Gross, R. (2008). *Fearless Symmetry: Exposing the Hidden Patterns of Numbers* (Revised Edition). Princeton University Press.

Azzouni, J. (forthcoming). What is PA + ~Con(PA) about and Where? *Synthese.*

Azcel, P. (1988). *Non-Well-Founded Sets. SLI Lecture Notes.* Vol. 14. Stanford, CA: Stanford University, Center for the Study of Language and Information.

Balaguer, M. (1995). A Platonist Epistemology. *Synthese, 103*, 303–325.

Balaguer, M. (2001). A Theory of Mathematical Correctness and Mathematical Truth. *Pacific Philosophical Quarterly, 82*, 87–114.

Balaguer, M. (2018). Fictionalism in the Philosophy of Mathematics. In E. N. Zalta & U. Nodelman (Eds.), *The Stanford Encyclopedia of Philosophy* (Spring 2023 Edition). Retrieved from https://plato.stanford.edu/archives/spr2023/entries/fictionalism-mathematics/

Bealer, G. (1999). A Theory of the A Priori. *Philosophical Perspectives, 13*, 29–55.

Bell, J. (1981). Category Theory and the Foundations of Mathematics. *British Journal for the Philosophy of Science, 32*, 349–358.

Bell, J., & Hellman, G. (2006). Pluralism and the Foundations of Mathematics. In K. Waters, H. Longino, & S. Kellert (Eds.), *Scientific Pluralism (Minnesota Studies in Philosophy of Science, Vol. 19).* University of Minnesota Press.

Benacerraf, P. (1965). What Numbers Could Not Be. *Philosophical Review, 74*, 47–73.

Benacerraf, P. (1973). Mathematical Truth. *Journal of Philosophy, 60*, 661–679.

Benardete, J. (1964). *Infinity: An Essay in Metaphysics.* Clarendon Press.

Bengson, John. [2015] "Grasping the Third Realm." Szabó Gendler, Tamar and John Hawthorne (eds.). Oxford Studies in Epistemology, Vol. 5. : Oxford University Press.

Bevan, M., & Paseau, A. C. (2020). Review of Morality and Mathematics. *Philosophia Mathematica, 28.*

Blackburn, S. (1984). *Spreading the Word.* Clarendon Press.

Blackburn, S. (2008). *The Oxford Dictionary of Philosophy.* Oxford University Press.

Boolos, G. (1971). The Iterative Conception of Set. *Journal of Philosophy, 68*, 215–231.

Boolos, G. (1999). Must We Believe in Set Theory? In *Logic, Logic, and Logic.* Harvard University Press.

Cameron, R. (2006). *Comment on 'Kripke's (Alleged) Argument for the Necessity of Identity Statements'.* Wo's Weblog. Retrieved from https://www.umsu.de/wo/archive/2006/08/09/Kripke_s__Alleged__Argument_for_the_Necessity_of_Identity_Statements

Cameron, R. (2008). Turtles all the way down: Regress, Priority and Fundamentality. *The Philosophical Quarterly, 58*, 1–14.

Carnap, R. (2001/1937). *The Logical Syntax of Language.* Routledge.

Carroll, S. (2010). *From Eternity to Here: The Question for the Ultimate Theory of Time.* Penguin.

Cartier, P., Dhombres, J., Heinzmann, G., & Villani, C. (2016). *Freedom in Mathematics*. Springer.
Cartwright, R. (1968). Some Remarks on Essentialism. *Journal of Philosophy, 65*, 615–626.
Chudnoff, E. (2013). *Intuition*. Oxford University Press.
Churchland, P. (2014). *Matter and Consciousness* (3rd ed.). MIT Press.
Clarke-Doane, J. (2012). Morality and Mathematics: The Evolutionary Challenge. *Ethics, 122*, 313–340.
Clarke-Doane, J. (2019). Modal Objectivity. *Noûs, 53*, 266–295.
Clarke-Doane, J. (2020). *Morality and Mathematics*. Oxford University Press.
Clarke-Doane, J. (2022). *Mathematics and Metaphilosophy (Elements in Philosophy of Mathematics)*. Cambridge University Press.
Clarke-Doane, J. (forthcoming). Replies to Rosen, Leiter, and Dutilh Novaes. *Philosophy and Phenomenological Research* (author meets critic symposium on *Morality and Mathematics*).
Coffa, A. (1982). Kant, Bolzano, and the emergence of logicism. *The Journal of Philosophy, 79*(11), 679–689.
Cohen, P. (1971). Comments on the Foundations of Set Theory. In D. Scott (Ed.), *Axiomatic Set Theory (Proceedings of Symposia of Pure Mathematics, Vol. XIII, Part I)*. American Mathematical Society.
Cohen, P. (2005). The Nature of Mathematical Proof. *Philosophical Transactions of the Royal Society, 363*, 2416.
Crowe, M. (1975). Ten 'Laws' Concerning Patterns of Change in the History of Mathematics. Reprinted in D. Gillies (Ed.), *Revolutions in Mathematics*. Oxford University Press.
Daniels, N. (2020). Reflective Equilibrium. In E. N. Zalta (Ed.), *The Stanford Encyclopedia of Philosophy* (Summer 2020 Edition). Retrieved from https://plato.stanford.edu/archives/sum2020/entries/reflective-equilibrium/
Devlin, K. (1977). *The Axiom of Constructability: A Guide for the Mathematician (Lecture Notes on Mathematics: 617)*. Springer Verlag.
Devlin, K. (1981). Infinite Trees and the Axiom of Constructibility. *Bulletin of the London Mathematical Society, 13*, 193–206.
Dixon, T. S. (2016). What Is the Well-Foundedness of Grounding? *Mind, 125*, 439–468.
Dummett, M. (1991). *The Logical Basis of Metaphysics*. Harvard University Press.
Easwaran, K. (2008). The Role of Axioms in Mathematics. *Erkenntnis, 68*, 381–391.
Edgington, D. (2004). Two Kinds of Possibility. *Aristotelian Society Supplementary Volume, 78*, 1–22.
Feferman, S. (1992). Why a Little Bit Goes a Long Way: Logical Foundations of Scientifically Applicable Mathematics. *Proceedings of the Philosophy of Science Association, 2*, 442–455.
Ferrier, E. (2019). Against the Iterative Conception of Set. *Philosophical Studies, 176*, 2681–2703.

Field, H. (1980). *Science without Numbers*. Princeton University Press.

Field, H. (1989). *Realism, Mathematics, and Modality*. Blackwell.

Forster, T. (forthcoming). *The Axioms of Set Theory*. Cambridge University Press. Retrieved July 4, 2018, from (in preparation): https://www.dpmms.cam.ac.uk/~tf/axiomsofsettheory.pdf

Fraenkel, A., Bar-Hillel, Y., & Levy, A. (1973). *Foundations of Set Theory (Studies in Logic and the Foundations of Mathematics, Vol. 67)*. Elsevier Science Publishers.

Frege, G. (1980/1884). *The Foundations of Arithmetic: A Logico-Mathematical Inquiry into the Concept of Number* (2nd rev. ed., J. L. Austin, Trans.). Northwestern University Press.

Friedman, H. (2002). *Philosophical Problems in Logic (Lecture Notes)*. Princeton University. Retrieved from https://cpb-us-w2.wpmucdn.com/u.osu.edu/dist/1/1952/files/2014/01/Princeton532-1pa84c4.pdf

Gibbard, A. (1975). Contingent Identity. *Journal of Philosophical Logic, 4*, 187–221.

Gibbard, A. (2003). *Thinking How to Live*. Harvard University Press.

Girle, R. (2017). *Modal Logics and Philosophy* (2nd ed.). McGill-Queens University Press.

Gödel, K. (1947). What is Cantor's Continuum Problem? *American Mathematical Monthly, 54*, 515–525.

Greene, J. (2013). *Moral Tribes: Emotion, Reason, and the Gap Between Us and Them*. Penguin.

Hamkins, J. D. (2012). The Set-theortic Multiverse. *Review of Symbolic Logic, 5*, 416–449.

Grünbaum, A. (1973). *Philosophical Problems of Space and Time*. Reidel.

Hanson, N. R. (1958). *Patterns of Discovery*. Cambridge University Press.

Harman, G. (1965). The Inference to the Best Explanation. *The Philosophical Review, 74*, 88–95.

Harris, M. (2012). Do Androids Prove Theorems in their Sleep? In Doxiadis and Mazur [2012].

Heller, M. (1986). *Questions to the Universe (Astronomy and Astrophysics Series, Vol. 14)*. Pachart Publishing House.

Huber-Dyson, V. (1991). *Gödel's Theorems: A Workbook on Formalization*. Teubner Verlag.

Huemer, M. (2005). *Ethical Intuitionism*. Palgrave Macmillan.

Incurvati, L. (2014). The Graph Conception of Set. *Journal of Philosophical Logic, 43*, 181–208.

Jensen, R. (1995). Inner Models and Large Cardinals. *Bulletin of Symbolic Logic, 1*, 393–407.

Kennedy, J., & van Atten, M. (2004). Gödel's Modernism: On Set-Theoretic Incompleteness. *Graduate Faculty Philosophy Journal, 25*, 289–350.

Kennedy, J., & van Atten, M. (2008). 'Gödel's Modernism: On Set-Theoretic Incompleteness,' Revisited. In S. Lindström, K. Segerberg, V. Stoltenberg-Hansen,

& E. Palmgren (Eds.), *Logicism, Intuitionism, and Formalism: What has Become of Them?* Springer.

Kilmister, C. W. (1980). Zeno, Aristotle, Weyl and Shuard: Two-and-a-Half Millennia of Worries over Number. *Mathematical Gazette, 64*, 149–158.

Koellner, P. (2006). On the Question of Absolute Undecidability. *Philospohia Mathematica, 14*(2), 153–188.

Kreisel, G. (1967). Observations on Popular Discussions of the Foundations of Mathematics. In D. Scott (Ed.), *Axiomatic Set Theory (Proceedings of Symposia in Pure Mathematics, Vol. XIII, Part I)*. American Mathematical Association.

Kripke, S. (1980). *Naming and Necessity*. Cambridge: Harvard University Press.

Leiter, B. (2009). *Moral Skepticism and Moral Disagreement in Nietzsche*. Public Law Working Paper no. 257. University of Chicago. Retrieved from http://ssrn.com/abstractpl315061

Lewis, D. (1983). New Work for a Theory of Universals. *Australasian Journal of Philosophy, 61*, 343–377.

Mackie, J. L. (1977). *Ethics: Inventing Right and Wrong*. Penguin.

Maddy, P. (1988a). Believing the Axioms: I. *Journal of Symbolic Logic, 53*, 481–511.

Maddy, P. (1988b). Believing the Axioms: II. *Journal of Symbolic Logic, 53*, 736–764.

Martin, D. A. (1998). Mathematical Evidence. In Dales and Oliveri [1998].

Mayberry, J. (2000). *The Foundations of Mathematics in the Theory of Sets*. Cambridge University Press.

Mazur, B. (2012). Visions, Dreams, and Mathematics. In Doxiadis and Mazur [2012].

Milgrom, M. (2002, August). Does Dark Matter Really Exist? *Scientific American, 287*(2), 42–46, 50–52.

Mill, J. S. (1882/2009). *A System of Logic, Ratiocinative and Inductive* (8th ed.). Harper & Brothers. Retrieved from www.guten-berg.org/files/27942/27942-pdf.pdf

Mogensen, A. (2016a). Disagreement in Moral Intuition as Defeaters. *Philosophical Quarterly*. Forthcoming.

Mogensen, A. (2016b). Disagreement in Moral Intuition as Defeaters. *Philosophical Quarterly, 67*(267), 282–302.

Mortensen, C. (2022). Inconsistent Mathematics. In E. N. Zalta & U. Nodelman (Eds.), *The Stanford Encyclopedia of Philosophy* (Winter 2022 Edition). Retrieved from https://plato.stanford.edu/archives/win2022/entries/mathematics-inconsistent/

Moss, L. S. (2018). Non-wellfounded Set Theory. In E. N. Zalta (Ed.), *The Stanford Encyclopedia of Philosophy* (Summer Edition). Retrieved from https://plato.stanford.edu/archives/sum2018/entries/nonwellfounded-set-theory/

Nagel, T. (1997). *The Last Word*. Oxford University Press.

Nelson, E. (1986). *Predicative Arithmetic (Mathematical Notes. No. 32)*. Princeton University Press.

Nelson, E. (2013). *Re: Illustrating Edward Nelson's Worldview with Nonstandard Models of Arithmetic*. Post on MathOverflow. October 31. Retrieved from https://

mathoverflow.net/questions/142669/illustrating-edward-nelsons-worldview-with-nonstandard-models-of-arithmetic

Nolan, D. (1997). Impossible Worlds: A Modest Approach Notre Dame. *Journal of Formal Logic, 38*, 535–572.

Nolan, D. (2005). *The Philosophy of David Lewis*. Routledge.

Poincaré, H. (1906). Les Mathematiques et la Logique. *Revue de Metaphysique et de Morale, 14*, 294–317.

Potter, M. (2004). *Set Theory and Its Philosophy*. Cambridge University Press.

Priest, G. (2008). *An Introduction to Non-Classical Logic: From If to Is* (2nd ed.). Cambridge University Press.

Pryor, J. (2000). The Skeptic and the Dogmatist. *Nous, 34*, 517–549.

Putnam, H. (2015). Intellectual Autobiography. In R. E. Auxier, D. R. Anderson, & L. E. Hahn (Eds.), *The Philosophy of Hilary Putnam (Library of Living Philosophers, Vol. XXXIV)*. Open Court.

Quine, W. V. O. (1937). New Foundations for Mathematical Logic. *American Mathematical Monthly, 44*, 70–80.

Quine, W. V. O. (1951). Two Dogmas of Empiricism. *Philosophical Review, 60*, 20–43.

Quine, W. V. O. (1958). The Philosophical Bearing of Modern Logic. In R. Klibansky (Ed.), *Philosophy in the Mid-Century, I* (pp. 3–4). La Nuova Italia.

Quine, W. V. O. (1969). *Set Theory and Its Logic* (rev. edn). Harvard University Press.

Quine, W. V. O. (1986). Reply to Charles Parsons. In L. E. Hahn & P. A. Schilp (Eds.), *The Philosophy of W. V. Quine (Library of Living Philosophers, Vol. 18)*. Open Court.

Quine, W. V. O., & Ullian. (1978). *The Web of Belief* (2nd ed.). McGraw-Hill Education.

Rabin, G. O., & Rabern, B. (2016). Well Founding Grounding Grounding. *Journal of Philosophical Logic, 45*, 349–379.

Rachels, J. (Ed.). (1998) Introduction. In *Ethical Theory 1: The Question of Objectivity (Oxford Readings in Philosophy)*. Oxford University Press.

Rawls, J. (1974). The Independence of Moral Theory. *Proceedings and Addresses of the American Philosophical Association, 47*, 5–22.

Rickles, D. (2016). *The Philosophy of Physics*. Polity Press.

Rieger, A. (2011). Paradox, ZF, and the Axiom of Foundation. In D. DeVidi, M. Hallet, & P. Clark (Eds.), *Logic, Mathematics, Philosophy, Vintage Enthusiasms: Essays in Honour of John L. Bell (The Western Ontario Series in Philosophy of Science)* (pp. 171–187). Springer.

Rosen, G. (2001). Nominalism, Naturalism, and Epistemic Relativism. *Philosophical Perspectives, 15*, 69–91.

Russell, B. (1903/2022). *The Principles of Mathematics*. Cambridge University Press. Retrieved from http://people.umass.edu/klement/pom/

Russell, B. (1906). On some Difficulties in the Theory of Transfinite Numbers and Order Types. *Proceedings of the London Mathematical Society, 4*, 29–53.

Russell, B. (1907). The Regressive Method for Discovering the Premises of Mathematics. In D. Lackey (Ed.), *[1973] Essays in Analysis, by Bertrand Russell.* Allen and Unwin Ltd.

Russell, B. (1918). The Philosophy of Logical Atomism. *The Monist, XXVIII*, 495–526.

Russell, B. (1952/2023). *Excerpt from "Remarks on my 80th Birthday".* Retrieved from https://www.maa.org/press/periodicals/convergence/quotations/russell-bertrand-1872-1970-13

Russell, B., & Whitehead, A. N. (1910/1961). *Principia Mathematica, Vol. 1* (2nd ed.). Cambridge University Press. Retrieved from http://www.uhu.es/francisco.moreno/gii_mac/docs/Principia_Mathematica_vol1.pdf

Schechter, J. (2010). The Reliability Challenge and the Epistemology of Logic. *Philosophical Perspectives, 24*, 437–464.

Schroeder, M. (2010). *Being For: Evaluating the Semantic Program for Expressivism.* Oxford University Press.

Shapiro, S. (1997). *Philosophy of Mathematics: Structure and Ontology*. Oxford University Press.

Shelah, S. (2003). Logical Dreams. *Bulletin of the American Mathematical Society, 40*, 203–228.

Shoenfeld, J. (1977). The Axioms of Set Theory. In J. Barwise (Ed.), *Handbook of Mathematical Logic* (pp. 321–344). North-Holland.

Sider, T. (2011). Writing the Book of the World. *Oxford: Oxford University Press.*

Silk, A. (2015). How to Be an Ethical Expressivist. *Philosophy and Phenomenological Research, 91*, 47–81.

Singer, P. (1994). Introduction. In P. Singer (Ed.), *Ethics*. Oxford University Press.

Steel, J. (2014). Gödel's Program. In J. Kennedy (Ed.), *Interpreting Gödel: Critical Essays* (pp. 153–179). Cambridge University Press.

Stratton-Lake, P. (2020). Intuitionism in Ethics. In E. N. Zalta (Ed.), *The Stanford Encyclopedia of Philosophy* (Summer 2020 Edition). Retrieved from https://plato.stanford.edu/archives/sum2020/entries/intuitionism-ethics/

Warren, J. (2020). *Shadows of Syntax*. Oxford University Press.

Weyl, H. (1918/1994). *The Continuum. A Critical Examination of the Foundation of Analysis* (S. Pollard & T. Bole, Trans.). Dover.

Williamson, T. (2005). The Presidential Address: Armchair Philosophy, Metaphysical Modality and Counterfactual Thinking. *Proceedings of the Aristotelian Society, 105*, 1–23.

Williamson, T. (2012). Logic and Neutrality. *The Stone (New York Times).* May 13. Retrieved from https://opinionator.blogs.nytimes.com/2012/05/13/logicand-neutrality/

Williamson, T. (2013). *Modal Logic as Metaphysics*. Oxford University Press.

Wilson, M. (1983). Why Contingent Identity is Necessary. *Philosophical Studies, 43*, 301–327.

Woodin, H. (2010). Strong Axioms of Infinity and the Search for V. In R. Bhatia, A. Pal, G. Rangarajan, V. Srinivas & M. Vanninathan (Eds.), *Proceedings of the International Congress of Mathematicians 2010 (ICM 2010) (In 4 Volumes)*. World Scientific.

Zeilberger, D. (2004). "Real" Analysis is a Degenerate Case of Discrete Analysis. In B. Aulbach, S. N. Elaydi, & G. Ladas (Eds.), *Proceedings of the Sixth International Conference on Difference Equations*. CRC Press.

4

Arealism, Thin Realism, and the Problem of Extrinsic Evidence

Jeffrey R. Schatz

1 The Problem of Extrinsic Evidence

In recent years, set theory has been uniquely interesting for philosophers due to the presence of non-proof-based forms of reasoning not often found in other areas of mathematics. While many areas of "core" mathematics take their axioms as given, and focus on proving theorems from these axioms, in set theory the matter of which axioms should be accepted remains a live research question. Historically, much of this effort focused on identifying and justifying the now-standard theory ZFC, or its widely accepted extension $ZFC + LCs$; today, the aim of axiom selection has shifted to finding new, defensible axioms capable of settling independent questions like the Continuum Hypothesis. Given that traditional proof methods cannot be any help in defending new axioms candidates—since axioms must be accepted prior to any proving—an

Thanks to all of the participants in the UNICAMP-Konstanz Philosophy of Set Theory Seminar in 2021 for their feedback on an early presentation of this material. In addition, special thanks to Sorin Bangu, Neil Barton, Kevin Cahill, Leon Commandeur, Evelyn Erickson, Ole Thomassen Hjortland, Torfinn Thomesen Huvenes, Penelope Maddy, Chris Mitsch, Tore Fjetland Øgaard, Sindre Olaussen Søderstrøm, and Evan Sommers for their comments on drafts of this chapter. Finally, thanks to the two anonymous reviewers for their many helpful suggestions.

J. R. Schatz (✉)
Department of Philosophy, University of Bergen, Bergen, Norway

C. Antos et al. (eds.), *The Palgrave Companion to the Philosophy of Set Theory*,
Palgrave Companions, https://doi.org/10.1007/978-3-031-62387-5_4

important epistemological question is what methods are used to justify axiom choices in set theory, and whether those methods are rationally defensible.

An important early work toward understanding how axiom candidates are, in fact, defended was Maddy's "Believing the Axioms."[1] In these articles, Maddy collects and describes the wide variety of arguments which had been proposed in defense of the axiom candidates of ZFC, as well as stronger extensions of this theory. Surveying the wide scope of justificatory stories on offer, Maddy notes that there is no monolithic sort of evidence that encompasses all of the defenses provided for axiom candidates. Instead, most axiom candidates are defended through distinct arguments of two basic types: intrinsic and extrinsic justifications.[2] Roughly, a justification of an axiom candidate is intrinsic if it suggests that an axiom is self-evident or otherwise follows immediately from a satisfactory understanding of the basic concepts of set theory.[3] Alternatively, a justification is extrinsic if it supports an axiom candidate through the fruitfulness or desirability of its consequences for some mathematical purposes.[4] We will briefly consider each of these types of justification in turn.

With regard to the former, intrinsic justifications attempt to defend axiom candidates by showing that they in some way follow immediately from the very nature of sets or of set theory. Kurt Gödel thought that such considerations could justify the acceptance of axioms extending ZFC, namely, what are now known as the small large cardinal axioms:[5] "the axioms of set theory by no means form a system closed in itself, but…the very concept of set on which they are based assert the existence of still further iterations of the operation 'set of.'"[6] Though such axioms would follow directly from the nature of set theory as a discipline, Gödel thought that this connection need not be obviously apparent, instead requiring sophisticated logical machinery to draw out: "there

[1]See Maddy (1988a) and Maddy (1988b). We will rely heavily on Maddy's account of the justification of $ZFC + LCs$ in this section.

[2]This distinction is originally due to Zermelo (1908a), and was subsequently popularized by Gödel (1983), though Russell's discussion of the regressive method of axioms in Russell (1973) provides an important precursor to this distinction.

[3]As Maddy puts it simply: an intrinsic justification is one which claims an axiom is "self-evident, intuitive, [or] part of the 'concept of set'" (Maddy, 2011, p. 47).

[4]Again, put simply: an extrinsic justification claims an axiom is "effective, fruitful, [or] productive" (Maddy, 2011, p. 47).

[5]A large cardinal axiom asserts the existence of a stage in the ordinal hierarchy that goes beyond those that provably exist in ZFC: these axioms typically imply that there is a set-sized model of ZFC within the class-sized set-theoretic universe. A large cardinal axiom is considered small if it is consistent with the theory $ZFC + V = L$ (more on this theory below). Alternatively, a large cardinal axiom implies $V \neq L$.

[6]See Gödel (1983), p. 476

may exist…other (hitherto unknown) axioms of set theory which a more profound understanding of the concepts underlying logic and mathematics would enable us to recognize as implied by these concepts."[7] Nonetheless, the distinguishing feature of an intrinsic justification is that it relies on the internal properties of the axiom itself, without making reference to its broader consequences for the theory of sets as a whole.

As a paradigmatic example of an intrinsic justification, consider the axiom of replacement. Replacement is an axiom schema which asserts that, for any function, if a collection can be generated by applying the function to a set, then the generated collection is itself also a set.[8] In Zermelo's original axiomatization of set theory, there was no version of the axiom of replacement.[9] However, after Fraenkel and Skolem independently realized the necessity of replacement for proving the existence of certain ordinal and cardinal numbers,[10] Zermelo accepted replacement as a happy amendment to his theory, leading to the now-standard axiomatization ZFC. While many arguments have been put forward for replacement, one historically important example relies on the notion of limitation of size. By this argument, it is part of the concept of 'set' that any collection which is not so large as to be inconsistent is in fact a set.[11] In this light, replacement can be defended as asserting that any collection which is 'no bigger than' a genuine set (in the sense of being a functional image of that set) must also fail to be inconsistently big, and so also forms a set. While there are a number of important consequences of replacement for set theory, which would make it quite difficult to live without the axiom schema of replacement, note that this argument does not make any reference to these consequences;[12] instead, it defends the schema as following directly from the (suitably understood and interpreted) concepts underlying contemporary set theory.

While many philosophers have traditionally preferred intrinsic arguments for set-theoretic axioms as either more reliable or providing greater illumination of the axiom candidate, it is worth noting that this preference seems to

---

[7] See Gödel (1983), p. 477.

[8] Formally, if a class F is a function, then for any set X there exists a set $Y = F(X) = \{F(x)|x \in X\}$.

[9] See Zermelo (1908b).

[10] For example, the ordinal $\omega \times \omega$ or the cardinal $\aleph_\omega$. In contrast to large cardinals, these are fairly small infinite numbers, and do not imply the existence of models of set-sized set theory.

[11] See Hallett (1986), especially ch. 4 and 5, for a detailed examination of the history of limitation of size arguments in set theory, as well as a defense of the importance of such justifications for understanding set theory today.

[12] See Maddy (1988a) pp. 489–90. In particular, replacement is necessary for transfinite recursion and for Borel determinacy. It is often defended with alternative extrinsic arguments on this basis.

be in stark contrast to the norms within set-theoretic practice itself. As Maddy notes, this philosophical preference seems to be tied to the misplaced hope that working set theorists would be in the business of providing a foundationalist epistemology for all of mathematics and logic from the axioms of set theory.[13] For example, Mary Tiles, in her classic survey of the philosophy of set theory, argues that it would be circular to defend set-theoretic axioms in terms of their consequences, only then to claim these axioms underlie all of logic: "...the conception of set theory as providing a logical foundation for mathematics. To claim this status for set theory it is necessary to claim an independent and intrinsic justification for the assertion of set-theoretic axioms. It would be circular indeed to justify the logical foundations by appeal to their logical consequences, i.e., by appeal to the propositions for which they are going to provide the foundations."[14] Regardless of whether such arguments establish the priority of intrinsic evidence for mathematical foundations understood in such a strong epistemological sense, Maddy notes that this simply is not what actual set theorists are after. Instead, set theorists pursue a wide variety of mathematical goals independently of the hopes of epistemic foundationalists.[15] For these more properly mathematical foundational roles, there seems to be no reason to prejudge the matter and prefer intrinsic justifications to any other sort of justificatory argument.[16]

On the other hand, extrinsic justifications seek to defend an axiom candidate broadly in terms of the desirability of their consequences for set theory as a whole, or for other areas of mathematics. In contrast to intrinsic justifications, extrinsic ones make full reference to the role of an axiom candidate in set theory against the backdrop of the collection of already accepted axioms. After discussing his hope for intrinsically justified axiom candidates going beyond ZFC, Gödel goes on to suggest that the open questions of set theory might instead be decided only by extrinsically justified axiom candidates: "There might exist axioms so abundant in their verifiable consequences, shedding so much light upon a whole field, and yielding such powerful methods for

[13] See Maddy (2011), part V section 4 for a detailed treatment of such preferences for intrinsic justifications.

[14] See Tiles (1989).

[15] See Maddy (2017) for a detailed taxonomy of the different roles that set theory is put to as a foundation of mathematics. As Maddy notes, only the purported foundational task of "epistemic source" requires such a strong preference for intrinsic justifications: see p. 291 especially.

[16] Maddy (2011) argues for a stronger claim, namely that extrinsic justifications are preferable for set-theoretic axioms, and intrinsic justifications attain only a defeasible, heuristic status. While the author finds this argument compelling, for present purposes no more is needed than that intrinsic and extrinsic justifications are on a relatively even footing. For this reason, we will remain agnostic about this further question.

solving problems...that, no matter whether or not they are instrinsically necessary, they would have to be accepted at least in the same sense as any well-established physical theory."[17] Gödel notes that such an axiom candidate should be accepted on this basis "even in case it had no intrinsic necessity at all."[18] This suggestion has proven somewhat prophetic, as set theorists have since developed axiom candidates, even the strongest advocates of which do not pretend to have any serious intrinsic support, but which are capable of settling just the open questions that Gödel had in mind.[19]

It will be instructive to consider two important examples of extrinsic justifications in the history of axiom selection debates. First, we consider Zermelo's defense of his axiom of choice, the event which proved the impetus for axiomatizing set theory in the first place.[20] Originally, Zermelo introduced the axiom of choice to prove that every set has a well-order.[21] However, there was significant backlash over Zermelo's use of choice in the proof, with many fearing that the axiom was an improper assumption, if not downright contradictory. As a result, Zermelo returned to the matter just a few years later, producing a new proof of the well-ordering principle, as well as providing and defending the first explicit axiomatization of set theory, including the axiom of choice.[22]

In this latter work, Zermelo is refreshingly explicit about what he takes the justifications for choice to be. First noting that there may be some claim to the intrinsic self-evidence of the principle—after all, many of the axiom's strongest critics unknowingly utilized the axiom throughout their own work[23]—Zermelo quickly shifts focus to extrinsic justifications for the axiom. Regardless of whether the appeal to self-evidence is accepted, Zermelo claims, the axiom of choice simply cannot be rejected due to its importance for many key theorems of set theory, analysis, and algebra. For this reason, Zeremlo suggests that the axiom must be accepted as essential to mathematics, and

[17] See Gödel (1983), p. 477.

[18] Ibid.

[19] In particular, consider large cardinal axioms such as the existence of infinitely many Woodin large cardinals with a measurable cardinal on top, Woodin's $V = Ult(L)$ axiom, and the strong forcing axioms, such as MM^{++}. For the first, see Koellner (2014) section 4. For the $V = Ult(L)$ axiom, see Woodin (2017). For strong forcing axioms, see Magidor (2012).

[20] See Moore (1982) for a comprehensive account of this historical episode and its aftermath.

[21] A well-order on a set is an ordering such that every non-empty subset has a smallest member. See Zermelo (1904) for this proof. In fact, the axiom of choice and the existence of well-orders on every set are equivalent in ZF.

[22] See Zermelo (1908a) for the new proof of the well-ordering principle; see Zermelo (1908b) for the presentation of his axiomatization.

[23] See Moore (1982) chapters 3.1 and 3.4.

thereby justified through its good works: "the question that can be objectively decided, whether the principle is necessary for science, I should now like to submit to judgment by presenting a number of elementary and fundamental theorems and problems that, in my opinion, could not be dealt with at all without the principle of choice."[24] Since so many results in set theory and wider mathematics relied on the axiom of choice, rationality required accepting the somewhat surprising axiom candidate, instead of ejecting the prior successful mathematics: "principles must be judged from the point of view of science, and not science from the point of view of principles fixed once and for all."[25] When in 1938 Gödel proved the consistency of choice relative to the axioms of ZF, thereby showing conclusively that there is no greater risk of paradox from accepting choice, ZFC became the de facto standard for set-theoretic research.[26] So strongly was this evidence taken to be that when, in 1962, the axiom of determinacy was proposed, it was almost immediately rejected for contradicting the axiom of choice.[27]

In the case of Zermelo's own argument for the axiom of choice, it was found that the axiom was necessary for important theorems and results already utilized by the mathematical community at the time. This is in marked contrast to another important extrinsic justification which has been offered for choice, that it leads to a structure of the cardinal numbers which is orderly and well behaved. Without choice, we cannot know that there is a linear ordering on the cardinal numbers: prior to Zermelo's statement of choice, it was a live possibility that the cardinality of the real numbers simply was not representable as an aleph at all.[28] Instead, there are models of $ZF + \neg AC$ which have incomparable cardinalities, with the resulting structure on the cardinals capable of becoming quite wild and unruly.[29] While in principle this need not prevent systematic study of the cardinal numbers, in practice the lack of AC has proved a quite significant barrier for the development of any such study: without the general facts about the relationships between cardinalities provable with choice, there simply is no satisfactory starting point for a theory of the structure of the cardinal numbers. On the other hand,

[24] See Zermelo (1908b), pp. 187–8.

[25] See Zermelo (1908b) p. 189.

[26] See Gödel (1940).

[27] See Maddy (1988b), section 5. Instead of considering the axiom of determinacy, set theorists sought out restrictions of the axiom consistent with ZFC.

[28] See Moore (1982) section 2.1 for more on this fact, and its relationship to König's attempted refutation of the Continuum Hypothesis.

[29] See Jech (1973) for more on the cardinal numbers without choice.

Zermelo's axiom of choice led to the development of the strong general theory of the cardinals which underlies much of contemporary set theory. In this way, choice is in effect a necessary prerequisite for any unified and tractable theory of the cardinal numbers.

Given the importance of the study of cardinalities for set theory as a whole, and the practical necessity of choice for this study, we find a different sort of justificatory story supporting choice. In particular, this extrinsic argument suggests that since many of the goals of set theory require a useful theory of cardinal numbers, and choice provides just such a theory, choice should be accepted in our theory of sets.[30] That is, choice should be accepted since it opens up a fruitful and important avenue of future inquiry that would otherwise remain hopelessly clouded and obscure. In stark contrast to the implications of choice for areas of core mathematics which were identified by Zermelo himself, the structure theory of the cardinals under choice did not precede the acceptance of choice by the set-theoretic community. That is, choice did not vouchsafe some antecedent part of mathematical practice. Instead, choice was seen as permitting work on good, new mathematics, and the desirability of this further mathematics (compared to the alternative) encouraged set theorists to accept choice and pursue this new facet of set-theoretic research. There simply were not prior theorems about the cardinal numbers that required the acceptance of choice; the systematic study of this field of set theory followed after choice had been accepted. Without the axiom of choice implying that every set has *some* cardinality, even the basic theorems of cardinal arithmetic and cardinality comparison would become prohibitively onerous to pursue.[31]

These two cases reveal an important difference between two sorts of extrinsic justificatory arguments. On the one hand, there are arguments like Zermelo's argument of the necessity of choice for core mathematics, which aim to show that an axiom candidate is already central to mathematical practice as it exists. Similarly, there are arguments which support an axiom candidate on the basis of it providing simpler proofs of already established theorems, or of providing unified proofs of previously disparate and unconnected mathematical facts.[32] Given the focus of such arguments on the relation of the axiom candidate to prior mathematical activity, we will refer to such arguments

[30]In particular, the goal of providing a measure of the consistency strength of mathematical theories requires a tractable theory of the cardinal numbers. See Maddy (2017) for more on this goal.

[31]See Maddy (1988a) Section 1.7 for a brief treatment of the extrinsic arguments in support of choice. See Moore (1982) Ch. 4 for an extended treatment of these arguments in their historical context.

[32]See Martin (1998) for examples of extrinsic arguments of these sorts.

as *backwards-looking* extrinsic justifications. On the other hand, arguments like the desirability of choice for the study of cardinal numbers have no such connection to prior mathematical results; instead, these sorts of arguments aim to justify an axiom candidate on the basis that it is more apt to lead to fruitful and productive mathematics in the future. Given the focus of such arguments on future avenues of mathematical research, we will refer to such arguments as *forward-looking* extrinsic justifications. While the differences between these categories are hopefully apparent, it is worth noting that there are wide differences between the sorts of arguments even within one of these categories. The scope of extrinsic justifications is wide and varied, including a plethora of distinct argumentative strategies.

With this background in place, we are now prepared to introduce the main focus of the current chapter, the problem of extrinsic evidence. Put informally, the problem is how the acceptance of an axiom candidate on the basis of its desirability can possibly be rationally acceptable. In other words, what vouchsafe do we have against the possibility that, in spite of how necessary some axiom proves to be for mathematical science, it is unfortunately the case that set theorists are rationally compelled to endorse instead some less appealing axiom: unlucky us! Compounding this concern is the apparent fact that extrinsic justifications seem unreliable in other domains besides mathematics. Even after it was established how useful the atomic theory was for late nineteenth-century physics and chemistry, with every sort of extrinsic justification applying to the theory, these concerns were not considered sufficient for establishing the existence of atoms; instead, full endorsement of the atomic theory was withheld until Perrin's experiments were able to establish direct support for the theory.[33] If extrinsic arguments did not suffice for the atomic theory, then why are they enough for the existence of large cardinals? Even worse, there is a seeming parallel between extrinsic justifications and obviously defective arguments like William James's argument from the subjective necessity one might have for God's existence to the existence of God; no matter how fruitful God's existence might be for the life of a believer, this just doesn't seem to be the sort of evidence that could establish a substantive fact about divinity.[34] The problem of extrinsic evidence demands an explanation of how extrinsic justifications can be acceptable in set theory, while similar arguments in other areas seem to be merely misguided bouts of wishful thinking.

---

[33] See Maddy (1997), pp. 135–43 for this historical case study.

[34] See James (1979) for the classical origin of this argument. See Misak (2016) for a contemporary examination of this argument.

We must be somewhat careful with precisely how the problem of extrinsic evidence is expressed. One natural proposal would be to express the problem as explaining how extrinsic evidence can be a reliable guide to set-theoretic truth; in fact, this is precisely how Maddy first formulated the problem.[35] Unfortunately, this formulation is a bit problematic: it is an open question in the metaphysics of set theory whether there is a notion of truth in set theory, and, if so, how this notion of truth relates to the truth concept in other areas of inquiry. Arealists, nominalists, and fictionalists, among others, all reject the notion of truth in set theory. Furthermore, the problem of extrinsic evidence is just as pressing for those who reject set-theoretic truth as for those who accept it, if not more so: Without an objective notion of mathematical truth, is there anything that constrains axiom selection, or does radical subjectivity reign supreme? For this reason, it would be best to formulate the problem in a way that does not beg the question against any of the live options for the metaphysics of sets.

Instead we will formulate the problem in a metaphysically neutral way, as comprising two related questions. The first, epistemological question asks what the general structure of an extrinsic justification is, and how these justifications functionally provide support for a particular theory of sets. Ideally, an answer to this question would make clear the criteria which could distinguish a good extrinsic justification from a specious one, perhaps even providing a rational basis for deciding between the very different sorts of extrinsic arguments offered in favor of strong axiom candidates like $V = Ult(L)$ and the forcing axioms. The second, metaphysical question asks what sets are like such that extrinsic justifications suffice to motivate theory selection, unsupplemented with further, more 'direct' sorts of evidence. An answer to this question should make clear what the difference between mathematics and empirical science is such that extrinsic justifications are given greater weight in the former than the latter. A wholly satisfactory solution to the problem would make it clear what limited domain extrinsic justifications are reliable in, explaining why they should not be used in other contexts (the metaphysical component), as well as providing guidelines for how to evaluate proposed extrinsic arguments (the epistemological component). A solution to the problem of extrinsic evidence would therefore combine a general theory of extrinsic justifications with a compatible account of the metaphysics of sets.

---

[35] See Maddy (2007), p. 366: "To put the problem more generally: granting that selecting for maximization generates theories we like, what reason do we have to think it likely to generate theories that are true?".

Only if a theory of extrinsic justifications meets both of these criteria can it be considered an acceptable response to the problem of extrinsic evidence.

Note that in the present chapter we are not concerned with the question of whether extrinsic justifications are acceptable in set theory. Instead, for present purposes we take as an operating assumption that the sorts of extrinsic justifications which have historically led to successful mathematics are in fact reliable. With this assumption in place, we then focus instead on seeking whether there is a naturalistically acceptable account of how these justifications work. If there were no solution to the problem of extrinsic justifications as here stated, one might naturally worry about the reliability of extrinsic arguments. In the remainder of this chapter, we aim to show that a satisfactory solution can be provided, representing a further reason to defend the use of extrinsic justifications in set theory.

2 Three Recent Responses to the Problem

With this background in place, we are now prepared to consider the various possible responses to the problem of extrinsic evidence. As it so happens, there have been three distinct recent attempts to solve this problem. The first, by Barton, Ternullo, and Venturi, attempts to explain extrinsic justifications as a sort of broadly abductive form of reasoning, more or less continuous with the use of inference to the best explanation in empirical science.[36] The second, by Soysal, attempts to solve the problem through a novel account of analyticity inspired by the late Carnap.[37] Finally, the third and original response, by Maddy, seeks to explain extrinsic justifications as a sort of practical means-ends reasoning, leading to a surprising account of the nature of sets.[38] In this section, we will consider each of these responses in turn, before proposing a novel response to the problem in the following section.

Let us first consider the response given by Barton, Ternullo, and Venturi. In this account, Barton et al. argue that there is no epistemically significant distinction between intrinsic and extrinsic justifications, but instead all justification of set-theoretic axioms takes a single form: inference to the best explanation. As they summarize the view, "justification of axioms in set theory consists of finding the best explanations for relevant mathematical data"

[36] See Barton et al. (2020).
[37] See Soysal (2024).
[38] See Maddy (2011).

where mathematical data consists mostly of "the body of axioms and theorems accepted by that community at that time."[39] For example, Zermelo's argument for the axiom of choice on the basis of its necessity for important theorems in core mathematics can be seen as proposing an explanation (the axiom of choice in its full generality) for a disparate range of mathematical data (each of the theorems in algebra and analysis which require a restricted choice axiom, even if not the full general axiom). On this approach to the problem, then, extrinsic arguments are a rational basis for theory choice in set theory in just the same way that an inference-to-the-best-explanation can be in natural science. Furthermore, Barton et al. claim that these sorts of extrinsic considerations are broadly acceptable in other areas of inquiry. Much of the rest of this account attempts to lay out a notion of explanation suitable to mathematical contexts, but we will pass over these complications at present. Let us call this proposed solution the *abductive response.*

There seem to be two distinct challenges to an abductive response to the problem of extrinsic evidence faces. The first applies to the epistemological question, which asks what the nature of extrinsic justifications is. Here, Barton et al. have a clear answer: An extrinsic argument for an axiom is an argument that the axiom best explains some prior collection of mathematical axioms and theorems. But, while it is clear how some extrinsic arguments fit this mold—in particular what we have called the backwards-looking extrinsic arguments—it does not seem plausible that all extrinsic justifications have this form. For forward-looking extrinsic arguments, there simply is no antecedent mathematical work that can be cast as the data to be explained by the axiom candidate. Instead, in the case of the argument for choice from the desirability of the theories of cardinality it makes possible, the acceptance of choice was necessary prior to mathematical work proceeding in this field at all! Instead of following and providing a more general account of historical theorems, the forward-looking argument sought to open a hitherto unexplored frontier of research and proof-activity. It frankly does not seem remotely plausible to claim that this argument for choice was an attempt to explain a body of theorems about cardinal numbers that would take fifty years to develop. Thus, the abductive response faces a clear dilemma with respect to the epistemological question: It must either find a way of arguing that forward-looking extrinsic justifications are nonetheless proposed explanations for data that will exist in the future, or make the (decidedly non-naturalistic) move of rejecting these

[39] See Barton et al. (2020) p. 14–15. They add that mathematical data might also include "non-theorem-like observations," but we leave this additional complication to the side.

sorts of extrinsic justifications as improper, despite their widespread use and acceptance by the working set theorists engaging in axiom selection debates.[40] Neither response seems particularly promising.

Nonetheless, even if proponents of an abductive response are able to account for forward-looking extrinsic justifications (as defenders of abduction in empirical science might well believe they can!), there remains a further, more serious challenge. The abductive response seeks to account for extrinsic justifications in set theory through an analogy with abductive inferences in the natural sciences; but in stark contrast to the situation in set theory, abductive inferences are *not* always seen as a sufficient basis for theory choice in natural science. The efficacy of abductive inference in natural science is a central question in contemporary philosophy of science, with realists defending inference to the best explanation as providing only weak and defeasible support for a theory, and critics arguing instead that abductive inference does not provide *any* justification for a theory whatsoever.[41] Furthermore, this resistance to using abductive justification alone as a basis for theory choice can also be found in the history of physics: with the nineteenth-century theory of atoms, as well as subsequent theories of subatomic particles, direct evidence for the existence of microscopic entities was seen as necessary for the acceptance of these theories, in spite of all the prior abductive evidence in support of these theories.[42] Thus, there is a strong disanology between inference to the best explanation in empirical science and extrinsic justification in set theory, with only the latter being seen as sufficient by itself to justify accepting a theory. Thus, the question remains for Barton et al. of why abductive inference is so much stronger in set theory than in other areas of inquiry, and why this increased reliance on abduction is rationally permitted. In effect, what is needed is a response to the metaphysical component of the problem of extrinsic evidence, explaining what it is about set theory that justifies the comparatively

---

[40] It is worth noting that not all modern set theorists are in fact engaged in the question of axiom selection. First, multiverse theorists such as Joel Hamkins believe that no axioms beyond ZFC—or perhaps even less—should be accepted wholesale, and instead that practitioners should study all of the various models of a weak theory of sets. See in particular Hamkins (2015), as well as Hamkins (2012), for this sort of position. Secondly, some areas of current set-theoretic research, such as cardinal characteristics of the continuum, do not need axioms beyond ZFC to be pursued. Neither of these two sorts of set theorist would necessarily be motivated to pursue axiom selection. Nonetheless, many set theorists today do pursue exactly these sorts of questions, and this chapter may be considered an attempt to determine how best to choose axioms for those mathematicians and philosophers who work on axiom selection. Thanks to an anonymous reviewer for stressing this point.

[41] For tempered realist defenses of abduction, see Psillos (1996) and Lipton (1993). For critiques of even a restricted role for abductive inference in rational theory choice, see van Fraassen (1980), van Fraassen (1983), and Stanford (2010).

[42] See Maddy (1997) chapter II.6 and Maddy (2007) section IV.1.

lofty status of inferences to the best explanation in this domain. Without such an account of the nature of sets, a sizable explanatory gap remains in accounting for the use of extrinsic justifications in set theory compared to other areas of inquiry.

In contrast to the abductive response, the account of extrinsic evidence given by Soysal focuses on the nature of set-theoretic language. Soysal argues that the key to understanding extrinsic justifications lies in the process of how the concept of 'set' gets its meaning. Assuming metasemantic descriptivism, the meaning of 'set' includes any and all properties that the vast majority of the linguistic community of set theorists would assent to it having.[43] In particular, this includes the property of 'extrinsic justifications are evidence for truths about X.' That is, it is simply part of the conceptual apparatus shaping set theory that extrinsic justifications are reliable; furthermore, Soysal notes that this fact is actually an analytic truth about the words 'set' and 'membership.' On this account, in fact, seemingly all undisputed mathematical claims are analytic truths in virtue of the intersubjective agreement of the set-theoretic community. Intrinsic justifications, then, show that some set-theoretic claim follows from the meaning of 'set' and 'membership' in some particular, determinate way; extrinsic justifications, on the other hand, claim that an axiom is an opaque analytic truth, following from the meanings underlying set theory in some-way-we-know-not-how.[44] For example, on this response Zermelo's defense of choice as required for core mathematics would really amount to a claim that (enough of) the mathematical community implicitly assumes 'all sets have a choice set,' with these beliefs combining to be partially constitutive of what it means to be a set. Let us call this proposed solution the *analyticity response*.

As with the abductive response, there seem to be two distinct serious challenges to the analyticity response. Again, the first challenge relates to the epistemological component of the problem of extrinsic evidence. A satisfactory solution to the problem should provide at least some informal criteria for distinguishing between good and bad extrinsic arguments: after all, the history of set theory is littered with examples of rejected arguments for particular axiom candidates.For example, consider the intricate and complex extrinsic

[43]This approach to language and analyticity is directly inspired by the later Carnap, especially Carnap (1947). It is worth noting that this approach to language is markedly different from that of the logical positivist view of Carnap (1937). In particular, the former has a much more open-ended approach to analytic truths, in contrast to the strict formal notion of analyticity found in the latter.

[44]See Soysal (2024), p. 19. Soysal suggests that this may in effect collapse the intrinsic/extrinsic distinction, leaving all set-theoretic arguments as intrinsic justifications.

justification for $\neg CH$ found in Woodin (2010), and the subsequent arguments against this justification found in Woodin (2017). In particular, we note that this example shows that there is basis for rational disagreement about whether an extrinsic justification is successful even without disagreement between distinct mathematicians. Unfortunately, the analyticity response does not provide any epistemological criteria with which to evaluate a proposed extrinsic justification; instead, the only seeming way to evaluate such a proposal is to put it to a vote in the set-theoretic community, and see whether it achieves a sufficient threshold to demonstrate its hidden relationship to the meanings of 'set' and 'membership.' This is problematic on two counts. First of all, there is little insight to be gained from this account of extrinsic evidence: even if the analyticity account is true, a separate description of how extrinsic evidence is structured in more explicitly mathematical terms would be required for a further understanding of set-theoretic methodology.[45] Secondly, one might naturally fear that such an account leads to set theory being a wholly unobjective discipline, with the set-theoretic community constrained by nothing but its own arbitrary will. Such a radically subjective account of set theory strongly contrasts with the sense within set theory of objective constraints on successful arguments. A naturalistically acceptable response to the problem of extrinsic evidence should therefore be extremely reticent to accept such a view, unless compelled to by the force of evidence.

Nonetheless, even if this concern can be overcome, and metalinguistic descriptivism is accepted as the correct account of set-theoretic language, the analyticity response faces a further challenge. Regarding the metaphysical component of the problem of extrinsic evidence, Soysal is unfortunately silent. There is no account of what the subject matter of set theory is that makes extrinsic justification so much more powerful than in other areas of scientific inquiry; instead, it is taken as a brute and opaque fact that this methodology is rationally acceptable in set theory, since it is in fact accepted by set theorists. As described above, this is in stark contrast to the standards of empirical inquiry in the sciences. But not just any collection of meaning-constitutive claims can be combined to generate a coherent and useful scientific concept: What is it about set theory that accords with a strong acceptance of extrinsic justifications in contrast to other areas of inquiry? One might naturally hope that a full response to the problem of extrinsic evidence would explain the source of the strong disanalogy between empirical science and set-theoretic methodology, if not providing an explicit defense of the methods of set theory as following

[45] See Maddy's response to Soysal (2024) in the same volume for a closely related critique.

from its objects and its aims. That is, as with the abductive response, the analyticity response calls for supplementation with an account of the nature of sets that explains set theory's unique acceptance of extrinsic justification without requiring some further sort of direct evidence.

Thus, abstracting from the particular accounts of the nature of extrinsic justification given by the abductive and analyticity responses, an explanatory gap remains to be filled. In either case, while an epistemological account of extrinsic evidence is provided, there is no accompanying response to the metaphysical component of the problem. As a result, any acceptance of either account must remain conditional, requiring further work to establish whether there is an account of the nature of set theory compatible with the lofty status that extrinsic justifications are given. Answering this further challenge is certainly a non-trivial task. Furthermore, given the additional challenges the accounts face even regarding the epistemological portion of the challenge, there is reason for skepticism regarding whether these accounts can be successfully developed to meet this further challenge. For this reason, we will now put these two accounts to the side, and consider a third response to the problem of extrinsic evidence. Unlike the two alternatives already considered, this next response directly addresses both the epistemological and metaphysical components of the problem of extrinsic evidence.

Maddy's response to the problem of extrinsic evidence was developed over an extended period of time, with a number of important steps in its development: in particular, elements of Maddy (1997, 2011), and Maddy (2017) each play an important role in generating a multifaceted account of the nature of extrinsic evidence in set theory.[46] Maddy begins with an account of set theory as a human activity, with theories, methods, and tools being developed to serve concrete mathematical purposes. As a foundational subject in mathematics, many of these purposes relate to the field of mathematics as a whole. Among these goals are shared standard (to develop a single theory from which counterparts of all mathematical theorems can be proven), generous arena (to provide a single arena where counterparts for the objects of the many subfields of mathematics can all be studied together), and risk assessment (to permit the measurement and comparison of the consistency strengths of various theories).[47] Importantly, these goals are all inherently mathematical,

[46] In the next few paragraphs, we will develop an account of Maddy's response to the problem of extrinsic evidence from mainly these three works. While the account differs in focus and in emphasis on some points, it is our hope that this reconstruction of the view is largely in keeping with the spirit of Maddy's view.

[47] See Maddy (2017), esp. section 1.

independent of any philosophical projects that might call for a foundation for mathematics. In this way, set theory is just one branch of mathematics among many others, each defined by its particular goals.

With this background understanding of set theory in place, much of set-theoretic practice can be seen as taking the form of means-ends reasoning: Given some goal(s) of set theory, what axioms generate the best theory for accomplishing said goal(s)? That is, set-theoretic inquiry is not best understood as a primarily descriptive discipline, seeking to articulate the truths of some distant mathematical realm, but instead a thoroughly practical branch of mathematics, developing the best tools for particular mathematical tasks, and then actively applying those methods to the task at hand.[48] The role of extrinsic justifications in set theory, then, is to highlight the suitability of some axiom candidate for the goals of set theory, as they are found in both historical and contemporary practice.[49] That is, the general form of an extrinsic justification of some axiom candidate is an argument that, for some particular set-theoretic goal or goals, that axiom candidate is the best means for achieving those ends. For example, Zermelo's argument for choice on the basis of its necessity for proving theorems in core mathematics can be understood as a proposal that choice is required for the goal of shared standard: without choice, our chosen theory of sets would simply not be strong enough to prove important theorems of analysis and algebra. Similarly, the argument for choice on the basis of the desirability of the orderly theory of cardinal numbers it permits can be understood as a proposal that choice is required for the goal of risk assessment: a well-behaved linear ordering of the cardinal numbers is an essential prerequisite for systematic comparison of the consistency strengths of various theories through the large cardinals. Given this analysis of extrinsic justifications, the criteria for evaluating such an argument is clear: An extrinsic justification is good just in case it successfully establishes an axiom candidate as the proper means for achieving some mathematical goal. Extrinsic justifications should therefore be evaluated according to the canons of practical reason, and not those of theoretical reasoning. Since this account considers extrinsic justifications as practical bits of means-ends reasoning, let us call this proposed solution the *practical response*.

With Maddy's response to the epistemological component of the problem of extrinsic evidence in place, we now turn to the metaphysical portion of this problem: What must sets be like such that means-ends reasoning is sufficient

[48] See Maddy (1997), esp. section III,4.

[49] See Maddy (2017) for an analytical/historical approach to understanding the goals of set theory, and see Kant and Džamonja (2019) for a more sociological/empirical approach to identifying set theory's goals.

to motivate the choice of a particular axiomatization? Maddy first notes what this epistemology implies set theory cannot be. If means-ends reasoning does not require supplementation with further, more direct sorts of evidence—as current set-theoretic practice suggests it doesn't—then set theory cannot be a descriptive discipline in analogy with physics, which seeks to describe truths of an independently existing realm of set-theoretic objects.[50] Without such additional evidence, the possibility remains open that the set-theoretic reality is simply radically unconducive to our particular mathematical goals. Thus, such a robust form of realism is not compatible with the analysis of extrinsic justifications given by the practical response; either the metaphysical account or the epistemological analysis must be rejected. As a naturalist, seeking to preserve the mathematics as much as possible, Maddy chooses to reject robust realism. But then what *could* the nature of sets instead be?

Maddy concludes that there are two possible answers to this question: a thin form of realism and arealism. While on the surface these two accounts of the metaphysics of sets seem wildly different—after all, the former holds that there are sets, while the latter does not—Maddy posits that there is in fact no substantive disagreement between the two views. Before articulating these two positions, it will therefore be instructive to consider how this could possibly be so. With that in mind, we must make a few remarks on the nature of language as a human activity.

It is an unfortunately appealing temptation to assume that our well-worn linguistic concepts have an unlimited range of applicability. Presented with a wild story of a possible world far different from the actual world we inhabit, it can be natural to assume that there must be some fact of the matter about how our words apply there. Similarly, our most trusted concepts, fruitfully applied to medium-sized dry goods over the centuries, can appear to apply to the domains of the very small or very large without issue, if only we reflect sufficiently on our linguistic intuitions. Alas, this assumption of conceptual preparedness—that our concepts must be applicable in a uniquely correct manner in whichever real or imagined situation we encounter—has come under scrutiny from philosophical and scientific quarters.

Conceptual unpreparedness and its implications are a common theme in the writings of J.L. Austin. In "The Meaning of a Word," Austin presents a rather compelling example of the limitations of our words.[51] Given Austin's way with prose, it is worth presenting this argument in its fullness:

---

[50] See Maddy (2011), p. 58.

[51] See Austin (1979).

> Suppose that I live in harmony and friendship for four years with a cat: and then it delivers a philippic. We ask ourselves, perhaps, 'Is it a real cat? Or is it *not* a real cat?' 'Either it *is*, or it *is not*, but we cannot be sure which.' Now actually, that is not so: *neither* 'It is a real cat' *nor* 'It is not a real cat' fits the facts semantically: each is designed for situations other than this one: you could not say the former of something which delivers philippics, nor yet the latter of something which has behaved as this had for four years...there is no short description which is not misleading: the only thing to do, and that can easily be done, is to set out the description of the facts at length. Ordinary language breaks down in extraordinary cases.[52]

Austin's point here is that nothing in our established usage of the concept 'cat' can suffice to determine whether such a philippic delivering creature can be correctly described as a cat. Instead, it must be accepted that 'words fail us,' and that any decision on whether or not to make a cat attribution will be arbitrary.[53] In the face of a one-off occurrence, practical wisdom would suggest simply describing all of the facts, and refraining from any decision on how the word 'cat' applies in this context. However, if philippic delivering cats were to become a more common occurrence, it would be natural to instead make some decision on whether such beings should be called cats or not. While it is not clear what factors could be brought to bear on such a decision—perhaps whether the philippic cats and non-philippic cats could be cleanly differentiated, or how common these philippic occurrences came to be—it should be clear by Austin's argument that no facts of prior usage can settle this question. Instead, a genuinely new decision on linguistic usage would be required.[54]

Austin's focus remained firmly fixated on conceptual unpreparedness regarding ordinary language. An additional reason for not considering cases of scientific language, however, was Austin's belief that unpreparedness would only be a problem for the rough-and-ready languages of everyday life. In contrast, Austin argues that a scientific language, like that of physics, would be a maximally conceptually prepared language: "Now no doubt an ideal language would not break down, whatever happened. In doing physics, for example, where our language is tightened up in order precisely to describe complicated

[52] See Austin (1979) pp. 67–8.

[53] See Austin (1979) p. 68.

[54] If the example of a philippic delivering cat is too fanciful for one's tastes, consider instead the example of amorphous ice. While ice is commonly defined in scientific texts as a crystalline structure of solid water, amorphous ice—created by rapidly cooling water—is solid water, without the typical crystalline structure. See Wilson (2006) p. 55, and Maddy (2011) pp. 106–7 for Maddy's presentation of this case.

and unusual cases concisely, we *prepare linguistically for the worst.*"[55] While such optimism regarding scientific languages was perhaps understandable when Austin wrote, at the height of efforts to apply formal logic to clarify scientific theories, work since Austin's time has raised doubts about the conceptual preparedness of even well-understood scientific languages, like those of classical mechanics. In particular, Wilson's *Wandering Significance* serves to establish that Austin's concerns regarding ordinary languages can also be applied to the languages of science.[56] Perhaps even more relevant for our present purposes, Shapiro has recently suggested that Waismann's notion of open texture predicates—in our terms, predicates that exhibit conceptual unpreparedness—can be found in mathematical language.[57]

Returning our discussion to set theory and the problem of extrinsic evidence, Maddy's metaphysical account of the nature of set theory begins from the claim that mathematics presents a situation of conceptual unpreparedness for our concepts of 'truth' and 'existence.' We quite naturally assume that our practices of ascribing truth to statements, and affirming the existence of spatiotemporal objects, must uniquely determine whether the 'truth' concept applies to our best justified mathematical theories, and whether the objects those theories speak about exist. Nonetheless, Maddy argues, this is not the case.[58] There are many analogies between the objects of the empirical world and mathematical objects, but there are just as many disanalogies, for example, sets are nonspatiotemporal and acausal. Should existence be attributed to such objects—emphasizing the similarities between sets and ordinary empirical objects—or rejected—emphasizing the stark differences? Maddy's point is that there simply is no fact of the matter in our prior patterns of usage of 'truth' and 'exists' as to whether they should be extended to this context or not. Any account of set theory as discipline will have to acknowledge both the similarities and the dissimilarities between set theory and empirical science, explicitly accounting for the myriad ways that set theory is unlike any other area of investigation. Once these features are accounted for, nothing of substance hangs on the decision of whether to bestow the honorific 'true' on mathematical theories, or whether to give mathematical objects the lofty title of 'existing.' It is a purely arbitrary choice as to whether or not to extend our

[55] See Austin (1979) p. 68.

[56] See Wilson (2006).

[57] See Waismann (1951), as well as Shapiro (2006), Shapiro and Roberts (2021), and Zayton (2022) for recent work on open texture predicates in mathematics. Many thanks to a second anonymous reviewer for pointing out this connection.

[58] See Maddy (2011) ch IV.5 for this argument.

language and use the old words in this new context. If the former decision is made, thin realism results; if the latter, then arealism.

Thin realism, resulting from a decision to extend the use of the 'truth' concept into the realm of mathematics, asserts that the current best theory of sets is our current best theory of what is true of the set-theoretic universe. In this way, it endorses the pre-theoretical assumption that set theory is in the business of discovering true statements. Nonetheless, one must be on guard to avoid assuming too much about this sort of set-theoretic truth; unlike the truths of empirical science, they are not discovered directly. Furthermore, unlike robust realism, thin realism does allege that the realm of set-theoretic objects is unique, in that only its objects can be studied in these particular ways.[59] Instead, the truths of set theory seem to be rather closely connected to the goals of the set-theoretic community, as revealed by the widespread acceptance of extrinsic evidence. Furthermore, set-theoretic methods seem to be the only way to discover truths about sets, with no need for or even possibility of an independent sort of verification. So long as all of these qualifications are kept in mind, Maddy suggests that there is no barrier to speaking of mathematical truth as a thin realist.[60]

While we agree with Maddy that in theory there is no problem with speaking of truth as a thin realist, unfortunately the situation in practice is a bit less clear. Historically, philosophers of mathematics have often fallen into the temptation to overemphasize the similarities between mathematics and other areas of inquiry. As a result, there has been an unfortunate tendency in the philosophy of set theory for thin realists to implicitly—and perhaps unknowingly—drift toward robust realism, leading to the differences between math and science being ignored. Wittgenstein begins his 1939 lectures by noting just this point about the philosophy of real numbers: "misunderstandings...arise from a tendency to assimilate to each other expressions which

[59] See Maddy (2011) p. 62.

[60] Note that both Barton et al. and Soysal do directly respond to Maddy on the problem of extrinsic evidence. However, both focus on the thin realist version of the response. In the case of the latter, arealism is explicitly excluded from consideration at the outset. In the case of the former, Barton et al. challenge the notion of 'mathematical depth.' The question of depth and its objectivity goes well beyond the scope of the present chapter, but we merely note in passing that depth is only essential for Maddy's thin realism: "what if depth should turn out not to be 'in the math,' but only 'in us'? I say this without pretending to a defense, but my inclination is to think that in the imagined eventuality, the ground beneath thin realism would collapse. We would be left with a version of arealism as the only viable option remaining" (Ernst et al., 2015, p. 248). Thus, Barton et al.'s objections to Maddy, if successful, only provide further support for versions of arealism. In neither case do these objections effect an arealist solution to the problem of extrinsic evidence, like that offered below.

have very different functions in language."[61] In order to guard against this common psychological tendency to rely too much, or too quickly, on analogies between disparate concepts, Wittgenstein proposes to speak in such a way that the differences are immediately apparent in the surface grammar of his theorizing: "Hence I will have to stress the differences between things, where ordinarily the similarities are stressed."[62] Given the problems of a shift into robust realism as a result of overemphasizing the similarities between math and empirical science—and the ensuing Gordian knot of epistemological puzzles such a shift generates—we posit that philosophers of mathematics are best served by guarding against any possible conflation between their metaphysical position and robust forms of realism.

For just this reason, we also find it wise to conduct our theorizing in a way that makes the differences between set theory and empirical sciences immediately apparent. In particular, we propose that theorizing about the metaphysics of sets should be done, at least preliminarily, without making any use of the 'truth' concept. Speaking as an arealist, with the concepts of ordinary empirical inquiry restricted just to the empirical world, we are forced to come to terms with the nature of set theory directly in its own terms. In this way, we prevent ourselves from inadvertently smuggling in non-trivial assumptions about the similarities between mathematics and empirical science. Once we have responded to the problem of extrinsic evidence, and a theory has been articulated and defended without relying on notions of 'truth' or 'existence,' only then should we permit an expansion of these concepts into the mathematical realm. That is, though we agree with Maddy that there is no *theoretical* reason to choose arealism over thin realism, we nonetheless suggest that—at least while engaged in philosophy of mathematics—there are good *pragmatic* reasons to prefer arealism. Since this position proposes arealism as a useful framework for studying the philosophy of set theory, and not as the ultimate metaphysical position, let us refer to this approach as *methodological arealism*, in contrast to metaphysical arealism, which would hold that there are rational reasons to accept the truth of arealism, contra thin realism.[63] In the following section we will use this approach, articulating a response to the

---

[61]Wittgenstein (1989), p. 15.

[62]Wittgenstein (1989), p. 15. There are close connections between the version of arealism advanced below and Wittgenstein's philosophy of mathematics as a whole. While it goes beyond the scope of the present chapter to pursue these connections, see Bangu (2019) and Bangu (2023) for a potentially compatible account of the later Wittgenstein's broader approach to mathematics.

[63]It is our hope that Maddy would consider this a friendly amendment to her view of the relation between thin realism and arealism, expecially in light of the remarks on the necessity of depth for thin realism in Ernst et al. (2015).

metaphysical component of the problem of extrinsic evidence while speaking as an arealist. With this response in place, we will then demonstrate how to reformulate it as a thin realist position, combining the utility of theorizing as a methodological arealist with the ease of speaking about set theory as a thin realist.

3 An Arealist Solution to the Problem

In contrast to thin realism, the arealist rejects the extension of the concepts of 'truth' and 'existence' into the realm of mathematics. For this reason, the arealist finds that set theory is not the study of sets—there are no such things!—and its methods are neither reliable nor truth-seeking. It is important to note, however, that this rejection of mathematical truth is of a markedly different nature than that of the mathematical fictionalist. For the fictionalist, there is a substantial notion of mathematical truth that *could* obtain, but simply doesn't; mathematical theories articulate what it would be like if their claims were true, and mathematical objects did exist. Arealism, in contrast, denies that there is any meaningful notion of 'truth' or 'existence' applicable to abstract, mathematical objects. In rejecting the claim that ZFC is true, the arealist is not denying that some property or theoretical virtue is rightfully attributed to the theory: there simply is no sense to the claim that "ZFC is true." By the same token, there is no meaningful property of existence that sets lack, but instead it just does not make sense to ascribe existence to objects like sets.

If set theory is not about sets, then what is it about for the arealist? It is first and foremost the study of a wide range of methods and tools for achieving a particular collection of mathematical tasks. As Maddy puts it, for the arealist "Set theory is the activity of developing a theory of sets that will effectively serve a concrete and ever-evolving range of mathematical purposes."[64] While the history of set theory is full of varied purposes to which set theory has been applied, there is also a fairly consistent core of mathematical pursuits that have characterized set theory from Cantor and Zermelo up to today: elucidating objects from core mathematics with formal set-theoretic counterparts, providing relative interpretations of mathematical theories, finding shared assumptions underlying proofs in disparate mathematical fields, etc. Such tasks can be understood wholly separate from any notion of a set-

[64] Maddy (2011) p. 89.

theoretic universe, or notion of set-theoretic truth. In this way, set theory is a strongly syntactic discipline, with theories and their manipulations forming the centerpiece of the field, and any apparently semantic reference to particular objects or structures only secondary to the formal tasks at the heart of the field. The arealist therefore finds set theory to be a rich field characterized by its goals and traditional tricks and tools with which it approaches its myriad puzzles. In this way, the arealist view of set theory shares somewhat more similarities with branches of engineering than with theoretical physics.

Does the arealist view amount to anything more than a brute version of formalism? This certainly depends on how formalism is defined. In its strongest versions, though, formalism amounts to the claim that mathematics is nothing more than the manipulation of meaningless symbols in proofs. Understood this way, there are two important differences between the arealist and the brute formalist. First, the arealist would deny that the symbols of mathematics—even a branch of pure mathematics, like set theory—are in any interesting way meaningless. While the symbols of set theory do not attain their meaning by directing at particular objects, there are other ways for symbols and words to become meaningful: through their structured usage in a pattern of language which comes into contact with other bits of human language in regular ways. The language of set theory is a more complicated language game than many areas of empirical science, but nonetheless represents a functional language game, connected to the world through the various mathematical goals it pursues. Secondly, the arealist is focused on mathematical activity broadly considered, and not just proof. The arealist is just as interested in the work of developing new mathematical definitions, evaluating axiom candidates, and non-deductive forms of mathematical reasoning, as well as that of formal and informal proof. Nonetheless, if these two nuances of the brute formalist position are dropped, then there are broad similarities between the arealist and formalist positions on the nature of mathematics.

With the arealist position made clear, we can now return to the explanatory challenge posed by the metaphysical component of the problem of extrinsic evidence: What must set theory be like that extrinsic justifications are rationally compelling? Given the practical response to the problem, this becomes the question of what set theory is like that makes means-ends reasoning itself sufficient for theory choice in mathematics. For the arealist, importantly, no theory of sets is metaphysically special: all theories, regardless of their extrinsic support (or lack thereof), are just as metaphysically 'good' as any other. There is therefore no need for—or even possibility of—demonstrating that a particular desirable axiomatization is connected to the world in some special way, or is uniquely correct in its description of the mathematical world. *Any* theory of sets could be used for our mathematical purposes if we so desired; the question, then, is whether it is practically wise to do so. Given the wide scope of possible choices for our foundational theory, then, the only considerations that can be brought to bear for or against it are practical considerations regarding how useful the theory can be. But these sorts of practical considerations are precisely the sorts of features brought to bear in the means-ends reasoning of extrinsic justifications! Since all of the possible set-theoretic tools are equally acceptable from a metaphysical point of view, there is nothing more to say for or against an axiomatization than how well the tool is suited for the tasks at hand. The only way to rationally discriminate between mathematical theories is through their utility toward our ends.

It is worth noting that on this arealist view, it is something of a misnomer to refer to extrinsic justifications as a sort of evidence. In the empirical sciences, evidence is often understood as a sort of data which serves to indicate the presence of some more observationally distant objects, or the truth of some further less directly verifiable claims. On the arealist view, however, extrinsic justifications do not indicate that any additional, separate property is true of the realm of mathematics, that is, the fruitfulness or desirability of an axiom candidate shows us nothing about the axiom than that it is fruitful and desirable! However, this is not a deficiency of extrinsic justifications, since there simply is nothing else that could be asked for from a mathematical theory. All mathematical theories are equally good (or bad, as it were) at describing the world, so there is no way to differentiate them in terms of their metaphysical accuracy. All that remains is the question of which theories are best suited to our particular, concrete aims and projects. For the arealist, then, extrinsic 'evidence' is really nothing more than an argument for the usefulness of a particular mathematical tool; but, again, there is nothing further that could be coherently asked for from a defense of a set-theoretic theory.

While this arealist response to the problem of extrinsic evidence hopefully seems at least somewhat compelling, one might still feel a bit queasy about the rejection of mathematical truth. After all, as Soysal notes, it is fairly standard in set-theoretic practice to speak of true and false theories, and to speak of the universe of sets; it would be something of a burden if the philosopher were committed to ending this sort of speech as metaphysically improper.[65] Thankfully, however, the methodological arealist has no need—or even desire—to require such a shift in how mathematicians speak. With the response to the problem in place, and free from inadvertent assumptions of

[65] See Soysal (2024), p. 9.

robust notions of mathematical truth, there is no reason not to reintroduce a notion of truth into mathematical discourse. Reformulating the metaphysical response to the problem into thin realist discourse, we find that extrinsic evidence indicates the truth of a mathematical theory because of the very thin notion of truth at play here. There is nothing more to the truth of a theory of sets than that it has the desirable virtues indicated by extrinsic arguments: truth is not a separate property beyond the theoretical virtues, but simply is a concise and pithy way of describing the presence of these virtues. But so long as this is kept in mind, it does no harm to our philosophical research to speak about math in the ordinary ways. In doing so, the philosopher of set theory can work as a methodological arealist, bracketing off any improper suppositions of similarities between mathematics and empirical science, but then go through everyday life speaking about mathematics in the ordinary way. In this way, we hope that methodological arealism combines the philosophical advantages of arealism with the tractability of expression of thin realism.

In conclusion, in this chapter we outlined a central puzzle in the philosophy of set theory: the problem of extrinsic evidence. We then surveyed three recent responses to the problem, finding that only Maddy's practical response mounted a full response to both the epistemological and metaphysical components of the problem. We then outlined a slightly altered version of Maddy's response to the metaphysical problem, using a methodological prescription of arealism to ensure that our reasoning was free of unwanted presuppositions about the nature of mathematics; while this is different from Maddy's own formulation of the response, it is our sincere hope that this will be taken as a friendly amendment to the response. Finally, we demonstrated some of the implications of the practical response for our understanding of mathematics and its relation to human activity. In so doing, we might faintly hope that this chapter has been a stalking horse to establish two thoughts about the proper method for philosophy of set theory. First, we hope we have demonstrated the value of keeping in mind the role of set theory as something that mathematicians do, separate from any reflections on an independent mathematical realm. It is our belief that philosophical reasoning about mathematics can fall into traps and puzzles when it drifts too far from how mathematics is actually used. Secondly, we hope to have shown the usefulness of combining close epistemological study of the methods of set theory with broader reflections about the metaphysics of mathematics. We sincerely believe that both the epistemological and metaphysical work benefit from being done in concert.

References

Austin, J. L. (1979). The meaning of a word. In J.O. Urmson & G.J. Warnock (Eds.), *Philosophical papers*. Oxford University Press.

Bangu, S. (2019). Hard and blind: On wittgenstein's genealogical view of logical necessity. *Philosophy and Phenomenological Research, 102*(2), 439–458. https://doi.org/10.1111/phpr.12661

Bangu, S. (2023). Wittgenstein on proof and concept-formation. *Philosophical Quarterly*, pqad111. https://academic.oup.com/pq/advance-article/doi/10.1093/pq/pqad111/7468832

Barton, N., Ternullo, C., & Venturi, G. (2020). On forms of justification in set theory. *Australasian Journal of Logic, 17*(4), 158–200 (2020)

Carnap, R. (1937). *The Logical Syntax of Language*. K. Paul, Trench, Trubner.

Carnap, R. (1947). *Meaning and necessity: A study in semantics and modal logic*. University of Chicago Press.

Ernst, M., Heis, J., Maddy, P., McNulty, M. B., & Weatherall, J. (2015). Afterword to special issue on mathematical depth. *Philosophia Mathematica, 23*(2), 242–254. ISSN 0031-8019. https://doi.org/10.1093/philmat/nkv002

Gödel, K. (1940). *The consistency of the continuum hypothesis*. Annals of Mathematics Studies (No. 3). Princeton University Press.

Gödel, K. (1983). What is cantor's continuum problem? In P. Benacerraf & H. Putnam (Eds.), *Philosophy of mathematics: Selected readings* (2nd ed., pp. 470–485). Cambridge University Press.

Hallett, M. (1986). *Cantorian set theory and limitation of size*. Clarendon Press.

Hamkins, J. D. (2012). The set-theoretic multiverse. *Review of Symbolic Logic, 5*(3), 416–449. https://doi.org/10.1017/s1755020311000359

Hamkins, J. D. (2015). Is the dream solution of the continuum hypothesis attainable? *Notre Dame Journal of Formal Logic, 56*(1), 135–145. https://doi.org/10.1215/00294527-2835047

James, W. (1979). *The will to believe: And other essays in popular philosophy*. Cambridge University Press.

Jech, T. (1973). Chapter 11 cardinal numbers in set theory without choice. In T. Jech (Ed.), *The axiom of choice*. Studies in Logic and the Foundations of Mathematics (Vol. 75, pp. 151–166). Elsevier.

Kant, D., & Džamonja, M. (2019). Interview with a set theorist. In S. Centrone, D. Kant, & D. Sarikaya (Eds.), *Reflections on the foundations of mathematics* (pp. 3–26). Springer.

Koellner, P. (2014). Large cardinals and determinacy. In E. N. Zalta (Ed.), *The stanford encyclopedia of philosophy*. Metaphysics Research Lab, Stanford University, (Spring 2014 Ed.).

Lipton, P. (1993). Is the best good enough? *Proceedings of the Aristotelian Society, 93*, 89–104.

Maddy, P. (1988a). Believing the axioms. I. *Journal of Symbolic Logic, 53*(2), 481–511. https://doi.org/10.2307/2274520

Maddy, P. (1988b). Believing the axioms. II. *Journal of Symbolic Logic, 53*(3), 736–764. https://doi.org/10.2307/2274569

Maddy, P. (1997). *Naturalism in mathematics*. Oxford University Press, 1997.

Maddy, P. (2007). *Second philosophy: A naturalistic method.* Oxford University Press.

Maddy, P. (2011). *Defending the axioms: On the philosophical foundations of set theory*. Oxford University Press.

Maddy, P. (2017). Set-theoretic foundations. In A. E. Caicedo, J. Cummings, P. Koellner, & P. B. Larson (Eds.), *Foundations of mathematics: Logic at harvard essays in honor of W. Hugh Woodin's 60th birthday*. American Mathematical Society.

Magidor, M. (2012). Some set theories are more equal. *Talk given at the Harvard EFI Project in the Spring of 2012.*

Martin, D. A. (1998). Mathematical evidence. In H.G. Dales & G. Oliveri (Eds.), *Truth in mathematics*. Oxford Science Publications.

Misak, C. (2016). *Cambridge pragmatism: From peirce and james to ramsey and wittgenstein*. Oxford University Press.

Moore, G. H. (1982). *Zermelo's axiom of choice*. Springer. https://doi.org/10.1007/978-1-4613-9478-5

Psillos, S. (1996). On van fraassen's critique of abductive reasoning. *Philosophical Quarterly, 46*(182), 31–47. https://doi.org/10.2307/2956303

Russell, B. (1973). The regressive method of discovering the premises of mathematics. In D. P. Lackey (Ed.), *Essays in analysis* (pp. 272–283). Allen & Unwin.

Shapiro, S. (2006). Computability, proof, and open-texture. In A. Olszewski, J. Wolenski, & R. Janusz (Eds.), *Church's Thesis After 70 Years* (pp. 420–455). Ontos Verlag.

Shapiro, S., & Roberts, C. (2021). Open texture and mathematics. *Notre Dame Journal of Formal Logic, 62*(1), 173–191. https://doi.org/10.1215/00294527-2021-0007

Soysal, Z. (2024). A Carnapian solution to the puzzle of extrinsic evidence. In S. Arbeiter & J. Kennedy (Eds.), *Outstanding contributions to logic: Penelope maddy*. Springer.

Stanford, K. (2010). *Exceeding our grasp*. Oxford University Press.

Tiles, M. (1989). *The philosophy of set theory: An historical introduction to Cantor's paradise.* Dover Publications.

Waismann, F. (1951). Verifiability. *Journal of Symbolic Logic, 19*(1), 117–44. https://doi.org/10.2307/2267243

Wilson, M. (2006). *Wandering significance: An essay on conceptual behavior.* Clarendon Press.

Wittgenstein, L. (1989). *Wittgenstein's Lectures on the Foundations of Mathematics, Cambridge, 1939.* University of Chicago Press.

Woodin, W. H. (2010). *The axiom of determinacy, forcing axioms, and the nonstationary ideal.* De Gruyter Series in Logic and its Applications (Vol. 1). Walter de Gruyter

GmbH, (Revised edition, 2010). ISBN 978-3-11-019702-0. https://doi.org/10.1515/9783110213171

Woodin, W. H. (2017). In search of Ultimate-L: the 19th Midrasha Mathematicae Lectures. *Bulletin of Symbolic Logic, 23*(1), 1–109. ISSN 1079-8986. https://doi.org/10.1017/bsl.2016.34

van Fraassen, B. (1980). *The scientific image*. Oxford University Press.

van Fraassen, B. (1983). Glymour on evidence and explanation. In J. Earman (Ed.), *Testing scientific theories* (pp. 165–176). University of Minnesota Press.

Zayton, B. (2022). Open texture, rigor, and proof. *Synthese, 200*(4), 1–20. https://doi.org/10.1007/s11229-022-03842-4

Zermelo, E. (1904). Beweis, daß jede menge wohlgeordnet werden kann. *Mathematische Annalen, 59*(4), 514–516. ISSN 1432-1807. https://doi.org/10.1007/BF01445300

Zermelo, E. (1908a). Neuer beweis fÜr die möglichkeit einer wohlordnung. *Mathematische Annalen - MATH ANN, 65*(1), 107–128. https://doi.org/10.1007/BF01450054

Zermelo, E. (1908b). Untersuchungen über die grundlagen der mengenlehre. I. *Mathematische Annalen, 65*(2), 261–281. ISSN 1432-1807. https://doi.org/10.1007/BF01449999

5

The Hidden Use of New Axioms

Deborah Kant

1 Introduction

Philosophers of set theory are very interested in new axioms of set theory, typically related to the question of axiom adoption. In this chapter, I analyse part of set-theoretic practice by presenting the hidden use of new axioms, a specific way new axioms are used by set-theoretic practitioners. This relates to the question of axiom adoption, and I argue that the hidden use of a new axiom provides extrinsic reasons in support of this axiom via the idea of verifiable consequences, introduced by Gödel (1947). This perspective is especially relevant for set-theoretic practitioners with an absolutist view who search for new justifiable axioms that can extend the standard theory ZFC. However, a focus on axiom adoption has two shortcomings. First, there is strong evidence that the view that extrinsic justification is valid reasoning for the truth of new axioms is not the default among set-theoretic practitioners. Rather, a part of the community is conclusively accepting ZFC and sees no need for further axiom adoption; they endorse a pluralist view. Second, a substantial proportion of set-theoretic practitioners are aiming for ZFC-proofs and are reluctant to explicitly using new axioms in proofs. One might conjecture that new axioms simply do not appear in the practice of these set theorists. A philosophical focus on axiom adoption disregards these two areas of the reality of practising set theorists.

D. Kant (✉)
Department of Mathematics, University of Hamburg, Hamburg, Germany

C. Antos et al. (eds.), *The Palgrave Companion to the Philosophy of Set Theory*, Palgrave Companions, https://doi.org/10.1007/978-3-031-62387-5_5

In a recent study of the set-theoretic community, I analysed the practice of set-theoretic practitioners from various research areas and backgrounds. One advantage of my methodology is that it is able to retrieve information on set-theoretic practices that is otherwise unavailable (i.e. not contained in published research), allowing this to be analysed for philosophical purposes. Some of these results provide the empirical basis of this chapter. The specific practice presented here is called *hidden use*, because the new axioms are eliminated from the proof. Besides the epistemic significance for set-theoretic practitioners with an absolutist view, such hidden use has pragmatic significance for set-theoretic practitioners with a pluralist view—*ZFC is enough and no new axioms should be adopted*—and for set-theoretic practitioners whose fundamental interest is in ZFC-proofs. The analysis, therefore, also sheds light on the parts of set-theoretic practice that are usually disregarded in philosophical discourse about the roles of axioms. This includes, for example, refuting the conjecture that new axioms do not appear in the practice of set-theoretic practitioners interested in ZFC-proofs. Moreover, from a social-epistemological perspective, the pragmatic significance is also epistemic, because it contributes to the extension of set-theoretic knowledge.

The hidden use of new axioms is a two-step procedure resulting in a ZFC-proof of some statement S. In the first step, some new axiom believed to be consistent with ZFC is used as a source of 'extra power' to prove S that is believed to be decidable in ZFC alone. At this point, set theorists learned that, if S is indeed decidable in ZFC, then S rather than its negation, $\neg S$, is provable. The second step involves an attempt to eliminate the new axiom, and if this is successful, the set-theoretic practitioner ends up with a ZFC-proof of S.

While every consistent new axiom can be used in this way, in this chapter, I focus on the hidden use of *large cardinal axioms*, because their use is widespread. Therefore, a large amount of data is available, and I believe initially focusing on a restricted class of axioms aids clarity. An overly general approach can also lead to some fuzziness about the detail. Large cardinal axioms are part of set-theoretic practice but they are not considered to be part of the standard axioms. I provide a summary of the data related to large cardinal axioms, which enables a proper embedding of the hidden use in the realm of different set-theoretic practices using large cardinal axioms.

The analysis is based on information gathered in an explorative interview study between 2017 and 2019 on set-theoretic independence with 28 set-theoretic practitioners. The interviewees, who work in various research areas and have fundamentally different views, were asked about a number of research-related topics. To guarantee anonymity, the source interviews for interview quotations are not indicated. This measure is necessary, because readers of this chapter probably know the interviewees and might identify them even with a small amount of information. More details on the method and important parts of the results are included in my dissertation.[1]

The chapter is structured as follows: Sect. 2 gives a brief review of relevant literature on the roles of axioms in mathematical practice. Section 3 summarises relevant data from the study to provide an overview of set-theoretic practices involving large cardinal axioms. The discussion in Sect. 4 focuses solely on the hidden use, especially of large cardinal axioms. It includes a conceptualisation of hidden use, provides two examples from published research literature, and describes the significance of the hidden use for set-theoretic practitioners. Section 5 concludes the chapter and raises some open questions.

2 Studying the Roles of Axioms in Mathematical Practice

In the literature on the philosophy of mathematics, the roles of axioms are seldom considered from a purely practical point of view. Most questions relate to the issue of their justification in one way or another. While an orthodox view sees axioms as self-evident truths, this has been rejected by many philosophers. Here, Maddy highlights this development:

> [A]ssumptions once thought to be self-evident have turned out to be debatable, like the law of the excluded middle, or outright false, like the idea that every property determines a set. Conversely, the axiomatization of set theory has led to the consideration of axiom candidates that no one finds obvious, not even their staunchest supporters. (Maddy, 1988a, p. 481)

Self-evidence is a kind of intrinsic justification of axioms, and despite some problems, intrinsic justification is not off the table; today, it is usually related to an informal conception of mathematical objects, such as the iterative conception of set (Boolos, 1971).

Easwaran proposes four necessary conditions different from self-evidence for the adoption of an axiom: An axiom should (1) be widely acceptable, (2) be useful in proving interesting consequences, (3) avoid philosophical

[1] Pragmatic Insights into Set-theoretic Independence, to be published in *Studies in Theoretical Philosophy*, Vittorio Klostermann (2025).

problems, and (4) be independent from the other axioms (Easwaran, 2008, p. 387). His main claim is that the adoption of axioms is a social practice based on the resolution of some philosophical problems while bracketing other philosophical disagreements. His points (1), (2), and (4) seem uncontroversial to me. Point (3) is more interesting. According to Easwaran, mathematical practitioners with different philosophical views and possibly for different reasons may accept the same axioms, thereby bracketing their disagreement on philosophical issues. Easwaran emphasises, moreover, that his analysis fits mathematical practice:

> [I]t seems that axioms are not chosen because they are inherently certain and let us make an uncertain result certain-they can certainly play this role, but that is not how or why they are chosen. Rather, they are uncontroversial and we use them to make a controversial result uncontroversial. I would like to suggest that this is the real role of axioms in mathematics-to stop arguing about our disagreements, and just work together on proving theorems. (Easwaran, 2008, p. 385)

Clearly, Easwaran objects to the view that axioms should be seen conclusively as self-evident statements. I agree with him on this point, as well as on the requirement that a philosophical analysis of the roles of mathematical axioms should be consistent with mathematical practice.

In the philosophy of set theory, the most prominent defender of a shift towards mathematical practice is Maddy, who reconstructs from mathematical practice the methodological principles, including the adoption of axioms, which govern mathematics (see, e.g., Maddy, 1997, 2011). My view regarding the focus on mathematical practice is completely in line with those of Easwaran and Maddy, but I go further still. While they are both mainly interested in the question of axiom adoption, my interest in the roles of axioms in mathematical practice is much wider and includes uses such as in the discovery process of mathematical proofs. This aligns with a social-epistemological perspective investigating the mechanisms within a scientific community that produce scientific knowledge.

That axioms play important roles in mathematical practice beside their adoption is supported by Schlimm (2013). In his systematisation of these roles, he distinguishes between three dimensions of axiom systems—Presentation, Role, and Function—and argues that while the presentation (language and consequence relation) is fixed, role and function are usually dependent on the user and not inherent to the axiom system:

> [T]he power of axiom systems stems from the possibility of changing our perspective and using them in different ways. ...Putting forward an axiomatization does not commit mathematicians to one particular perspective. (Schlimm, 2013, p. 81)

I strongly agree with Schlimm's viewpoint, and aim to show how even the significance of very specific practices like the hidden use of new axioms varies according to the user.

3 Data on Large Cardinal Axioms in Set-Theoretic Practice

This section gives an overview of the relevant data included in the study regarding the roles of large cardinal axioms in set-theoretic practice. Some of the following data are probably known to people with set-theoretic expertise and are available in the research literature. Therefore, this section is brief and I then focus on data that are probably less well known. Before presenting some results of the study, I describe the participants who were interviewed. The reader might want to skip this part initially and only come back when questions about methodological details arise.

3.1 Sample Set

This subsection gives some quantitative evaluations of the sample of twenty-eight professional set theorists to show that the sample possibly represents more than 8% of the community, and that the sample is diverse in terms of research area, age, gender, location of home institution, and view on the possibility of extending ZFC.

First, regarding the size of the current set-theoretic community, there are no conclusive data on the total number of professional set theorists. But a preliminary hint at the community's size is given by the 'list of homepages of set theorists' managed by the set theorist Jean A. Larson, which lists 323 set theorists. I spoke to 8.6% of them (by adding one person to the list), and I invited 45 (13.6%) of them to participate in the study (by adding two people to the list).[2] If one extrapolates that per 43 set theorists, two are not listed, one

[2] See the web page https://people.clas.ufl.edu/jal/set-theory-homepages/#I, accessed Aug 12, 2020.

Table 5.1 Distribution of research areas

Combinatorics	13
Descriptive set theory	11
Ergodic theory	4
Inner model theory	8
Forcing axioms	8
Large cardinals and forcing	8
Forcing	8
Set-theoretic and general topology	5
Cardinal characteristics	4
Determinacy and large cardinals	3
Recursion theory	3
Class forcing	2
Set theory of the reals (forcing)	2
Small research areas (very specific)	4

obtains a total number of 338 set theorists, and 28 out of 338 is still 8.3%. As noted, this is a preliminary evaluation owing to the lack of conclusive data.

Second, regarding the specific research areas of the interviewees, all main research areas in set theory are represented (see Table 5.1). Each interviewee indicated between one and five research areas, including some additional smaller ones.

Regarding their age, the interviewees are all professional set theorists with a (past) permanent position as a professor of mathematics with a research focus on set theory. One shortcoming of the study in this respect is that the views of the younger generation are not represented. However, the study still represents different generations: Six of the interviewees obtained their PhD before 1980, four between 1980 and 1989, nine from 1990 to 1999, and nine after 1999. The years of obtaining the PhD were taken from the *Mathematics Genealogy Project*.[3]

With regard to gender, four of the interviewees (14%) are women according to my evaluation (the interviewees were not explicitly asked about their gender). The sample does not seem to be biased in this respect, because the majority of set theorists are male.

Regarding the location of their home institution, fifteen of the interviewees are affiliated to a European university, eleven to a university in the USA, and two to a university outside Europe and the USA.[4] There are groups outside Europe and the USA that were not included in the interviews, but there is no

[3]See https://mathgenealogy.org, accessed Nov 11, 2020.

[4]Specific locations are not indicated because of anonymity.

obvious corresponding bias, because most set-theoretic research seems to be concentrated in these Western regions.

Regarding their views on independence, eleven of the interviewees have an absolutist view and another eleven a pluralist view. Hence, the sample is not biased in this respect. The remaining six interviews do not contain a coding in one of these two categories.

These data show that the sample is broad, especially according to the range of different research areas, but also that it is limited in certain respects, for example, it captures mainly the European and US-American research context. As such, it is not biased in some obviously misleading way. All in all, these data provide evidence that a broad cross-section of views are represented in the study.

3.2 Results

The interview data suggest that a majority, but not every set theorist, uses large cardinal axioms in their research. Twenty-three of the twenty-eight interview partners indicate that they use large cardinal axioms. This is more than those who report using forcing axioms and determinacy principles, which supports the assumption that large cardinal axioms are the most widely used new axioms in set theory.

The additional results summarised in this subsection identify different uses of as well as research questions about large cardinal axioms, and suggest consistency beliefs, occasional reluctance to use them, and partial acceptance.

Uses of Large Cardinal Axioms Large cardinal axioms are used by set-theoretic practitioners as tools to achieve some mathematical goal. Besides the insight into set-theoretic practices, the usefulness of new axioms is philosophically relevant, and even cited as evidence in favour of some new axioms (see, e.g., Viale, 2019). The following uses were each mentioned by one or more interviewees in the study. (The order of the list is arbitrary.) Large cardinal axioms are used:

- for certain *forcing arguments*, because they enable forcing constructions that are otherwise impossible. The research area called 'large cardinals and forcing' is dedicated to this research. An important research question in this context is how one preserves the large cardinal property.
- to *learn about independent statements*. Via their consequences, large cardinal axioms organise independent statements. This use is shared with other new axioms.

- *outside set theory*, which does not seem to be widespread, but even the rare cases are relevant. An example is the use of Vopěnka's principle in algebraic topology.
- for *consistency proofs* of other new axioms and principles: A theory is usually considered consistent if it is proven consistent relative to large cardinal axioms.
- as a *consistency measure* of new axioms and principles. As one interviewee expressed this:

> *The really surprising thing is that almost every concept that we come up with, whose consistency strength is beyond that of ZFC, aligns exactly with some large cardinal concept in its consistency strength. And this is just phenomenally bizarre. Somehow, this large cardinal concept ends up being a measuring stick for consistency strength.*[5]

- for "*methodological guidance*". A few interviewees noted that they do not use large cardinal axioms directly in their work but consider them to be useful as guiding principles. This guidance, one interviewee stated, consists in the assumption that if a statement in their research area is true assuming a large cardinal, it is probably true without. This use is discussed extensively as a case study below.

Research Questions About Large Cardinal Axioms Large cardinal axioms are not only considered to be tools to solve problems but are also investigated as objects of study in themselves. The interviewees mentioned that set theorists investigate large cardinal axioms by asking about:

- the *structure* of large cardinals, which are, for instance, important in their use for forcing constructions.
- the *consequences* of large cardinal axioms, which are, among other things, used to evaluate the large cardinal axioms. Interviewees explained that the fact that the existence of Woodin cardinals implies generic absoluteness is itself an interesting fact about Woodin cardinals; and that the theorem, that an inner model with a supercompact cardinal would also contain all larger large cardinals, is a fascinating fact about supercompact cardinals.[6]

---

[5]Interview quotations are written in small italics.

[6]The most prominent results in this context are probably the theorems establishing a deep link between large cardinal axioms and determinacy principles. This was a major surprise at the time (in the 1980s). For instance, Larson describes this development as follows:

- the existence of a *canonical inner model* with some large cardinal in it. The existence of such an inner model is cited as the most important evidence in favour of the consistency of large cardinal axioms. The research area of 'inner model theory' is devoted to the construction of these models.
- the *order* of large cardinals. The linearity of the large cardinals is also quoted as evidence in favour of their consistency.[7]
- *equivalent formulations* of large cardinal axioms, which are important for their use in proofs, because some equivalent formulations are better suited to certain applications. Moreover, it is interesting for the evaluation of large cardinal axioms in terms of plausibility or even acceptability. This research question is also posed in relation to other new axioms.

Consistency Beliefs and Reluctance to Use Large Cardinal Axioms Using large cardinal axioms in the ways described presupposes some belief in their consistency. The data indeed suggest that set theorists generally believe in the consistency of large cardinal axioms, but that they, nevertheless, sometimes prefer to avoid their use.

To give more detail, no interviewee revealed any doubt about the consistency of large cardinal axioms. Some of them explained, moreover, why they think proof of inconsistency is very improbable. For example, if these axioms were inconsistent, one would obtain weird results because everything would be provable from them, one interviewee argued. Another pointed to the amount of time that has passed without any inconsistency being found. One may interpret this argument as suggesting an inductive justification for their consistency beliefs. This person concluded: *"I don't think that one can really say that large cardinal assumptions are on some sort of shaking ground or something."* One peculiar observation is that although the existence of inner models is referred to as evidence in favour of the consistency of large cardinal axioms, consistency

> In a dramatic development, the hypotheses for these results [determinacy hypotheses] would be significantly reduced through work of Woodin, Martin and John Steel. The initial impetus for this development was a seminal result of Matthew Foreman, Menachem Magidor and Saharon Shelah which showed, assuming the existence of a supercompact cardinal, that there exists a generic elementary embedding with well-founded range and critical point ω_1. (Larson, 2020, p. 6)

These results were quoted extensively by the interview partners as a major breakthrough. Set theorists were amazed that large cardinal assumptions have such neat consequences, to the extent that some of them became convinced of their truth. The quoted paper by Larson (2020) gives a detailed overview of the specific results.

[7] Please note that there are exceptions to the linearity phenomenon.

beliefs do not seem to be weaker for supercompact cardinals (for which no inner model has been constructed to date) than for smaller ones.

In contrast to these consistency beliefs in the set-theoretic community, the data also show that set theorists sometimes prefer to avoid the use of large cardinal axioms. One interviewee, for example, noted: "*Of course, if you prove something using some of these [large] cardinals, the question is always: Is this necessary?*"

But the data also show that a subtle evaluation takes place. For instance, one can observe that the usefulness of large cardinals works against the reluctance to use them, and that there may be more reluctance about using the large cardinal axioms than about using some of their consequences. This is expressed in the following quotation:

> *Every time I use the existence of a measurable, the existence of a supercompact, it's a big jump in faith, so to say. For me, it is like an extra effort. But it's an extra effort to the moment in which I realise that this assumption is giving me this combinatorics for free. And I feel happy to work with the given combinatorics, irrespective of where it comes from.*

Partial Acceptance of Large Cardinal Axioms The use of large cardinal axioms in set-theoretic practice requires a belief in their consistency, but it does not require a belief in their truth or, formulated less strongly, any genuine acceptance of large cardinal axioms. Still, the data suggest that a substantial proportion of the community genuinely accept large cardinal axioms; some of the interviewees explicitly expressed their belief in them. The following quotation illustrates this observation:

> *Among those that are, let's say, Platonist, or have a point of view which is not too dissimilar to Platonism, I think, there [are] not [many] questions about the truth of large cardinal axiom[s].*

These results have provided the reader with some background on large cardinal axioms in set-theoretic practice. The next section is dedicated to an analysis of the use of large cardinal axioms for 'methodological guidance'. Although 'methodological guidance' is a suitable name for this use, I call it *hidden use*, because this name highlights the specific characteristic of interest: New axioms are used to prove a theorem, but they are eliminated afterwards, and therefore do not appear in the final ZFC-proof.

4 Discussion: The Hidden Use of Large Cardinal Axioms

The use of large cardinal axioms for methodological guidance is only suitable in certain specific areas of set theory that are mostly interested in pure ZFC results (or results in even weaker systems). In these areas, the use of large cardinal axioms should be hidden: They should not appear in the final proof.

These areas are typically descriptive set theory, forcing on the reals, set-theoretic and general topology, or cardinal characteristics. In descriptive set theory, for example, interviewees gave very clear statements like: *"For descriptive set theory, as you know, we typically prove results that hold in ZF even."* or:

> *Most of the time when I'm working on something, I have built-in faith that it's not independent from ZF+DC. But that's just where I tend to work. Every so often, something comes up that's further out and then I don't know so much. And then, occasionally, something comes up where one can just see pretty quickly that there's some independence going on. But I would say 95% of the time I go on thinking that there is no chance that there is any independence phenomenon there, and I can't recall being wrong.*

However, one major conclusion of this chapter is that, nevertheless, new axioms that go beyond ZFC are used by set-theoretic practitioners in these areas. One of my interview questions asked about axioms used besides the ZFC axioms, here posed to a descriptive set theorist:

> *I: Which axioms do you use in your work apart from ZFC axioms? …*
> *IP: So, I suppose, the Axiom of Determinacy plays a role, also weaker versions like Projective Determinacy. And occasionally Martin's Axiom comes in, …either because it allows you to generalise that result that you have without ….*[8] *Or, it can come in on a rare occasion, I think, in my research more, as opposed to my results. For what can happen sometimes is that in sort of desperation to try to get traction on a problem, …I might say to myself that, well, suppose now, we had Martin's Axiom, and I could actually meet these κ many dense sets—and I'm well aware that I've now moved past /,*[9] *I want to prove something that should be just a ZFC theorem—but, in desperation, I might be looking for the extra power. And then, if it is a good idea,*

[8]Taken out of the quote: *"perhaps [about] countable trees, or this could be about sets that are described using countable trees. And then, you might be able to generalise them to trees of higher cardinality, or sets described by trees of higher cardinality."*

[9]The sign '/' marks that the sentence was not finished.

you might end up realising, well, you never really needed to meet more than countably many sets. So, it actually was a ZFC theorem.

In their remark, this person elaborated on the use of Martin's Axiom, but they added that *"the extra power is more typically something like determinacy."*

So, although I focus on the hidden use of large cardinal axioms, forcing axioms and determinacy principles can also be, and indeed are, used in the same way.

Another interviewee working in descriptive set theory who is also an expert on forcing reflected on their use of large cardinal axioms:

Certainly, large cardinal axioms are always there. They may not be present in the work that I produce, but they at least serve as a methodological guide. I think it's extremely useful to have them; at least to have them as a methodological guide. Of course, if you actually use them in your work, then non-set theorists will take a dim view of it. So, it's better not to do it. But they're helpful on a methodological level.

A third interviewee, working in set-theoretic and general topology, expressed a similar idea.

What is it that set-theoretic practitioners learn by using the *extra power* of a new axiom that can then be eliminated? If they assume that the statement in question is very probably decidable within ZFC, why don't they try to find a ZFC-proof directly? Let me present an illustrative quote about this stage of the proof-finding procedure:

If I start new at a problem, I really have to do this, we call it the Magidor strategy because he advocates it: On the even calendar day, you try 'yes', and on the odd day, you try 'no'. You should not waste time on one direction because the opposite might be true. And as long it's far from intuition, one really has to proceed both [ways]. Sometimes, I also have to give up after certain months or so. Then, I say 'I can't afford any more to work in vain on such and such problem.' I don't give up forever but sometimes for some years or so.

At this stage of the proof-finding procedure, set-theoretic practitioners are tackling the question of whether some statement is probably true or not, and if they use an additional axiom to prove the statement or its negation, they have learned quite a lot. They can now abandon the *Magidor strategy*, because they know in which direction they should proceed. If they can prove some statement S using a new axiom, and they believe that S is not independent of ZFC, they can now try to find a proof of S without wasting any more thought on the possibility that the negation might be true.

4.1 Conceptualisation

The idea of hidden use is that assuming that some statement S is not independent of ZFC ($\text{ZFC} \vdash S \vee \text{ZFC} \vdash \neg S$), one first uses the *extra power* of an additional new axiom to prove either S or its negation, say S: $\text{ZFC} + \text{NA} \vdash S$ or, equivalently: $\text{ZFC} \vdash \text{NA} \rightarrow S$.[10] Subsequently, because it is assumed that S is not independent, set theorists try to eliminate the use of the new axiom in the proof, and if they succeed, they provide a ZFC-proof: $\text{ZFC} \vdash S$.

If, for some statement S, set-theoretic practitioners assume that S is decidable in ZFC, then the following scenario is possible and conceptualises the hidden use of new axioms:

Assumption:	$\text{ZFC} \vdash S \vee \text{ZFC} \vdash \neg S$	*Question*: Is S true or false?
Step 1:	$\text{ZFC} + \text{NA} \vdash S$	*Conjecture*: S is true.
Step 2:	$\text{ZFC} \vdash S$	*Proof of the conjecture and confirmation of the assumption.*

Note that hidden use does not reveal any logical connection between the new axioms and the statement S. For, if S was proven in ZFC, then it trivially holds that $\text{ZFC}+\text{NA} \vdash S$. The characteristic of hidden use is rather that there is a time when this proof is non-trivial because the new axioms are directly used.

One further remark on the assumption in the framework: In some cases, as described in the quotes above, bringing their extensive experience to bear, set-theoretic practitioners are sure that the statement S is not independent. Subsequently, they indeed succeed in finding a ZFC-proof of either S or its negation. In other cases, the assumption is rather a second question: Is S independent or not? This question can be answered in the negative by showing either that the use of the new axiom was actually necessary or that the negation of S is consistent. When discussing the examples in the next subsection, I come back to this possibility.

[10] Wherever I refer to proofs, please note that I mean what De Toffoli introduced as *simil-proofs*, i.e. "arguments that look like proofs to the relevant experts" (De Toffoli, 2020, p. 824).

4.2 Exemplary Proofs

The idea of the hidden use of a new axiom is that it is used in the discovery process of a theorem but not in the final proof and that it guides the search for a proof but can be eliminated in the end. Described like this, of course, it is very hard to find instances of hidden use, unless one asks practitioners for examples. Because the final proof does not mention the new axiom, as a reader of a final ZFC-proof, one does not learn about the new axiom's use in earlier stages. Because of this methodological difficulty, I draw on instances in which the first proof was published too. The two examples are Borel Determinacy and Cichoń's maximum, which were both first proved using large cardinal axioms.[11] They are from different times and research areas, and are about different kinds of statements. The ZFC-proof of Borel Determinacy was published in 1975. It is a plain statement about a property of sets of reals: *All Borel sets are determined.* The ZFC-proof of Cichoń's maximum was published in 2022. It is a consistency statement about separating the cardinal characteristics of the continuum.

4.2.1 Borel Determinacy

In this section, I summarise the historical development of Borel Determinacy. In 1953, Gale and Stewart studied two-player games on sets of reals and asked: "[H]ow pathological must the set [of reals] be for the game to be indeterminate?" (Gale & Stewart, 1953, p. 246).[12] They showed that sets on the first level of the Borel hierarchy, open sets, are determined. Some years later, in 1970, Martin proved that analytic sets are determined using a large cardinal axiom in his proof:

> We assume the existence of a measurable cardinal and prove that every analytic set is determinate. Our proof is fairly simple and makes a very direct use of the large cardinal assumption (we present it in terms of a Ramsey cardinal) and the fact that open games are determined. (Martin, 1970, p. 287)

[11] I owe the first example to Philipp Schlicht.

[12] I assume that the readers of this volume do have some set-theoretic expertise. But for comprehensiveness, a few words on determinacy. Determinacy is a property of sets of reals and of the related games. Let us take a set of reals and imagine two players who play a game in which they alternately choose natural numbers. This game results in an infinite sequence of natural numbers, which in turn refers to a real number. Now, player I wins the game if this real number is an element of the given set, otherwise player II wins. If one of the two players has a winning strategy, then we call the game as well as the set of reals *determined.*

Since every Borel set is analytic, Borel Determinacy holds too on the assumption of a measurable cardinal.

At the same time, Friedman (1971) showed that Borel Determinacy is independent of a fragment of ZFC that he calls Z, which excludes in particular the replacement schema.[13] A few years later, Martin showed that Borel Determinacy is provable in ZFC, so the measurable cardinal is not necessary. With reference to Friedman's work, he noted:

> Borel Determinacy is probably then the first theorem whose statement does not blatantly involve the axiom of replacement but whose proof is known to require the axiom of replacement. (Martin, 1975, p. 364)

In terms of our conceptualisation, in 1970, there was a proof of Borel Determinacy assuming a large cardinal axiom, but only a few years later, in 1975, there was a pure ZFC-proof of Borel Determinacy. In the meantime, set theorists were elaborating on details of the determinacy of sets on the Borel hierarchy and developing proving techniques. The question remained: Is the large cardinal assumption necessary to prove Borel Determinacy? Friedman showed that Z is not sufficient to prove it, so it had to be something stronger. And finally, Martin found a ZFC-proof. A proof making direct use of a large cardinal axiom could be replaced by a proof in ZFC.

It is plausible to assume that, in this case, the question of whether or not Borel Determinacy is independent remained open throughout the process. Martin told me in a private communication: "When I learned about Friedman's theorem, I began a long attempt to find out whether or not Borel Determinacy is provable in ZFC. If my memory is correct, I mainly tried to prove that the answer is yes. I spent little time trying to show that Borel Determinacy could not be proved in ZFC." So, he conjectured that Borel Determinacy might be provable but was probably still aware that it might also be independent. If we consider the development regarding the stronger statement of Analytic Determinacy, then we move on to the alternative case, in which the use of some large cardinal strength was indeed found to be necessary. In 1978, Harrington (1978) proved that Analytic Determinacy implies $0^\sharp$. Hence, in parallel to solving the question of whether Borel Determinacy is true or not, the question of whether or not Borel Determinacy, respectively Analytic Determinacy, is indepedent from ZFC was solved.

[13] Martin's and Friedman's articles were written around the same time. Each refers to the other's work as unpublished.

4.2.2 Cichoń's Maximum

In the case of proving Cichoń's maximum, a proof making direct use of large cardinal axioms could also be replaced by a proof in ZFC:

> It is consistent that all entries of Cichoń's diagram are pairwise different (apart from add($\mathcal{M}$) and cof($\mathcal{M}$), which are provably equal to other entries). However, the consistency proofs so far required large cardinal assumptions. In this work, we show the consistency without such assumptions. (Goldstern et al., 2022, p. 3951)

In an earlier paper from 2019, Goldstern et al. (2019) assumed the existence of four compact cardinals to prove the consistency of the following statement: $\aleph_1 < \mathrm{add}(\mathcal{N}) < \mathrm{cov}(\mathcal{N}) < \mathfrak{b} < \mathrm{non}(\mathcal{M}) < \mathrm{cov}(\mathcal{M}) < \mathfrak{d} < \mathrm{non}(\mathcal{N}) < \mathrm{cof}(\mathcal{N}) < 2^{\aleph_0}$.[14] They used the technique of forcing, and the compact cardinals provide Boolean ultrapower embeddings, which enable the construction of the respective forcing notions. The question of Cichoń's maximum is whether the cardinal characteristics of Cichoń's diagram can be separated simultaneously (respecting the two ZFC-provable equalities in the diagram), and the above-mentioned authors proved that they can.

In their first proof, the large cardinals are directly used for the forcing constructions. But the authors asked: "Can we prove the result without using large cardinals?" (Goldstern et al., 2019, p. 139), and (revealing the temporal incoherence between knowledge and published knowledge) the authors referred to a draft paper in which they had proved that this is indeed possible. In this paper, published in 2022, Goldstern, Kellner, Mejía, and Shelah "introduce a new method to control cardinal characteristics .... This method can replace the Boolean ultrapower embeddings in previous constructions, so in particular [they] can get Cichoń's maximum without assuming large cardinals" (Goldstern et al., 2022, p. 3953).

The example of Cichoń's maximum is another prototype for the hidden use of large cardinal axioms. It is interesting to note, moreover, that the authors directly referred to this use when they conjectured that the large cardinal assumptions should be eliminable in their proof:

---

[14] Again, for comprehensiveness, a few words on the content of this statement. The cardinal numbers in this inequality denote *cardinal characteristics*. In ZFC, it is provable that all of them are greater than $\aleph_0$ and less than or equal to $2^{\aleph_0}$. In research on cardinal characteristics, many forcing notions have been found that can show that certain pairs of cardinal characteristics can be separated. This question was subsequently generalised by trying to separate simultaneously more than two cardinal characteristics.

> It seems unlikely that any large cardinals are actually required; but a proof without them would probably be considerably more complicated. It is not unheard of that ZFC results first have (simpler) proofs using large cardinal assumptions. (Goldstern et al., 2019, p. 116)

The authors considered it a common pattern that a ZFC result is sometimes first proved assuming large cardinal axioms.

4.3 Significance of Hidden Use for Set-Theoretic Practitioners

The set-theoretic community is not a homogeneous group of scholars with similar research interests and framework beliefs on the nature of mathematics. Therefore, I present the various perspectives of different sub-groups of the set-theoretic community on the relevance of the hidden use of new axioms. The three sub-groups to be considered are set-theoretic practitioners with an absolutist view, set-theoretic practitioners with a pluralist view, and set-theoretic practitioners whose fundamental interest is in ZFC-proofs. We have already heard several voices from this latter sub-group above. The sections that follow start by characterising each of the first two sub-groups, and then present the significance of the hidden use for them.

4.3.1 Extrinsic Justification of New Axioms

Hidden use has an epistemic significance in the context of axiom justification. I discuss this claim and argue that a successful hidden use is a case of the kind of verification that is suggested by Gödel. From the perspective of set-theoretic practitioners with an absolutist view, this is important. They believe that ZFC should be extended by new axioms, and these axioms should typically be justified by convincing, extrinsic reasons.

Absolutist Views of Set-Theoretic Practitioners The convictions of set-theoretic practitioners with an absolutist view are varied. Some believe in the existence of a set-theoretic universe, some believe that set theory being a mathematical foundation implies that every sentence must be either true or false, and others simply believe that the independence phenomenon is surmountable by adding new axioms to ZFC. Examples of these views are presented below.

An absolutist view is often related to a realist position, for instance when set-theoretic practitioners talk about a *"real set-theoretic universe"* or describe set-theoretic research as follows:

> *We are actually describing a reality, which is the mathematical world or set-theoretic world, which is real, as real as the physical world. We are acquiring real knowledge about something, which we don't know quite yet what it is but....*

With regard to independence proofs (comprising two consistency proofs),[15] set-theoretic practitioners with an absolutist view often consider searching for new axioms to be more valuable than proving consistency results: *"The most interesting consistency results have been proved and in my perspective it's better to try to find new arguments to choose among the many possibilities which is the right one"*. They are *"more interested in searching for new axioms and the like than in finding out what is consistent with ZFC"*. This research goal is consistent with the view that ZFC axioms are not sufficient: *"We know [ZFC] doesn't tell you enough; there's not enough to settle important problems like CH and so on"*.[16]

Gödel's Idea of Verification and Extrinsic Justification Gödel proposed that new axioms may be justified not only intrinsically but also extrinsically via their consequences:

> [D]isregarding the intrinsic necessity of some new axiom, and even in case it had no intrinsic necessity at all, a decision about its truth is possible also in another way, namely, inductively by studying its 'success,' that is, its fruitfulness in consequences and in particular in 'verifiable' consequences, i.e., consequences demonstrable without the new axiom, whose proofs by means of the new axiom, however, are considerably simpler and easier to discover, and make it possible to condense into one proof many different proofs. The axioms for the system of real numbers, rejected by the intuitionists, have in this sense been verified to some extent owing to the fact that analytical number theory frequently allows us to prove number theoretical theorems which can subsequently be verified by elementary methods. A much higher degree of verification than that, however, is conceivable. There might exist axioms so abundant in their verifiable consequences, shedding so much light upon a whole discipline, and furnishing such powerful methods for solving given problems ...that quite irrespective of

---

[15]To prove that a statement A is independent of ZFC, one must prove that ZFC is consistent with A and also with $\neg A$.

[16]CH denotes the famous continuum hypothesis.

their intrinsic necessity they would have to be assumed at least in the same sense as any well established physical theory. (Gödel, 1947, p. 521)

There are two points worth noting in this quotation. First, for Gödel, proofs of "verifiable consequences" that use new axioms are much simpler, but these consequences can be demonstrated without the new axiom. Second, his examples are number theoretical consequences of the theory of real numbers, which were later proven by elementary means.

The first point directly applies to the hidden use of large cardinal axioms. A first and simpler proof that uses some large cardinal axiom was later replaced by a somewhat more complicated proof without assuming large cardinal axioms. Recall the quotation in the example of Cichoń's maximum: "It is not unheard of that ZFC results first have (simpler) proofs using large cardinal assumptions" (Goldstern et al., 2019, p. 116). Gödel's example is also analogous to our examples: Consequences of a stronger theory were later verified by the weaker theory.

Philosophers of set theory interested in the extrinsic justification of axioms have tried to explicate the notion of the fruitfulness or success of an axiom. In this context, verifiable consequences support the notion that an axiom is fruitful. Therefore, people tried to explicate *verification*. In an analysis of Gödel's works on independence, van Atten and Kennedy explicated verifiable consequences as arithmetically verifiable consequences (van Atten & Kennedy, 2009, p. 341). In the above quotation, the example refers to such consequences. Hidden use, however, would not correspond to this conception, because ZFC-provable statements are not necessarily arithmetically verifiable. The examples given above do not count as such arithmetical consequences. Borel Determinacy is about sets of reals, and Cichoń's maximum is about cardinal characteristics, which are both set-theoretic subject matters. But it does not seem necessary to restrict the notion of verifiable consequences conclusively to arithmetical consequences and non-set-theoretic subject matters. Gödel does not reject notions of verifiability that go beyond arithemetical consequences, and van Atten and Kennedy also refer to Borel Determinacy as an example:

> Borel Determinacy is a 'verifiable consequence,' in Gödel's sense of the phrase here, i.e., it was proved without using measurables, and the measurables in turn were verified by their having led to the 'correct' result. (van Atten & Kennedy, 2009, p. 343)

Hence, in their explication, van Atten and Kennedy do not actually restrict verifiable consequences to arithmetical consequences. In conclusion, I consider it a reasonable proposal to explicate 'verifiable consequence' as 'ZFC-provable consequence'. One final observation is that it is only the consequences of an axiom that can be verified and never the new axiom itself (if it really goes beyond ZFC).

The upshot of this subsection is a proposal for the clarification of an important aspect of extrinsic justification. I argue that the hidden use of large cardinal axioms identifies verifiable consequences of large cardinal axioms and that 'verifiable consequences' of an axiom are the ZFC-provable consequences. In other words, ZFC is considered to be a mathematical theory that is suitable for the verification of more uncertain components of a possible theory extension. Every such verifiable consequence of a large cardinal axiom is considered to be an extrinsic reason in favour of this axiom.

Hidden Use Compared to Martin's Mathematical Evidence Similar observations concerning justification and verification are provided by Martin (1998).[17] In his chapter, Martin presents two mathematical examples which, according to him, "count[ ] as evidence for mathematical truth" (Martin, 1998, p. 215) of determinacy principles such as Projective Determinacy. Similar to the hidden use of new axioms, the determinacy principle is used in addition to ZFC to prove some general statement. In Martin's examples, specific instances of this more general statement are verified, and so, Martin argues, the examples provide a case of prediction and confirmation. The determinacy principle predicts a general statement and all specific instances of it that were 'tested' are confirmed.[18] Martin's first example is the Cone Lemma, which states that if AD(PD/BD) holds, then every (projective/Borel) set of Turing degrees either contains a cone or its complement does. Martin comments:

> When I discovered the Cone Lemma, I became very excited. I was certain that I was about to achieve some notoriety within set theory by deducing a contradiction from AD. In fact I was pretty sure of refuting Borel Determinacy. I had spent the preceding five years as a recursion theorist, and I knew many sets of degrees. I started checking them out, confident that one of them would ...give me my contradiction. But this did not happen. For each set I considered, it was not hard to prove, from the standard ZFC axioms, that it or its complement contained a cone. ...

[17] Thanks to the anonymous referee who pointed me towards this reference.

[18] In the first version of this chapter, I said 'prediction-use' instead of 'hidden use'. I changed my mind, because although one characteristic of hidden use is indeed that the truth of S is predicted, the more important characteristic is that the new axiom can be eliminated.

I take it to be intuitively clear that we have here an example of prediction and confirmation. (Martin, 1998, p. 224)

The Cone Lemma was proved in 1968, and as the quote makes clear, there is a close relation to the developments concerning Borel Determinacy. Martin's attempts to refute Borel Determinacy, as described here, happened before he had proved that the existence of measurable cardinals implies Analytic Determinacy. But after his proof and Friedman's result (Friedman, 1971), he tried to prove Borel Determinacy. This reconstruction of events only strengthens the epistemic significance of the use of the measurable in proving Borel Determinacy in 1970, which seemed to have strongly supported the conjecture that Borel Determinacy might actually be provable in ZFC.

It is worth making a final note on Martin's examples and their relation to the hidden use of axioms. The examples are similar but not quite the same. The difference is that the use of the new axiom is not eliminated in Martin's examples as it is in hidden use. In the Cone Lemma, the conclusion is just as strong as the assumption. If one assumes AD, then the conclusion holds for every set of Turing degrees; if one assumes Projective or Borel Determinacy, then the conclusion only holds for Projective or Borel sets of Turing degrees. So, no eliminable extra power was used to prove the conclusion for Borel sets.

The characteristic of Martin's examples is rather that a general implication schema was proved, and that the conclusion could be proven for many specific instances without the assumption of some strong determinacy principle. This is what he desribes in the quote above. Hence, he argues that this verification of the specific instances provides evidence of the general stronger determinacy principle that predicted all specific instances. (Of course, since AD contradicts AC, Martin does not consider his examples to be evidence in favour of AD, but, with hindsight, in favour of PD/$AD^{L(\mathbb{R})}$.)

Consistency or Truth? This paragraph addresses the reviewers' requests for more clarity regarding the consistency or truth of axioms. I think that this can be clarified by again considering different sub-groups of the set-theoretic community. One of my main observations of set-theoretic practice is that the community is heterogeneous when it comes to the framework beliefs of its members. Regarding more foundational questions on set-theoretic axioms, it is simply disingenuous to make statements of the sort: 'Set-theoretic practitioners think this or that'. Such a statement presupposes a homogeneity that is not found in reality.

In this respect, an important distinction is the one between consistency beliefs about axioms on the one hand and actual beliefs in axioms or beliefs in the truth of axioms on the other. When I talk of *believing axioms* (in the same sense that Maddy, 1988a,b use the notion), this is actual belief in axioms and can be identified with believing in the truth of axioms. Scholars who prefer to avoid completely the notion of truth in the context of mathematical statements can use the notion of 'belief in axioms' instead.

If we now consider set-theoretic practitioners with an absolutist view, then we observe that they believe in the ZFC axioms, and most of them believe in large cardinal axioms and in Projective Determinacy. This is not only about believing in the consistency of these axioms but about believing them to be true. After all, the philosophical question of a justification of new axioms does not ask whether they are consistent—consistent new axioms would not solve the continuum hypothesis or any other question that is not answered in ZFC—it asks about the truth of these axioms. For members of this sub-group of the set-theoretic community, hidden use might have provided evidence in favour of the consistency of new axioms years ago when questions of consistency were more controversial. But, today, the question about new axioms for set-theoretic practitioners with an absolutist view is not about their consistency but about their truth, and, thus, hidden use is, as described above, a good way to provide support for the truth of axioms.

4.3.2 Pragmatic Significance

In this section, I describe the pragmatic significance of the hidden use of new axioms. Because the significance of hidden use in providing extrinsic evidence is not relevant for set-theoretic practitioners without an absolutist view, pragmatic significance here refers specifically to their view on hidden use. But, of course, pragmatic significance is relevant for set-theoretic practitioners with an absolutist view, too.

Hidden use is pragmatically relevant for set-theoretic practitioners with a pluralist view, but also, as described above, for set-theoretic practitioners who are interested in ZFC-proofs. These two sub-groups are not identical. Set-theoretic practitioners aiming for ZFC-proofs can have absolutist or pluralist views, or indeed neither. Let me present some observations from set-theoretic practitioners with a pluralist view.

Pluralist Views of Set-Theoretic Practitioners For set-theoretic practitioners with a pluralist view, the idea of a set-theoretic universe does not make sense. They say, for instance: *"I don't really know what is meant by it"*, and they think that *"we don't have a universe of ZFC anyway"*.

Typically, set theorists with a pluralist view are not annoyed by the independence phenomenon: *"The thing is, I guess, the independence phenomenon doesn't bother me"*; for them, independence is a *"fact of life"*. This acceptance of the current situation of set-theoretic independence as somehow settled is also reflected in their views on ZFC as the right and conclusive theory for sets—*"in some sense, you could say I believe in ZFC"*—that, according to them, does not need to be extended by further axioms: *"I don't really feel the need to sort of choose between axiom systems either.*[19] *They're all out there and you can study all of them, and I think that's all worthy of study."*

Set-theoretic practitioners with a pluralist view are usually convinced that no new axioms beyond ZFC will be accepted by the mathematical community: *"If I had a guess, I don't really expect any axioms beyond ZFC to be accepted and have this status in the mathematical community that the axioms of ZFC do"*.

Furthering Set-Theoretic Progress The pragmatic significance of the hidden use of new axioms is simply that it leads to new theorems that extend set-theoretic knowledge. The axioms are useful in this sense because they provide support during the proof-finding process. Other things are needed to end up with a finished proof. For example, the direct use of the new axioms in the first proof might inform the final ZFC-proof, but does not necessarily do so. In some cases, an essentially new proof idea is needed.[20] Of course, in the first case, the new axiom is even more useful.

Regarding consistency and truth, the pragmatic significance of the hidden use of new axioms does not relate to any question about the truth of new axioms, but it crucially involves consistency beliefs. Regarding the conceptualisation given in Sect. 4.1, if a new axiom is not taken to be consistent and $\mathrm{ZFC} + \mathrm{NA} \vdash S$ is proven in **Step 1**, then one cannot conjecture that S is true, because NA could be inconsistent and S could be false. In this case, the new axiom would not be of any help; only new axioms believed to be consistent are actually useful. The pragmatic significance shows how new axioms are useful even for set theorists who do not want to go beyond ZFC, but the new axioms are only as useful as far as they are believed to be consistent. In the case of large cardinal axioms, in Sect. 3, I presented some data supporting the view that set-theoretic practitioners do not doubt the consistency of large cardinal axioms, which in turn supports their pragmatic usefulness.

[19] Because set theorists work with various axiom systems that extend ZFC, the notion of *"axiom system"* in this quotation can be understood as referring to extensions of ZFC and not to axiom systems in general.

[20] Thanks to Benedikt Löwe for raising this point in discussion.

Regarding the epistemic value of hidden use, I want to add that pragmatic significance is also an epistemic significance from the social-epistemological viewpoint. Although for set-theoretic practitioners who do not believe in extrinsic justification, the hidden use of new axioms does not provide reasons justifying new axioms as true, it nevertheless leads to the extension of set-theoretic knowledge by answering valuable research questions among set-theoretic practitioners, because new true, valuable theorems are epistemically significant.

5 Conclusion

By analysing interview data on set-theoretic practices, this chapter provides insight into a novel role of axioms in mathematical practice. The case study of the hidden use of large cardinal axioms presents publicly unavailable information on mathematical practices. Moreover, this case study is relevant for the ever-growing research on mathematical practices as well as for the literature on the justification of mathematical axioms.

The set-theoretic community is rather heterogeneous when it comes to more foundational beliefs. Some set theorists have an absolutist view and accept the extrinsic justification of new axioms, while other set theorists have a pluralist view and do not accept extrinsic justification. A third relevant sub-group are set-theoretic practitioners who are aiming for ZFC-proofs. I have showed in this chapter that the hidden use of large cardinal axioms has significance for all of them. It can provide extrinsic reasons in favour of new axioms and it can be used in the discovery process of a ZFC-proof. Easwaran argued that mathematicians adopt axioms to bracket their philosophical disagreements, and I add the related observation that mathematicians agree on the usefulness of axioms while bracketing philosophical disagreements. The hidden use of new axioms is signficant either way. My analysis, moreover, illustrates Schlimm's thesis that using "an axiomatization does not commit mathematicians to one particular perspective".

Although the different perspectives on the significance of hidden use have been presented separately, the set-theoretic community should not be seen as being divided into disjoint sub-groups that each have their own individual perspective on the hidden use of new axioms. While the exploitation of the hidden use of new axioms in finding reasons in favour of these axioms is only open to people who endorse extrinsic justification, pragmatic significance is relevant to everyone, regardless of their beliefs on the question of axiom justification. One might plausibly assume that all set-theoretic practitioners by their nature are interested in furthering set-theoretic progress, and that pragmatic significance is valued and accepted by all of them.

I conclude with a few open questions and initial commentary on them as potential avenues to fruifully continue the work of this chapter:

More Mathematical Examples For a deeper insight into the hidden use of new axioms, informal examples of it would be necessary. This might be reports from practitioners in set theory. It would, moreover, be informative to know how many set theorists use which new axioms in this way. When does it work, when does it not work—how widely applicable is this use?

Is the Hidden Use of Large Cardinal Axioms Particularly Well Suited for Consistency Statements? Of the two examples cited of the hidden use of large cardinal axioms, one is a consistency statement and the other is not. Hence, hidden use can be successful in both cases. However, there might be differences according to the kind of statement. The two interviewees in the study who mentioned the hidden use of large cardinal axioms are both experts in forcing. Moreover, we have seen that large cardinals are especially useful in enabling certain forcing constructions. These two considerations might suggest that the hidden use of large cardinal axioms is probably more successful in cases where the statement of interest, S, is a consistency statement.

The Hidden Use of Other New Axioms Both determinacy principles and forcing axioms might play a similar role to some extent. Since many determinacy principles are implied by large cardinal axioms, their role would possibly be different within the details of the proof, but not in the logical strength of the assumptions. Regarding the axiom of determinacy and Martin's axiom, we have seen quotations confirming that set-theoretic practitioners working in descriptive set theory use forcing axioms and determinacy principles in the same way. It would be interesting to elaborate on the hidden use of these new axioms and see whether there are differences to the hidden use of large cardinal axioms. My conjecture is that the hidden use of other new axioms is as effective as that of large cardinal axioms, because they are all well-researched new axioms whose application is no longer difficult. I further conjecture that there are differences with regard to the research areas in which certain new axioms are applicable. One descriptive set theorist mentioned that they typically use AD, for instance. For forcing-related questions, large cardinal axioms are possibly used more. And forcing axioms seem to be often applied to topological questions.

Acknowledgments I am very grateful to the book's editors, Carolin Antos, Neil Barton, and Giorgio Venturi, for inviting me to contribute a chapter. I recall the seminar series organised as part of the preparation of this volume as a fruitful endeavour. I am very grateful for the detailed and thoughtful reports that I received from two anonymous referees. They helped to improve the chapter substantially. I am also very grateful for the discussions with the audiences when I presented this work at the Logic Colloquium 2023 in Milano, and at the Mathematical Logic Seminar in Hamburg.

References

Boolos, G. (1971). The iterative conception of set. *Journal of Philosophy, 68*(8), 215–231.

De Toffoli, S. (2020). Groundwork for a fallibilist account of mathematics. *The Philosophical Quarterly, 71*(4), 12.

Easwaran, K. (2008). The role of axioms in mathematics. *Erkenntnis, 68*(3), 381–391. https://doi.org/10.1007/s10670-008-9106-1

Friedman, H. M. (1971). Higher set theory and mathematical practice. *Annals of Mathematical Logic, 2*(3), 325–357. https://doi.org/10.1016/0003-4843(71)90018-0

Gale, D., & Stewart, F. M. (1953). Infinite games with perfect information. *Annals of Mathematics Studies, 28*, 245–266.

Gödel, K. (1947). What is Cantor's continuum problem? *The American Mathematical Monthly, 54*(9), 515–525.

Goldstern, M., Kellner, J., & Shelah, S. (2019). Cichon's maximum. *Annals of Mathematics, 190*(1), 113–143. https://doi.org/10.4007/annals.2019.190.1.2

Goldstern, M., Kellner, J., Mejía, D. A., & Shelah, S. (2022). Cichon's maximum without large cardinals. *Journal of the European Mathematical Society, 24*(11), 3951–3967.

Harrington, L. (1978). Analytic determinacy and $0^\sharp$. *The Journal of Symbolic Logic, 43*(4), 685–693. http://www.jstor.org/stable/2273508

Larson, P. B. (2020). A brief history of determinacy. In A. S. Kechris, B. Löwe, & J. R. Steel (Eds.), *Large cardinals, determinacy and other topics: The cabal seminar, volume IV*. Lecture Notes in Logic (Vol. 4, pp. 3–60). Cambridge University Press. https://doi.org/10.1017/9781316863534.002

Maddy, P. (1988a). Believing the axioms. I. *Journal of Symbolic Logic, 53*(2), 481–511.

Maddy, P. (1988b). Believing the axioms. II. *The Journal of Symbolic Logic, 53*(03), 736–764.

Maddy, P. (1997). *Naturalism in mathematics*. Oxford University Press.

Maddy, P. (2011). *Defending the axioms: On the philosophical foundations of set theory*. Oxford University Press.

Martin, D. A. (1970). Measurable cardinals and analytic games. *Fundamenta Mathematicae, 66*, 287–291.

Martin, D. A. (1975). Borel determinacy. *Annals of Mathematics, 102*(2), 363–371.

Martin, D. A. (1998). *Mathematical evidence* (pp. 215–231). Oxford University Press.

Schlimm, D. (2013). Axioms in mathematical practice. *Philosophia Mathematica, 21*(1), 37–92. https://doi.org/10.1093/philmat/nks036

van Atten, M., & Kennedy, J. (2009). Gödel's modernism: On set-theoretic incompleteness, Revisited. In S. Lindström, E. Palmgren, K. Segerberg, & V. Stoltenberg-Hansen (Eds.), *Logicism, intuitionism, and formalism: What has become of them?* (pp. 303–355). Springer Netherlands. https://doi.org/10.1007/978-1-4020-8926-8_15

Viale, M. (2019). Useful axioms, 2019. *arXiv, 1610.02832*, math.LO.

Part II

Formal Theories

6

The Iterative Notion of Function and the Iterative Notion of Set

Tim Button

Hilary Putnam once suggested that "the actual existence of sets as 'intangible objects' suffers…from a generalization of a problem first pointed out by Paul Benacerraf…are sets a kind of function or are functions a sort of set?"[1] Sadly, he did not elaborate; my aim here is to do so on his behalf.

There are well-known methods for treating sets as functions and functions as sets. But these do not raise any obvious philosophical or foundational puzzles (see Sect. 1). To raise such puzzles, we first need to provide a full-fledged function theory. I will supply one, which axiomatizes the iterative notion of function in exactly the same sense that **ZF** axiomatizes the iterative notion of set. Indeed, this theory is synonymous with **ZF** (see Sects. 2–3).

It might seem that set theory and function theory present us with rival foundations for mathematics, since they postulate different ontologies. But I will argue that appearances are deceptive. Set theory and function theory provide the very same judicial foundation for mathematics (see Sect. 4). They do not supply rival metaphysical foundations (see Sect. 5); indeed, if they supply metaphysical foundations at all, then they supply the very same metaphysical foundations (see Sect. 6).

[1] Putnam (2014, 13.Dec.2014).

T. Button (✉)
Department of Philosophy, University College London, London, UK
e-mail: tim.button@ucl.ac.uk

C. Antos et al. (eds.), *The Palgrave Companion to the Philosophy of Set Theory*,
Palgrave Companions, https://doi.org/10.1007/978-3-031-62387-5_6

1 Sets as Functions/Functions as Sets

There is a standard way to regard functions as sets. To avoid ambiguity, say that a set f is a $function_{\in}$ iff both (i) every member of f is a Kuratowski ordered pair, and (ii) if $\langle x, y\rangle \in f$ and $\langle x, z\rangle \in f$ then $y = z$. We then treat "$f(x) = y$" as an abbreviation of "$\langle x, y\rangle \in f$". Conversely, sets simply can be regarded as *characteristic* functions, that is, as functions whose values are always either 1 or 0. We then treat "$x \in y$" as an abbreviation for "$y(x) = 1$".

This simple observation suggests a formal equivalence. Consider the following re-axiomatization of **ZF**. Replace each atomic formula "$x \in y$" with "$@(y, x) = 1$", where "$@$" is a new two-place function symbol. We can read "$@(y, x)$" as "y applied to x". Next, add an axiom:

$$\forall y \forall x(@(y, x) = 0 \vee @(y, x) = 1)$$

Finally, treat 0 and 1 as objects such that: $@(0, x) = 0$ for all x; and $@(1, x) = 1$ iff $x = 0$. Call the result ZF_χ. Fairly trivially, ZF_χ is *synonymous* with **ZF**.

I should unpack this notion of synonymy. Two theories, **S** and **T**, are said to be *synonymous* when each interprets the other, and running the two interpretations consecutively gets you back where you began. More precisely, it is to say that there are translations, ∇ and Δ, which are *interpretations*, that is, theorems are translated to theorems, so that:

(1) if $\mathbf{T} \vdash \phi$ then $\mathbf{S} \vdash \phi^{\nabla}$, for any sentence ϕ in **T**'s language; and
(2) if $\mathbf{S} \vdash \phi$ then $\mathbf{T} \vdash \phi^{\Delta}$, for any sentence ϕ in **S**'s language

Moreover, composing these interpretations gets you back where you started, in the sense that:[2]

(3) $\mathbf{S} \vdash \phi \leftrightarrow \phi^{\Delta\nabla}$, for any formula ϕ in **S**'s language; and
(4) $\mathbf{T} \vdash \phi \leftrightarrow \phi^{\nabla\Delta}$, for any formula ϕ in **T**'s language

Now, **ZF** is a theory of sets, and ZF_χ is a theory of (characteristic) functions, but **ZF** and ZF_χ are *synonymous*, in the sense just defined. And something like this observation underwrote John von Neumann's extremely suggestive claim that the concepts *set* and *function* "are completely equivalent, since a function can be regarded as a set of pairs, and a set as a function that can take two values."[3]

---

[2] I write ϕ^{IJ} for $(\phi^I)^J$. Note that ϕ and $\phi^{\Delta\nabla}$ may be syntactically distinct, but they are provably equivalent. For more, see, for example, Friedman and Visser (2014, §§1–2), and Button and Walsh (2018, ch.5).

[3] von Neumann (1925, 221–2/396); see also (1925: 230–1/411–12, 1928: 676, 1929: 227, 231).

That conclusion is too quick. For one thing: there is nothing especially *set-theoretic* about that equivalence. Not just membership, but any relation, can be given this treatment: just read "$R(x_1, \ldots, x_n, y)$" as an abbreviation of "$@_R(y, x_1, \ldots, x_n) = 1$", and add an axiom:

$$\forall y \forall x_1 \ldots \forall x_n (@_R(y, x_1, \ldots, x_n) = 0 \vee @_R(y, x_1, \ldots, x_n) = 1)$$

This provides a perfectly general method for rewriting a theory in functional terms. But it is needlessly baroque: it just amounts to incorporating surrogates for truth-values, 0 and 1, into the theory's object language.

For another thing, the synonymy of ZF and ZF_χ would only suggest that the concept *set* is intimately connected with the concept *characteristic function*, not with *function* simpliciter. Now, to be fair to von Neumann, he was not considering the theory ZF_χ, but a much richer theory which encompassed functions which are *not* characteristic functions.[4] However, he also did not advance a synonymy result (the notion of *synonymy* had not yet been invented).[5] So I suspect that von Neumann's thought was roughly that the concepts *set* and *function* are equivalent, because his rich function theory can interpret (an unspecified)[6] set theory via characteristic functions, and (that same) set theory can interpret his function theory via functions$_\in$. (So we would have conditions (1) and (2) from above, but we would lack conditions (3) and (4).) But if that was indeed von Neumann's thought, it falls short of what he needed. Intuitively, (mere) mutual interpretability is insufficient to establish the equivalence of the relevant concepts; after all, PA and $\mathsf{PA}+\neg\mathrm{Con}(\mathsf{PA})$ are mutually interpretable, but they surely articulate very different conceptions of *number*.

All told, we do not *yet* have a good reason to agree with von Neumann's conclusion that the concepts *set* and *function* are "completely equivalent". Nonetheless, von Neumann was onto something; there *is* an important equivalence between these concepts. My aim in this chapter is to unpack this equivalence.

---

[4]So, rather than having an *axiom* like $\forall y \forall x(@(y, x) = 0 \vee @(y, x) = 1)$, he (1925: 229/403, 1928: 678, 684) stipulatively defines that the *sets* will be those (I-objects) y such that $\forall x(@(y, x) = 0 \vee @(y, x) = 1)$. In §2.2.1 of Chap. 6, Toby Meadows presents a theory, vNC, which is similar to von Neumann's. (I first encountered Meadows' work in March 2023, after writing this chapter.)

[5]That first occurs with Montague (1957, 56) and de Bouvère (1963).

[6]The theory NBG would do.

2 The Iterative Notion of *Function*

I must start by making my target concept of *function* more precise. In particular, and analogously with the iterative notion of *set*, I will consider the iterative notion of *function*.

Recall that we can introduce the iterative notion of *set* with the following story:[7]

> **The Set Story.** Sets are arranged in stages. Every set is found at some stage. At any stage **s**: for any sets found before **s**, we find a set whose members are exactly those sets. We find nothing else at **s**.

This Story articulates the bare-bones iterative notion of *set*. Adopting that conception of *set* is, of course, a choice.[8] But it is a common choice, and it has much to recommend it. I will say a bit more about its commendable qualities in Sect. 4.1; for now, I note that the iterative conception immediately blocks the threat of Russell's paradox.

Raphael Robinson noted that we can introduce an iterative notion of *function* with a similar story:[9]

> **The Function Story.** Functions are arranged in stages. Every function is found at some stage. At any stage **s**: we find all functions whose domains and ranges are exhausted by the functions we found earlier. We find nothing else at stage **s**.

Again, this Story lays down nothing more than the bare idea of an iterative notion of *function*. Furthermore, embracing the Story, and thereby adopting the iterative notion of *function*, is a choice.[10] But it is a good choice. Apart

[7]This formulation is from Button (2021a), but such stories are hardly novel to me! In that paper, I explain connections to, for example, Boolos (1971), Scott (1974), and Shoenfield (1977, 323). Note that we need no more (nor less) than this Story to introduce the (bare) idea of the iterative notion of *set*.

[8]We might instead have pursued Quine's **NF**.

[9]"Suppose that we have the idea 'function,' but nothing to use as arguments and values. We can start with the function 0 which is not defined for any arguments and hence does not need any values.... Then we construct a function 1 such that $1(0) = 0$, but $1(x)$ is undefined for $x \neq 0$...We proceed, always using as arguments and values functions already constructed" (Robinson, 1937, 29–30, notation slightly changed). Robinson then offered a deliberately simplified version of von Neumann's theory (which is quite different from my own $\mathsf{FLT}_{\mathsf{ZF}}$). I wrote this chapter in embarrassing ignorance of Robinson's paper; many thanks to Chris Menzel for bringing it to my attention.

[10]We could, instead, have developed a notion of *function* along the lines of Church's untyped λ-calculus or along the lines of Sant'Anna et al.'s (2020) theory Flow.

from anything else, it immediately blocks the threat posed by the paradoxes of naïve function theory. To illustrate, let d be this naïve diagonal function:[11]

$$\text{if } x(x) \neq x, \text{ then } d(x) = x$$

$$\text{if } x(x) = x, \text{ then } d(x) \text{ is undefined}$$

Contradiction follows swiftly: if $d(d) = d$ then $d(d)$ is undefined, which is absurd; so $d(d) \neq d$, and hence $d(d) = d$, a contradiction. But, given the Function Story, d obviously does not exist; its mapping behaviour requires it to be found after every function, which is absurd.

Two clarificatory comments might be helpful. The functions introduced in the Function Story are always *partial* functions: for example, $f(f)$ is undefined for every f. Moreover, the functions are *pure*: they only map from functions to functions. We could modify the Story to allow "urelements"—and ultimately we probably should—but for simplicity, and for parity of presentation with the case of pure sets, I will ignore them in this chapter.[12]

3 Functional-Level Theory

Of course, working within **ZF**, we can recursively define a system of (hereditary-)functions$_{\in}$ which would satisfy the Function Story. But I cannot rest content with what we can do within **ZF**. After all, I want to show that set-theoretic and function-theoretic approaches are on a par when it comes to mathematical foundations. To do this, I need to axiomatize the Function Story without relying upon some "prior" theory of sets (like **ZF**); I need to give you an axiomatic theory of functions which stands on its own two feet.

I will provide just such a theory in this section. Previous work makes this easier than it might at first seem. Elsewhere, I have shown how to axiomatize the Set Story,[13] and the same techniques can be used for the Function Story, with only minor adjustments. (For readability, I relegate most of the fiddlier details to the appendices.)

[11] Cf. von Neumann(1928: 676–7 fn5) and Robinson (1937, 31).

[12] Button (2021a, §§A–B) tackles the Set Story with urelements; the same techniques can be carried over to the Function Story.

[13] See Button (2021a).

3.1 Notational Preliminaries

I begin with some notation. Where σ and τ are any object-language terms, I write:

- "$\sigma(\tau)$" to indicate the value of applying σ to τ.
- "$\tau{\downarrow}$" to indicate that τ exists, that is, that $\exists y\ y = \tau$.
- "$\sigma \simeq \tau$" to indicate that $\sigma = \tau$ if either exists.[14]
- "$[\lambda x.\tau]$" for the (first-order) function which sends x to τ, if there is such a function.[15]

There are a few different ways to regiment these bits of notation more pedantically, but the exact choice of pedantry is relatively unimportant, so I relegate to Sect. B.1. I will continue with a key definition, which allows me to relate functions to "sets":

Definition 3.1 We introduce abbreviations as follows:[16]

$$g \mathrel{\varepsilon} f :\equiv (f(g){\downarrow} \vee \exists y\ f(y) = g)$$
$$g \Subset f :\equiv (\forall x \mathrel{\varepsilon} g) x \mathrel{\varepsilon} f$$

Say that f is a *set@* iff $(\forall x \mathrel{\varepsilon} f) f(x) = x$, i.e. f is any partial identity function. ◊

We might read "$g \mathrel{\varepsilon} f$" as "$g$ is in f's field" (i.e. in f's domain or its range). We might read "$g \Subset f$" as "f's field includes g's". And sets@ will serve as our function-theoretic surrogates for sets, easily allowing us to construct a set-hierarchy as an inner model of a function-hierarchy.[17]

---

[14] That is, $\sigma \simeq \tau$ means $((\sigma{\downarrow} \vee \tau{\downarrow}) \rightarrow \sigma = \tau)$; here, σ and τ are any object-language terms.

[15] That is, $[\lambda x.\tau](v) \simeq \tau[^v/_x]$, for all v.

[16] As usual, $(\forall x \lhd y)\phi$ abbreviates $\forall x(x \lhd y \rightarrow \phi)$, for any infix predicate $\lhd$.

[17] Specifically, as in Sect. 3.6, we will interpret **ZF**-sets as *hereditary-sets@*. But note that that interpretation does not witness the *synonymy* between **ZF** and $\mathsf{FLT}_{\mathsf{ZF}}$ (see the Main Theorem in Sect. 3.6). Under synonymy, *every* $\mathsf{FLT}_{\mathsf{ZF}}$-function "simulates" some **ZF**-set, and *every* **ZF**-set "simulates" some $\mathsf{FLT}_{\mathsf{ZF}}$-function (see footnote 30).

Many alternative definitions of sets@ would have worked equally well. With Robinson (1937, 31), we could have defined sets@ as functions obeying the principle that $\forall x(f(x){\downarrow} \rightarrow f(x) = 0)$. Or, with von Neumann (see Footnote 3) we could have defined sets@ as characteristic functions. Nothing much turns on this choice (although the idea of a characteristic function requires that there be at least two objects, which neither **FLT** nor **FST** guarantee). However, defining sets@ as partial identity functions gives an immediate link to category theory (see Theorem 3.4 and Sect. B.4).

3.2 Using Stages: FST

I now turn in earnest to the task of axiomatizing the Function Story. Since that Story mentions both stages and functions, I will start with a formal theory which does the same. My theory, then, has two distinct sorts of first-order variables, for *functions* (lower-case italics) and for **stages** (lower-case bold). It has these two primitives:

$<$: a relation between stages; read "$\mathbf{r} < \mathbf{s}$" as "$\mathbf{r}$ is before $\mathbf{s}$".
$\preceq$: a relation between a function and a stage; read "$f \preceq \mathbf{s}$" as "f is found at $\mathbf{s}$".

For brevity, let $f \prec \mathbf{s}$ abbreviate $\exists \mathbf{r}(f \preceq \mathbf{r} < \mathbf{s})$, that is, f is found before $\mathbf{s}$. Now let **FST**, for Functional Stage Theory, be the theory with these five axioms:

FunExt $\forall f \forall g\big(\forall x\, f(x) \simeq g(x) \rightarrow f = g\big)$
FunOrd $\forall \mathbf{r} \forall \mathbf{s} \forall \mathbf{t}(\mathbf{r} < \mathbf{s} < \mathbf{t} \rightarrow \mathbf{r} < \mathbf{t})$
FunStage $\forall f \exists \mathbf{s}\, f \preceq \mathbf{s}$
FunPri $\forall \mathbf{s}(\forall f \preceq \mathbf{s})(\forall x \,\varepsilon\, f) x \prec \mathbf{s}$
FunSpec $\forall P \forall \mathbf{s}\big(\forall x (P(x){\downarrow} \rightarrow (x \prec \mathbf{s} \wedge P(x) \prec \mathbf{s})\big) \rightarrow [\lambda x . P(x)] \preceq \mathbf{s}\big)$
with "P" a second-order function-variable

The first two axioms are analytic. FunExt says: functions are individuated by their mapping behaviour; I take it that this is a stipulative feature of the kind of functions we are considering. FunOrd says: *before* is a transitive relation on stage; I take it that this is required by the very idea of a notion of *before*.[18] Then the next three axioms are clearly given by the Story. So: FunStage says: every function is found at some stage. FunPri says: everything in a function's field is found (strictly) before the function itself is found. And FunSpec says: if P is a (second-order) map and everything in P's field is found before $\mathbf{s}$, then P determines a (first-order) function which is found at stage $\mathbf{s}$.

I just invoked second-order logic. I will do so throughout, assuming (impredicative) second-order Comprehension. That said, my use of second-order logic is mostly just for notational convenience; almost everything I

[18] Strikingly, there is no need to assume that *before* is a well-order. Our axioms allow us to prove well-foundedness. We cannot prove linearity, but this is unimportant, since the ordinal height of a stage determines exactly which functions are found at that stage. This all follows from Theorems 3.2–3.3, and it is characteristic of stage theories (cf. Button, 2021a: §5, 2021b: §4, 2022a: §4).

say can easily be first-orderized.[19] The only case where second-order logic is required is in the discussion of quasi-categoricity (see Sect. 3.4).

Clearly, **FST**'s axioms are all true of the Function Story, as told in Sect. 2. So this is an excellent first step in axiomatizing that Story. But it is just a first step. Recall my aims from the start of Sect. 3: I want to show that set-theoretic and function-theoretic approaches are on a par when it comes to mathematical foundations. Since we can axiomatize set theory without using a primitive **stage**-sort, I must next eliminate the **stage**-sort from function theory.

3.3 Eliminating Stages: FLT

With this goal in mind, let **FLT**, for Functional Level Theory, be the theory with these three axioms:

FunExt $\forall f \forall g \big(\forall x\, f(x) \simeq g(x) \to f = g\big)$

FunStrat $\forall a \exists s \big(Fev(s) \wedge a \Subset s\big)$

FunComp $\forall P \big(\exists f \forall x \big(P(x)\!\downarrow\ \to\ (x\ \varepsilon\ f \wedge P(x)\ \varepsilon\ f)\big) \to [\lambda x . P(x)] \text{ exists}\big)$
with "P" a second-order function-variable

As promised, these axioms do away with stages. But they require some discussion. We have already encountered FunExt in formulating **FST**. Then FunComp is a principle which turns (second-order) maps into (first-order) functions; in effect, FunComp plays the same role in **FLT** as FunSpec played in **FST**. The trickiest axiom to explain is FunStrat. This uses a new predicate, "*Fev*", and we should pronounce "*Fev*(s)" as "s is a function-level", or "s is a fevel" for short. Intuitively, **FLT**'s fevels serve as function-theoretic proxies for **FST**'s stages, so that FunStrat plays roughly the same role in **FLT** as FunStage played in **FST**. But, to ensure that the fevels behave this way, we must not take "*Fev*" as a primitive, but must instead define it.

This can be done. But the required definition is both cunning and ugly, and no one will be able to tell at a glance that it is the right definition. So I will not pause here to talk it through (see Definition B.2 for more). What matters here is just that *the definition works.*

The sense in which the definition *works* is made precise by considering three characteristic theorems for **FLT** (for technical discussion, see Sect. B; after announcing a Theorem, I list in brackets the axioms used in its proof):

---

[19] Specifically, FunSpec would become this scheme. If $\forall x \forall y \forall z\big((\phi(x, y) \wedge \phi(x, z)) \to (y = z \wedge x \preceq \mathbf{s} \wedge y \preceq \mathbf{s})\big)$, then $(\exists f \preceq \mathbf{s}) \forall x \forall y (f(x) = y \leftrightarrow \phi(x, y))$.

Theorem 3.2 **FST** $\vdash \phi$ *iff* **FLT** $\vdash \phi$, *for any* **FLT**-*sentence* ϕ.

Theorem 3.3 (FunExt + FunComp) *The fevels are well-ordered by* ε.[20]

Theorem 3.4 (**FLT**) *We have a category: the arrows are the functions; the objects and their identity arrows are exactly the sets*$_@$*; composition is functional composition.*

Theorem 3.2 says that **FLT** is equivalent to **FST**, in the sense that they make exactly the same claims about functions. Theorem 3.3 says that our stage-surrogates, the fevels, are well-ordered by ε. We can prove this result without using FunStrat; when we add FunStrat, we obtain the extremely powerful tool of ε-induction (see Definition B.3). Finally, Theorem 3.4 says that our functions constitute a category (see Sect. B.4).

3.4 FLT is Quasi-Categorical

The Function Story introduces the (bare idea of the) iterative notion of *function*. So, every cumulative hierarchy of functions answers to that Story. Since **FST**'s axioms are all true of that Story, every cumulative hierarchy of functions obeys **FST**, and hence also **FLT**, by Theorem 3.2. Hence, by Theorems 3.3 and 3.4, *every cumulative hierarchy of functions is a category which is arranged into well-ordered fevels*. This is a rather pleasant result.[21]

We can push this point further. The Function Story said nothing about the height of the hierarchy of functions. So, by design, **FLT** says nothing about that either. Still, as a second-order theory, **FLT** is quasi-categorical. Informally, we can spell out the point as follows: *Any two cumulative hierarchies of functions are structurally identical for so far as they both run, but one may be taller than the other.*

There are at least two ways to explicate this idea, but **FLT** is quasi-categorical on both explications: it is both *externally* and *internally* categorical. I will not go into details on this point here, since I have done so in detail elsewhere.[22] However, an immediate upshot is that the size of a standard model of **FLT** is

[20] That is: (1) if some fevel is F, then there is an ε-minimal fevel which is F; and (2) $s \mathrel{\varepsilon} t \lor s = t \lor t \mathrel{\varepsilon} s$, for any fevels s and t.

[21] Not least because the bare idea of the iterative notion of *function*—as explicated via the Function Story—made no explicit reference to either categories or well-ordering (cf. Scott, 1974: 210; Button, 2021a: §5, 2021b: §4, 2022a: §4; also footnote 18).

[22] See Button(2021a: §6, 2022a: §5).

determined entirely by the order-type of its fevels. Specifically, consider this function, p, from ordinals to cardinals:

$$p(1) := 1 \quad p(\alpha + 1) := (p(\alpha) + 1)^{p(\alpha)} \quad p(\alpha) := \sup_{\beta<\alpha} p(\beta) \text{ for limit } \alpha$$

For any ordinal $\alpha > 0$, there are (standard) models of **FLT** with (exactly) an α-sequence of fevels. And, given an α-sequence of fevels, there are $p(\alpha)$ functions.[23] Evidently p grows quite quickly: one fevel yields one function;[24] but with four fevels we already have a billion functions.

3.5 Adding More Functions

FLT provides us with a mathematical playground of functions, arranged in well-ordered fevels. However, since **FLT** allows that there can be only one fevel, and hence only one function, the vanilla theory of **FLT** allows our playground to be pretty desolate.

To ensure that our hierarchy of functions has a certain height—to populate the playground—here are some principles that we could add to **FLT**:

FunEndless $\forall s \exists t\ s \varepsilon t$
FunInfinity $\exists s(\exists p \varepsilon s)(\forall q \varepsilon s)\exists r(q \varepsilon r \varepsilon s)$
FunSupercomp $\forall P\big(\exists f \forall x\big(P(x)\!\downarrow\ \rightarrow x \varepsilon f\big) \rightarrow [\lambda x.P(x)] \text{ exists}\big)$

Against the background of **FLT**, we can characterize these principles as follows. FunEndless says: every fevel has a successor (compare **ZF**'s Powersets). FunInfinity says: there is an ωth fevel (compare **ZF**'s Infinity). And FunSupercomp strictly strengthens FunComp, saying: no function's field can be mapped unboundedly into the fevels (compare **ZF**'s Replacement).

3.6 Synonymy

Let $\mathbf{FLT}_{\mathbf{ZF}}$ be the theory **FLT** + FunInfinity + FunSupercomp. The following result explains the name:[25]

[23] *Proof Sketch.* By induction; the only interesting clause is $p(\alpha + 1)$. Here, note that there are exactly $|B|^{|A|}$ *total* functions $A \longrightarrow B$; so, allowing for the one extra possibility of being undefined rather than landing in B, there are exactly $(|B| + 1)^{|A|}$ *partial* functions $A \longrightarrow B$.

[24] Specifically, the null function, which is undefined for all inputs.

[25] I have formulated $\mathbf{FLT}_{\mathbf{ZF}}$ as a *second-order* theory; then the result is that $\mathbf{FLT}_{\mathbf{ZF}}$ is synonymous with *second-order* **ZF**. However, we also have synonymy between the *first-order* versions of both theories.

Main Theorem ZF *and* $\mathsf{FLT}_{\mathsf{ZF}}$ *are synonymous.*[26]

I outlined the notion of *synonymy* in Sect. 1. But here is a helpful, if rough, gloss of the Main Theorem: in *some* sense, ZF and $\mathsf{FLT}_{\mathsf{ZF}}$ are notational variants; they wrap the same content in different notational packaging.

This rough gloss is deliberately provocative, and it raises many philosophical questions. I will explore those questions in the next two sections. First, I should quickly sketch the proof strategy for the Main Theorem.

In ZF, we can regard functions as functions$_{\in}$, that is, sets of ordered pairs of a certain sort (see Sect. 1). This provides us with an interpretation $I : \mathsf{FLT}_{\mathsf{ZF}} \longrightarrow \mathsf{ZF}$. In particular, working in ZF, we can interpret $\mathsf{FLT}_{\mathsf{ZF}}$ as dealing with *hereditary-functions*$_{\in}$, where a set is a hereditary-function$_{\in}$ iff it is a function$_{\in}$, whose field comprises only functions$_{\in}$, whose field comprises only functions$_{\in}$, etc.[27]

Conversely, in $\mathsf{FLT}_{\mathsf{ZF}}$, we can interpret sets as sets$_{@}$ (see Definition 3.1). This provides us with an interpretation $J : \mathsf{ZF} \longrightarrow \mathsf{FLT}_{\mathsf{ZF}}$. In particular: working in $\mathsf{FLT}_{\mathsf{ZF}}$, we can interpret ZF as dealing with *hereditary-sets*$_{@}$, where a function is a hereditary-set$_{@}$ iff it is a set$_{@}$, whose members (in the sense of ε) are all sets$_{@}$, whose members are all sets$_{@}$, etc.[28]

We now have interpretations going in both directions. With a little effort,[29] we can show that they are *bi*-interpretations. By the Friedman–Visser Theorem,[30] it follows that ZF and $\mathsf{FLT}_{\mathsf{ZF}}$ are synonymous, as required.

---

[26] In fact, we can prove a much stronger result. Let FLT_{+} be the theory FLT + FunEndless, that is, the theory which arises by adding "there is no last stage" to the Function Story. Comparably, LT_{+} is the theory which arises by adding "there is no last stage" to the Set Story (see Sect. A for details; notably, LT_{+} is a strict sub-theory of ZF). Then the stronger version of the Main Theorem is this: LT_{+} *and* FLT_{+} *are synonymous.*

Since ZF is the "industry standard" of set theory, I will only discuss the case of ZF and $\mathsf{FLT}_{\mathsf{ZF}}$ in the main text. But I would much prefer to focus on the case of FLT_{+}; after all, focussing on the weaker theories only makes my argument stronger. (I revisit this in footnotes 34, 41, and 43.)

[27] Showing that this *is* an interpretation takes more effort than one might expect (particularly if the aim is to establish the results for LT_{+} and FLT_{+}; see footnote 26). The key steps are as follows: first, prove FunExt^{I} and FunComp^{I} within LT_{+}; then you can appeal to the I-translation of (all the results you used to establish) Theorem 3.3; these entail the good behaviour of the fevels^{I}, thereby delivering FunStrat^{I}. (Similar but simpler moves are made in Button, 2022a: §C.)

[28] Just as we can use transfinite recursion in ZF to define *hereditary-functions*$_{\in}$, we can use transfinite recursion in $\mathsf{FLT}_{\mathsf{ZF}}$ to define *hereditary-sets*$_{@}$. The proof that J *is* an interpretation is similar to the proof that I is an interpretation (see footnote 27).

[29] The key tools are: patience, $\in$-induction in LT_{+}, and ε-induction in FLT_{+}.

[30] That is, Friedman and Visser (2014, Corollary 5.5). Note that the interpretations I and J do not *themselves* bear witness to the synonymy. Rather, the Friedman–Visser Theorem shows that we can use I and J to define other interpretations which witness synonymy.

4 Function-Theoretic Judicial Foundations

Let me take stock. I began Sect. 2 with the Function Story, which introduced the bare idea of the iterative notion of *function*. The Story is quasi-categorically axiomatized by **FLT**; so the Story entails that functions constitute a category which is arranged into well-ordered fevels. Moreover, we can easily enrich **FLT** with various claims to ensure that the sequence of fevels runs for a rather long time. In particular, the theory $\mathsf{FLT}_{\mathsf{ZF}}$ arises just by insisting that the hierarchy of functions is extremely tall. And $\mathsf{FLT}_{\mathsf{ZF}}$ is synonymous with **ZF**.

This synonymy suggests that we might be able to replace set-theoretic foundations (as formalized in **ZF**) with function-theoretic foundations (as formalized in $\mathsf{FLT}_{\mathsf{ZF}}$). In this section, I will show that this is correct. Indeed, I will show that function-theoretic and set-theoretic foundations provide us with exactly the same (judicial) foundations.

4.1 Maddy on Judicial Foundations

Many things have been meant by talk of mathematical "foundations". So I must start by clarifying the kind of foundational project I have in mind. In brief: my interest is not in *methodological* foundations, nor (initially) in *metaphysical* foundations, but in (what I call) *judicial* foundations.

Methodological Foundations A mathematical theory might aim to supply us *methodological* foundations. Specifically, it might aim to provide a vocabulary and toolkit for dealing with situations that arise frequently across the motley of mathematics.

Category theory has some claim to providing methodological foundations; think of the unifying power of concepts like *functor*, *natural transformation*, or *adjunction*. Set theory, by contrast, has almost no such claim: to give a well-worn example, having shown that we can define the real numbers as sets of sets of sets of...sets, no practising mathematician ever digs back through that definition. Function theory has a similarly weak claim to providing methodological foundations, and for similar reasons. So I will not discuss methodological foundations any further.

Metaphysical Foundations Another possible foundational project is to determine what mathematical objects *are*.

Those who ask such questions are not usually satisfied with a grab-bag answer, along the lines of: *mathematical objects include numbers, groups, graphs, topologies, categories,...*. They will reply: *yes, but what exactly* are *these things?* And they will not be satisfied with a mathematical definition, like: *a category is a system of objects and arrows such that...*. They are in pursuit of something distinctively metaphysical.

Such people are asking for a *metaphysical foundation* for mathematics. I will set this metaphysical project aside for now, returning to it in Sect. 5, where we will be able to shed new light on it.

Judicial Foundations My focus in this section, instead, is on *judicial foundations*. The inspiration behind this phrase is in Penelope Maddy's suggestion that set theory is "a court of final appeal".[31] Here is her point, in detail:

> The force of set theoretic foundations is to bring (surrogates for) all mathematical objects and (instantiations of) all mathematical structures into one arena—the universe of sets—which allows the relations and interactions between them to be clearly displayed and investigated. Furthermore, the set theoretic axioms developed in this process are so broad and fundamental that they do more than reproduce the existing mathematics; they have strong consequences for existing fields and produce a mathematical theory that is immensely fruitful in its own right. Finally, perhaps most fundamentally, this single, unified arena for mathematics provides a court of final appeal for questions of mathematical existence and proof: if you want to know if there is a mathematical object of a certain sort, you ask (ultimately) if there is a set theoretic surrogate of that sort; if you want to know if a given statement is provable or disprovable, you mean (ultimately), from the axioms of the theory of sets.[32]

The crucial point is something like this.[33] Suppose you make a conjecture, in any area of mathematics. If someone can construct a set, in **ZF**, which provides a counter-example to the conjecture, then you lose. If you can show that the conjecture holds in (vanilla) **ZF**, then you win. And if you can show that the conjecture holds in **ZF** plus some extra large cardinal axiom, people might well be interested.

---

[31]Maddy (1997, 26). Maddy (2017, 296) has since expressed worries that the notion of "a final court of appeal" was "something of an exaggeration" and has outlined various nuanced notions in that ballpark. I would particularly emphasize the continuity between what I say about "judicial foundations" and Maddy (2017, 289–98) notions of a Meta-mathematical Corral, a Generous Arena, and a Shared Standard.

[32]Maddy (1997, 26).

[33]In addition to taking this from Maddy's writings, I would like to thank Michael Potter for impressing this point upon me many years ago.

This role for **ZF** is not, I think, a mere sociological quirk concerning mathematical practice. Instead, I think there are two very good reasons for regarding standard **ZF** as an excellent candidate for judicial foundations:

(a) it provides us with plenty of objects, which are richly structurable; and
(b) it is consistent and indeed coherent.

I will explain what I mean by "coherent" in Sect. 4.2. But let me quickly unpack these points, starting with (a). **ZF** tells us that there are plenty of objects. Moreover, the membership relation arranges these objects in an extremely tractable way. Using this given membership-structure, we have the resources to define many further relations on our objects, structuring and restructuring them to our heart's content. This is the point of (a). Indeed, this is *why* (in Maddy's phrase) we have a rich paradise of "set-theoretic surrogates" for other mathematical objects. Moreover, set theory gives us these objects, with their rich structure, without exploding under its own weight. That is the point of (b).

Features (a) and (b), then, are our two hallmarks for a decent judicial foundation. I will further elaborate on both features below, but my basic task for this section is to show that function theory possesses both features, to exactly the same extent as set theory. (Note: in Sects. 4.2–4.3, I will not really distinguish "set theory" from **ZF**, or "function theory" from $\mathsf{FLT}_{\mathsf{ZF}}$; I will explain why this is acceptable in Sect. 4.4.)

4.2 Consistency and Coherence

I begin by considering (b). I will argue that both set theory and function theory are consistent and coherent, starting with set theory.

Here is the strongest argument that I know of for the consistency of **ZF**:

($1_\in$) The Set Story of Sect. 2 introduces the bare idea of the iterative notion of *set*. We can augment that Story with two further claims: there is some limit stage, and no set can be mapped unboundedly into the sequence of stages. Call the result the Augmented Set Story.

($2_\in$) The Augmented Set Story is consistent. After all, the (unaugmented) Set Story is obviously consistent, and[34] the two further claims are also consistent.

($3_\in$) **ZF** quasi-categorically axiomatizes the Augmented Set Story.[35] So **ZF** inherits that Story's consistency.

I think this argument is a good one; indeed, it is why I believe that **ZF** is consistent. Moreover, the argument generalizes in an important way.

To explain the generalization, note that consistency is a rather weak property. Let PA_* be $\mathsf{PA} + \neg\mathrm{Con}(\mathsf{PA})$.[36] By Gödel's second incompleteness theorem, PA_* is consistent. But PA_* cannot be "trusted"; after all, PA_* "proves its own inconsistency", in the sense that $\mathsf{PA}_* \vdash \neg\mathrm{Con}(\mathsf{PA}_*)$. So, what we want is not just that our theory should be *syntactically* consistent, but that—speaking informally—our theory should have an *intended interpretation*.[37] Following Stewart Shapiro, I refer to this as the demand that the theory should be *coherent*.[38]

Fortunately, and as hinted, the argument ($1_\in$)–($3_\in$) can be strengthened. The Set Story is not presented as uninterpreted syntax; it is presented in English, with an associated intended interpretation. Indeed, we are convinced that the Story is consistent precisely because it strikes us as coherent.[39] So we can rerun the argument ($1_\in$)–($3_\in$), with "coherent" in place of "consistent".

[34]This is obviously the weakest point of this argument. How can we be sure that adding these two claims doesn't undermine consistency? If this really worries us, we can retreat from **ZF** to LT_+ (see footnote 26 and Sect. A). The theory LT_+ quasi-categorically axiomatizes the Set Story when it is augmented only with the claim that there is no last stage (see Button, 2021a), and I literally cannot entertain the idea that such a weakly augmented story is inconsistent. Comparably, FLT_+ arises by augmenting the Function Story (only) with the claim that there is no last stage. Moreover, LT_+ and FLT_+ are synonymous. So the entire argument of this subsection can be rerun in terms of LT_+ and FLT_+ rather than **ZF** and $\mathsf{FLT}_{\mathsf{ZF}}$, and the argument will be all the stronger for it.

[35]For this, see Button (2021a).

[36]Here, "Con(...)" indicates a canonical way for forming consistency sentences; I am assuming(!) **PA**'s consistency.

[37]I do not want to assume that the notion of "interpretation" must be *model-theoretic*, for two reasons. First, I would then have to decide whether to formalize model theory set-theoretically or function-theoretically. Second, for reasons I have described at length elsewhere (Button & Walsh, 2018, chs. 6–12; Button, 2022b), I would rather work within a second-order object language than within a first-order metalanguage.

[38]Shapiro (1997, 95–6, 133–6).

[39]The obvious question arises: why should we think there are (or could be) enough objects to provide us with this intended interpretation of **ZF**? Read a certain way, this question will swiftly lead us from judicial into metaphysical foundations, whereupon I do not know exactly how to respond (see Sect. 6). However, I do want to signal my agreement with Isaacson (2011, 29), that we take a big step towards showing that **ZF** is coherent when we show that **ZF** quasi-categorically axiomatizes the Augmented Story, thereby "developing our mathematical understanding of the subject matter of this theory through informal rigor".

This concludes my case that set theory is both consistent and coherent. With this to hand, I now want to show the same of $\mathsf{FLT}_{\mathsf{ZF}}$.

Here is a cute argument. Since **ZF** and $\mathsf{FLT}_{\mathsf{ZF}}$ are synonymous, they are equiconsistent. We just saw that **ZF** is consistent. So $\mathsf{FLT}_{\mathsf{ZF}}$ is consistent too. QED?

The cute argument has two weaknesses. First: it is not completely obvious that synonymy always preserves coherence (rather than mere consistency). Second: if this is our *only* argument for $\mathsf{FLT}_{\mathsf{ZF}}$'s consistency, then $\mathsf{FLT}_{\mathsf{ZF}}$'s consistency is epistemically dependent upon **ZF**'s, in which case, at least epistemically, set theory rather than function theory will serve as the "ultimate" court of appeal.

Fortunately, both issues can be dealt with. An almost identical argument, with exactly the same suasive force, establishes the consistency of $\mathsf{FLT}_{\mathsf{ZF}}$:

($1_@$) The Function Story introduces the bare idea of the iterative notion of *function*. We can augment that Story with two further claims: there is some limit stage; and no function's field can be mapped unboundedly into the sequence of stages. Call the result the Augmented Function Story.

($2_@$) The Augmented Function Story is consistent. After all, the (unaugmented) Function Story is obviously consistent, and the two further claims are consistent.

($3_@$) $\mathsf{FLT}_{\mathsf{ZF}}$ quasi-categorically axiomatizes the Augmented Function Story. So $\mathsf{FLT}_{\mathsf{ZF}}$ inherits that Story's consistency.

Moreover, as before, we can rerun the argument with "coherent" in place of "consistent".[40] So $\mathsf{FLT}_{\mathsf{ZF}}$ is both consistent and coherent.

4.3 Function-Theoretic Foundations

I now turn to (a). In Sect. 4.1, I explained the sense in which **ZF** provides us with a vast array of richly structurable objects. I will now show that the same holds of $\mathsf{FLT}_{\mathsf{ZF}}$.

Once again, synonymy supplies us with a cute argument to this effect. Specifically: there is (ultimately) a **ZF**-surrogate for an object iff there is (ultimately) a $\mathsf{FLT}_{\mathsf{ZF}}$-surrogate for an object; just apply the synonymy to the description of the surrogate. QED.

[40]With, mutatis mutandis, the clarifications of footnotes 34, 37 and 39. Indeed, this highlights part of the importance of **FLT**'s quasi-categoricity (see Sect. 3.4).

Again, one might worry that this cute argument threatens to diminish function theory's status. After all: if our *only* method for finding surrogate-objects involves first using set theory and then invoking synonymy, then set theory may seem to have a kind of primacy over function theory.

This time, though, the worry is wholly misguided: it confuses *judicial* foundations with *methodological* foundations. Granted, in searching for a surrogate-object, sometimes set theory will be more wieldy than function theory. Moreover, issues of (un)wieldiness are of real practical importance. But they are orthogonal to judicial foundations. In judicial foundations, the question is whether there are *ultimately* surrogates for objects or derivations, and the phrase "ultimately" is allowed to conceal all manner of unwieldiness. Indeed, from a purely judicial perspective, it would not matter if we presented function-theoretic foundations just by starting with our favourite set-theoretic foundations and then translating them into function theory using the interpretation $J : \mathsf{ZF} \longrightarrow \mathsf{FLT_{ZF}}$ from Sect. 3.6. The result will be less wieldy than set-theoretic foundations we started with; still, we can (ultimately) find function-theoretic surrogates iff we can (ultimately) find set-theoretic surrogates and that is the only thing which matters, judicially speaking.

(All that said, I should probably note that there is no need to make function-theoretic foundations "parasitic" upon set-theoretic foundations. We can present "autonomous" function-theoretic foundations, providing surrogates for various mathematical entities in a setting where our basic unit of currency is functions rather than sets. Especially enthusiastic readers may enjoy playing around with this possibility for themselves, and they can find some playful ideas in Sect. C. But my philosophical ambitions do not require me to force less enthusiastic readers to dwell on this point.)

4.4 Adding New Axioms

So far, I have argued that $\mathsf{FLT_{ZF}}$ meets conditions (a) and (b) exactly as well as **ZF**. But I have spoken as if *function theory* meets (a) and (b) exactly as well as *set theory*. It may not be immediately obvious that the conflation between **ZF** and set theory is harmless. After all: set theorists often consider extensions of **ZF**,[41] and presumably function theorists (if there were any) would often consider extensions of $\mathsf{FLT_{ZF}}$. So, for all I have said so far, one might worry that the following situation is possible: set-theoretic considerations push us to add

[41]Of course, set theorists also sometimes consider theories which are *weaker* than **ZF**. As per footnotes 26 and 34, I can rerun the arguments of the last few sections using just LT_+.

some new axiom, ϕ, to ZF; function-theoretic considerations push us to add some new axiom, ψ, to $\mathsf{FLT}_{\mathsf{ZF}}$; but $\mathsf{ZF}+\phi$ is *not* synonymous with $\mathsf{FLT}_{\mathsf{ZF}}+\psi$, and indeed our favoured set-theoretic foundations and our favoured function-theoretic foundations have come apart.

Roughly put, my reply to this worry is as follows. Considering synonymy will show that we never really have a case where set-theoretic and function-theoretic considerations pull us in different directions in the search for new axioms. To refine this reply, though, I will separately consider extrinsic and intrinsic arguments for new axioms.[42]

Extrinsic Arguments for New Axioms An extrinsic argument for ϕ is an argument that we should accept ϕ because doing so has better all-things-considered consequences for other areas of mathematics. I claim that synonymy always preserves extrinsic arguments.

To see why, suppose that the synonymy is witnessed by translations $\nabla : \mathsf{FLT}_{\mathsf{ZF}} \longrightarrow \mathsf{ZF}$ and $\Delta : \mathsf{ZF} \longrightarrow \mathsf{FLT}_{\mathsf{ZF}}$.[43] Then $\mathsf{ZF} + \phi$ is synonymous with $\mathsf{FLT}_{\mathsf{ZF}} + \phi^{\Delta}$, for every set-theoretic sentence ϕ. Similarly, $\mathsf{ZF} + \psi^{\nabla}$ is synonymous with $\mathsf{FLT}_{\mathsf{ZF}} + \psi$, for every function-theoretic sentence ψ. Now, synonymous theories (ultimately) have the same consequences for all areas of mathematics. So, using "best" as short-hand for "(all things considered) ultimately best for mathematics", we have two schematic biconditionals:

(1E) It is best to add ϕ to ZF iff it is best to add ϕ^{Δ} to $\mathsf{FLT}_{\mathsf{ZF}}$.
(2E) It is best to add ψ to $\mathsf{FLT}_{\mathsf{ZF}}$ iff it is best to add ψ^{∇} to ZF.

Now consider a situation where we are worried that set-theoretic and function-theoretic considerations are, for extrinsic reasons, pulling us in different directions. For example, and without loss of generality, suppose that Anne claims that it is best to add CH to ZF, and Bob claims that it is best to add (some statement which entails) $\neg\mathsf{CH}^{\Delta}$ to $\mathsf{FLT}_{\mathsf{ZF}}$. By (1E), Bob must accept that it is best to add $\neg\mathsf{CH}$ to ZF. But then it is clear that the disagreement between Anne and Bob has nothing per se to do with sets *versus* functions. Anne might have had essentially the same disagreement—about whether to accept or reject the continuum hypothesis—with a set theorist who has never heard of $\mathsf{FLT}_{\mathsf{ZF}}$. And exactly the same consequences, for all areas of mathematics, would have been at stake in the purely intra-set-theoretic formulation of this disagreement.

---

[42] See, for example, Maddy (1988a,b).

[43] As in footnote 26, this argument can be run using LT_{+} and FLT_{+}.

Intrinsic Arguments for New Axioms Having considered extrinsic arguments, let me now consider intrinsic arguments for new axioms. An intrinsic argument for ϕ is an argument that we should accept ϕ because ϕ is something like a conceptual truth. Again, I will claim that such arguments are preserved under synonymy, in the specific case of intrinsic arguments concerning the iterative notions of *set* and *function*.

To argue for this claim, I want to start by establishing these two conditionals:

(1i) If ZF intrinsically supports ϕ, then $\mathsf{FLT}_{\mathsf{ZF}}$ intrinsically supports ϕ^J.
(2i) If $\mathsf{FLT}_{\mathsf{ZF}}$ intrinsically supports ψ, then ZF intrinsically supports ψ^I.

Here, $I : \mathsf{FLT}_{\mathsf{ZF}} \longrightarrow \mathsf{ZF}$ and $J : \mathsf{ZF} \longrightarrow \mathsf{FLT}_{\mathsf{ZF}}$ are the translations mentioned in Sect. 3.6. So I renders $\mathsf{FLT}_{\mathsf{ZF}}$-functions as hereditary-functions$_{\in}$, and J renders ZF-sets as hereditary-sets$_{@}$.

To establish (1i), suppose we are working within $\mathsf{FLT}_{\mathsf{ZF}}$. Suppose, too, we read all mention of "sets", in the Augmented Set Story of Sect. 4.2, as concerned with sets^J, that is, hereditary-sets$_{@}$ (as defined in Sect. 3.6). Then we will agree that ZF^J is a quasi-categorical axiomatization of the Augmented Set Story (read in this way). So, whatever reasoning is supposed to show that ZF intrinsically supports ϕ, that same reasoning will show (to an exactly equal extent) that ZF^J supports ϕ^J. Finally, note that $\mathsf{FLT}_{\mathsf{ZF}}$ proves ZF^J. So, the intrinsic argument for ϕ becomes an intrinsic argument for ϕ^J, as required. An exactly similar argument establishes (2i).

Using the conditionals (1i) and (2i), I now want to turn them into biconditionals:

(1I) ZF intrinsically supports ϕ iff $\mathsf{FLT}_{\mathsf{ZF}}$ intrinsically supports ϕ^J.
(2I) $\mathsf{FLT}_{\mathsf{ZF}}$ intrinsically supports ψ iff $\mathsf{FLT}_{\mathsf{ZF}}$ intrinsically supports ψ^I.

The left-to-right direction of (1I) is just (1i). For the right-to-left direction, suppose $\mathsf{FLT}_{\mathsf{ZF}}$ intrinsically supports ϕ^J. Then ZF intrinsically supports ϕ^{JI}, by (2i). Since I and J comprise a bi-interpretation, $\mathsf{ZF} \vdash \phi \leftrightarrow \phi^{JI}$. So ZF intrinsically supports ϕ, delivering (1I). An exactly similar argument establishes (2I).

Armed with (1I) and (2I), we can now argue as we did concerning extrinsic arguments. For example, and without loss of generality, suppose that Anne claims that ZF intrinsically supports CH, and Bob claims that $\mathsf{FLT}_{\mathsf{ZF}}$ intrinsically supports $\neg\mathsf{CH}^J$. By (1I), Bob must accept that ZF intrinsically supports $\neg\mathsf{CH}$. And so, again, Anne and Bob's disagreement has nothing per se to do with sets *versus* functions.

To be clear: I am not denying that Anne may not find some heuristic value in thinking in terms of sets rather than functions (or vice versa, for Bob).[44] And I have taken no stance over which of Anne and Bob is *right*; I am not suggesting any method for *deciding* who is right; I am not even offering an account of what it might even *mean* to be right. My point is only that, by considering synonymy, we can see that Anne's and Bob's disagreement is not one where set-theoretic foundations rival function-theoretic foundations per se. There may be genuine rivalry over **CH**; but iff so, there is exactly the same rivalry over $\mathsf{CH}^{\triangle}$.

4.5 The Same Judicial Foundation

Allow me to recap the main points of this section. Set theory coherently presents us with an astonishingly fertile array of richly structurable objects; objects which can serve as surrogates for mathematical objects from almost anywhere else. This is the sense in which set theory provides us with a mathematical paradise.

Exactly the same, though, can be said of function theory. Function theory equally coherently introduces us to an equally fertile array of richly structurable objects. It presents us with an alternative but equally splendid vision of paradise.

Indeed, we can now appreciate that these are two visions of the *very same* (judicial) paradise. Set theory and function theory establish the same ultimate court of final appeal—they supply the very same judicial foundation—since all their verdicts are equally trustworthy (given Sect. 4.2) and ultimately identical (by Sects. 4.3–4.4). The difference between set-theoretic and function-theoretic judicial foundations amounts, not to varying the court, but simply varying the language within which the judicial proceedings are conducted.

5 Joint-Carving Metaphysical Foundations

So say I. But some will doubt that set-theoretic and function-theoretic foundations are really *identical*. They will insist come that set theory and

[44] Indeed, the above suggests a minor variant on the approach of seeking intrinsic arguments for new axioms: tell the Function Story (or something similar); axiomatize it with a theory (like $\mathsf{FLT}_{\mathsf{ZF}}$) which is synonymous with **ZF**; arguing that (some natural extension of) the Story intrinsically motivates ϕ; infer that ϕ-under-interpretation holds of the sets.

function theory postulate *rival ontologies*: a paradise of sets, on the one hand, or a paradise of functions, on the other. Addressing this idea will move us away from considering judicial foundations and back to metaphysical foundations (which I set aside in Sect. 4.1).

In this section, I will examine and reject *joint-carving realism* about metaphysical foundations. (In Sect. 6, I will consider whether there can be metaphysical foundations without joint-carving.) The notion of joint-carving that I have in mind is encapsulated by Ted Sider's call-to-arms:

> The world has a distinguished structure, a privileged description. For a representation to be fully successful, truth is not enough; the representation must also use the right concepts, so that its conceptual structure matches reality's structure. There is an objectively correct way to "write the book of the world".[45]

The case study of this chapter raises an interesting challenge for joint-carving realists (a challenge which builds on Putnam; see Sect. 7 for more). In brief: since set theory and function theory are synonymous, both can be made true if either can. But, according to joint-carving realists, for a theory "to be fully successful, truth is not enough"; a *fully* successful theory must "use the right concepts". Joint-carving realists therefore face a substantial question: is set-*membership* or functional-*application* one of "the right concepts"?[46]

In this section, I will argue that it is a disaster to think that this question has an answer. First, though, I should flag a general discomfort with joint-carving realism. Joint-carving realists ask whether certain concepts are fundamental, where concepts are the semantic analogues of *predicates* and *function-symbols*.[47] A priori, it is strange to think that this question should be asked. Perhaps an idealist should expect that the structure of the (phenomenal) world must be grammatically tractable, so that "the joints of nature" would align with a specially chosen sub-sentential vocabulary. But I am not an idealist, and nor are joint-carving realists like Sider. I see no reason whatsoever to think that reality must have a fundamental structure which mirrors the grammatical structure of recognizably human languages. A fortiori, I feel no compulsion whatsoever to ask whether set-theoretic *membership* or function-theoretic *application* is joint-carving.

---

[45] Sider (2011, i).

[46] Sider (2011, §7.13, §8.2.2, §8.3.3, ch.13n.2) himself often suggests that membership might be joint-carving. Given the possibility of function-theoretic foundations, he should reconsider whether, instead, application is joint-carving.

[47] A few lines below his call-to-arms, Sider (2011, i) writes: "a fact is fundamental when it is stated in joint-carving terms.... Questions about which expressions carve at the joints are questions about how much structure reality contains."

Still, the question *has* been posed. My aim is to argue that it should not be answered.

5.1 Monistic Joint-Carving and Its Surds

It will help to make things personal. Allow me to introduce two characters, setty-Seb and functiony-Fern. They are joint-carving realists who adopt rival positions on the question at hand. Here they are, in stereo:

Seb: Membership is joint-carving. Application is derivative, induced from membership by interpretation *I*.	**Fern:** Application is joint-carving. Membership is derivative, induced from application by interpretation *J*.

Seb and Fern are offering *rival* approaches, in the sense that their claims cannot consistently both be advanced in one and the same breath. The joint-carving realist must, then, hold that at most one of Seb and Fern is right.

Now, one might think that things have already gone wrong, on the grounds that all synonymous theories must be equally joint-carving. But that would be mistaken. Not only is there no good general argument that synonymy preserves joint-carving-ness; synonymy does not always preserve (intuitive) truth. This is because, whilst synonymy establishes that two theories are neatly inter-translatable, the translations witnessing synonymy might be *mis*translations. To illustrate: a child who systematically confuses the words "pug" and "French bulldog" may offer a theory which is synonymous with mine (just translate "pug" to "French bulldog" and vice versa); they still say something false when, pointing at my handsome hound, Ragnar, they say "that's a pug".[48]

That said, my firm insistence that Ragnar is a French bulldog (not a pug) has very little in common with Seb's insistence that membership (not application) is fundamental. For one thing, we can point at dogs, but not at sets or functions (let alone at the metaphysician's idea of *joint-carving*). For another, as we saw in Sect. 4.4, there is no *mathematical* disagreement between Seb and Fern. Their apparent disagreement is purely *metaphysical*. And, as I will now argue, it is absurd.

[48] Of course, it is not always a given that children intend to speak the same language as those around them; they may intend to speak a code. The point is: if they intend to speak the same language as their peers, then (ceteris paribus) the correct translation is homophonic, so they say something false.

To see that Seb and Fern's disagreement is absurd, let us start by considering how it might be settled. Very likely, Seb, Fern, and any other joint-carving realists will point us in the direction of *metaphysical virtues*. We will be told to evaluate matters on the basis of: explanatoriness; unificatoriness; simplicity; tractability; ideological/ontological parsimony; etc.

For the sake of argument, I will make two huge concessions to the joint-carving realist: first, that these virtues are well-defined; second, that greater virtue is a reliable indicator of greater joint-carving-ness.[49] Even making these concessions, Seb's and Fern's approaches score equally well as regards any metaphysical virtues.

To be sure, in different contexts, pragmatic considerations may favour one approach over the other. Perhaps the functions-first approach is more immediately hospitable to certain kinds of algebraic reasoning; perhaps the sets-first approach is more immediately amenable in other settings. But as we saw in Sect. 4, there is no real mathematical breathing-room between the two approaches. Consequently, it is hopeless to believe that metaphysical virtues will help to decide which of sets and functions is fundamental.

Seb can, of course, simply stamp his feet and insist that membership simply *is* more fundamental than application. (Fern can stamp her feet right back at him.) But he will have to admit that there is no *reason why* this is so;[50] it is just a *brute fact*, an inexplicable and unknowable *metaphysical surd*.[51] Positing such a surd is formally consistent, of course, but it is extremely unappealing. Indeed, in the cost/benefit game that metaphysicians often play,[52] such a surd is a substantial cost.

The costs continue to rise, as we move from considering metaphysics to metasemantics. To see why, let us try to imagine (without loss of generality) that Seb is right to stamp his feet, because membership rather than application is joint-carving. So, according to Seb: *there are a great many objects, fundamentally related by membership, and derivatively related by application.* Now, focus on the italicized sentence. To express what it needs to, our word "membership" must *somehow* pick out the joint-carving, fundamental relation of membership, $\in$, rather than a derivative copy of that relation (such as $\in^{IJ}$). Indeed: *somehow*. We must ask Seb: *how*? How did it come to do this? Indeed, how could it *so much as possibly* do this?

[49] For the record: I suspect the first is false and the second is nonsense.

[50] So Seb would be violating any version of the principle of sufficient reason; see Amijee (2020) for a discussion of its resurgence in modern metaphysics.

[51] I am deliberately using Putnam (1981, 46–8) phrase; see also Sect. 7.

[52] The idea is traceable to Quine (1951, 14), but Lewis (1986, 4) gave us the method for playing it.

To be clear: I am not suggesting that we should start to worry that "membership" picks out some relation other than membership.[53] My baffled and baffling questions for Seb arise out of his own commitments. Seb insists that membership *is* joint-carving, and Seb must believe *that* "membership" refers to membership; so Seb owes us and himself an explanation of *why* "membership" refers to a joint-carving relation.

Cutting to the chase: no good explanation is possible. Seb will end up saying that membership, in particular, and joint-carving relations, in general, are *referentially magnetic*.[54] That last phrase has gained currency in contemporary philosophy, but we should be clear that Seb might as well have said that some relations *cry out to be named*;[55] that explanation stops here. Seb has been forced to posit a *metasemantic surd*. Of course, there is no inconsistency in positing such a surd. But it is as absurd as anything I can imagine within mathematical metasemantics.

In the end, Seb's world-view has so many degrees of freedom that it can only be held together by some (ab)surd(ity). And what goes for Seb goes for Fern. We must conclude that neither Seb nor Fern is right.

5.2 Pluralistic Joint-Carving and Its Surds

This conclusion does not immediately entail that joint-carving realism is wrong. Perhaps joint-carving realists can simply agree with us that Seb and Fern are both wrong.

A terrible way to do this would be to claim that, in fact, Ψs are fundamental, rather than sets or functions. Whatever exactly these Ψs are claimed to be, this will amount to enriching our earlier dialogue with a third character, Sy, who offers Ψ-theoretic metaphysical foundations as a rival to set-theoretic and function-theoretic foundations.[56] Barring some shocking logico-mathematical genius on Sy's part, this will be painfully epicyclic.

The only remaining option for the joint-carving realist is to insist that neither Seb nor Fern is right, because *both* membership and application are joint-carving. On this view, set-theoretic and function-theoretic foundations

[53] In particular, I want to cancel any suggestion that I am advocating for a model-theoretically inspired scepticism about meaning. I believe that such scepticism is deeply incoherent (see Button, 2013, ch.7, 2016; Button & Walsh, 2018, ch.9).

[54] See, for example, Sider (2011, 23–33).

[55] Again, I am echoing Putnam's rhetorical criticism of magical theories of reference. I am also implicitly drawing upon arguments like Button (2013, ch.12) and Wrigley (2018).

[56] Perhaps Sy endorses the Complemented Story, and $\mathsf{BLT}_{\mathsf{ZF}}$, as presented in Button (2022a).

present us with *distinct* ontologies, but these ontologies are not *competitors*. Instead, we have two disjoint but perfectly fundamental hierarchies—a set-hierarchy and a function-hierarchy—sitting happily alongside each other in mathematical heaven.

Making things personal again, let Pearl advocate this pluralistic version of joint-carving realism. In putting sets and functions on a par, Pearl has taken a step in the right direction. Alas, her continued adherence to joint-carving means that her position is still untenable, for both metaphysical and metasemantic reasons.

On the *metaphysical* front: Pearl's position is approximately as bad as Seb's and Fern's, but for slightly different reasons. Recall that Seb (and Fern) posited an inexplicable and for that reason unknowable fact: that membership rather than application (or vice versa) was fundamental. Pearl deliberately and explicitly avoids this, but in doing so she incurs the cost of profligacy at the fundamental level. After all, as we saw in Sect. 4, the synonymy result (my Main Theorem) shows that anything we can do with sets, we could instead do with functions (and vice versa), so there is no *need* to treat both as fundamental; either alone would do.[57]

On the *metasemantic* front: Pearl inherits all of Seb's and Fern's problems, unchanged. Exactly like Seb: Pearl will require that, *somehow*, our word "membership" picks out the joint-carving membership relation, as found in the fundamental set-hierarchy, rather than (for example) the J-translation of that relation, as found in the fundamental function-hierarchy. And exactly like Fern: Pearl will require that, *somehow*, our word "application" picks out the joint-carving application function , as found in the fundamental function-hierarchy, rather than (for example) the I-translation of that function, as found in the fundamental set-hierarchy. As before: my point is not that we should worry *what* "membership" or "application" refer to, but that Pearl cannot possibly explain *why* these words refer to (what she insists are) joint-carving stuff. She must posit a metasemantic surd; this is absurd.

Evidently, we must abandon Pearl's pluralism, just as we abandoned Seb's and Fern's monisms. And with this, joint-carving realism has run out of options. We must give up on the idea of joint-carving.

[57] Cf. the competition between Austerity and Reasons discussed in Button (2023a, §8).

6 Metaphysical Foundations Without Joint-Carving?

Having given up on the idea of joint-carving, must we also abandon the very idea of providing metaphysical foundations for mathematics? Candidly: I don't know, but *maybe* not. In this extremely speculative section, I will sketch an approach to metaphysical foundations which might just survive the demise of joint-carving realism. (For ease of readability, I will write as if I *endorse* this approach; but I want to emphasize that I am just *exploring* it.)

Suppose that, even after the discussion of Sect. 5, we still want set theory to supply us with metaphysical foundations. Given everything we have learned, we will have to accept that function theory also supplies us with metaphysical foundations. Moreover, we will eschew the idea that mathematical heaven has natural joints to carve. So we cannot say that some of the denizens of mathematical heaven are sets "first" (or "fundamentally") and functions "second" (or "derivatively"). Instead, we will be forced to say that they are sets *just as much* as they are functions. In slightly more detail, we will be forced to say something like this:

> Mathematical heaven comprises a great many objects, which are richly structurable (see (a) of Sect. 4.1). Moreover, these objects are sets, arranged in an iterative hierarchy; and the fact that they are so arranged convinces us that our glimpses of mathematical are not mere mirages (see (b) of Sect. 4.1). Equally, though: these very same objects are functions, arranged in an iterative hierarchy. That is to say: the objects of mathematical heaven can be described and conceptualized equally well in at least two different (but synonymous) ways. They are sets, and they are functions, and they are not one "before" they are other.

That is the approach. But it invites a puzzled question: *How can anything be both a set and a function?* This question is important, and I will spend the rest of this section grappling with it.[58]

We sometimes speak of sets and functions in cutesy terms: of sets as carrier bags full of objects (their members); of functions as things which, sea-cucumber-like, suck in arguments and spit out values. If we earnestly think about mathematical objects like this, then we will certainly be confused to be told that sets are also functions (carrier bags don't spit out objects). But,

[58]Thanks to Chris Menzel and Chris Scambler for really pressing on this question.

once we have explicitly spelled out this possible source of puzzlement, we can immediately set it aside. After all, such cutesy, carrier-bag-style imagery should never have been treated very seriously.[59]

We might instead think of sets and functions more *structurally*. If we do that, it becomes totally unremarkable to be told that sets are also functions. Comparably: we do not even blink if we are told that someone is both a conductor and married to one of the orchestra's violinists. And note that this way of dissolving the puzzlement does not require that we think of sets and functions in *wholly* structural terms: if we want, we can allow that sets and functions have some intrinsic features; we just need to accommodate their structural features when we characterize them *as* sets or as functions.

We might also think of sets and functions *presentationally*. Comparably: we would be perplexed to be told that something is both a duck and also a rabbit, but we are all familiar with *duck–rabbits*, that is, things which, viewed one way, are pictures of ducks and, viewed another, are pictures of rabbits. Inspired by duck–rabbits, we might say: mathematical heaven is a set–function-hierarchy: a bunch of entities which, viewed (or described) one way, are a hierarchy of sets and, viewed (or described) another, are a hierarchy of functions.

The *structural* and the *presentational* approaches need not be competitors. Indeed, they are very naturally complementary. In drawing a duck–rabbit, you lay down a system of geometric shapes; the spatial relations between these shapes make the two viewpoints possible. Comparably: the structure of mathematical heaven makes possible its presentational multiplicities.

Of course, the analogy between duck–rabbits and the (envisaged) set–function-hierarchy is limited in several ways, but one limitation merits particular comment. Presented with a duck–rabbit, we can describe it in *duck*-terms ("its beak points left") or in *rabbit*-terms ("its ears point left"); but we can also describe it more *neutrally*, exhaustively characterizing the underlying spatial structure by describing the various geometric shapes on the page (perhaps in terms of vector graphics). Now, we can certainly describe the set–function-hierarchy in *set*-terms (via the Set Story) or in *function*-terms (via the Function Story); but I very much doubt that the set–function-hierarchy can be adequately characterized in *neutral* terms.

This is not to say that there is anything *inadequate* about the non-neutral descriptions that we might offer, that is, our descriptions of the hierarchy in terms of sets or functions (or anything else). It is just to say that we cannot describe mathematical heaven without selecting some notational conventions.

---

[59] See Oliver and Smiley (2018, 396–7).

In turn: that is not (yet) to say that mathematical heaven is in any sense "constituted by" or "dependent upon" our conventional choices. It may just amount to the observation that our descriptions of mathematical heaven always bear our trace.

I do not pretend that this is the last word on the question: *How can anything be both a set and a function?* In the end, fully answering that question requires fully articulating this approach to metaphysical foundations. I have only offered a sketch of this approach here; I hope that this sketch is sufficient to show that the approach is worth pursuing.

7 Conclusion, and Putnam Revisited

I began this chapter with a quote from Putnam. That quote implicitly referred back to a comment he had offered nearly four decades earlier:

> A mathematical example: you have a theory according to which mathematical heaven consists of objects called sets. The theory says that some of these sets are functions—in fact it says that those sets which are sets of ordered pairs satisfying a certain functionality condition are functions, and then it proceeds to translate your favorite textbook on calculus. The other theory says that mathematical heaven consists of things called functions, even zero, 1, 2, 3 turn out to be functions, and it says that some of these functions are sets—in fact those functions which take on only the values zero and 1 are sets. Now, these are formally incompatible[. ...But they] are *equivalent descriptions*[...]. A sophisticated realist should not be bothered by the collapse of the 'One True Theory' version of realism.[60]

Putnam's remarks are suggestive, but somewhat opaque. What Putnam called "the 'One True Theory' version of realism" is just what I have called *joint-carving realism.*[61] Moreover, Putnam held that joint-carving realism collapses, because set theory and function theory are "*equivalent descriptions*". Unfortunately, he did not specify the relevant formal set theory, the relevant

[60] Putnam (1979a, 287–8).

[61] Putnam(1981: xi, 49, 73, 143, 210, 1989: 352) frequently described his opponents as believing that there is just 'One True Theory', during his internal realist period (and cf. Button, 2013, 9, 201). Note, though, the capital "T" on "True" (and cf. Field, 1982, 553–4). As flagged at the start of Sect. 5, Sider's position is explicitly not that there is only one true theory, but that there is only one theory whose "conceptual structure matches reality's structure".

formal function theory, or the relevant notion of an *equivalent description*.[62] And whilst Putnam stated that a "sophisticated realist" could acknowledge that we are dealing with "equivalent descriptions", he did not tell us how or what such sophisticated realism might amount to.[63]

This chapter can be read as a reconstruction (and partial vindication) of Putnam's remarks. I introduced the bare idea of an iterative concept of *function*, via the Function Story (see Sect. 2). By stages, I turned this Story into a quasi-categorical, categorial, formal theory, **FLT**. And an extension of this theory has all the power of **ZF**; indeed, **ZF** and $\mathsf{FLT}_{\mathsf{ZF}}$ are synonymous (see Sect. 3).

This perfectly precise equivalence shows that set theory and function theory provide exactly the same judicial foundations for mathematics, subject only to a change in notation (see Sect. 4). Moreover, when this point is fully appreciated, it leads to the collapse of joint-carving metaphysical foundations (see Sect. 5). But we may yet be left with the possibility of a "sophisticated realism", according to which set theory and function theory present us with the very same metaphysical foundation (see Sect. 6).

Acknowledgments In late 2019, I developed both **FST** and **FLT**, and sketched the proof of my Main Theorem (see Sect. 3.6). I sent this to Thomas Forster and Albert Visser at the time, and I would like to thank them both for early discussion and inspiration: the chapter was originally motivated by their suggestion that I should try to offer a level-theoretic axiomatization of multisets. (This is indeed possible; see Button, 2023b.)

In June 2021, I presented this chapter at the seminar series which gave rise to this volume, covering the formal theories, the proof-strategy for the Main Theorem, and the significance of all this for joint-carving realism. I wish to thank that audience for helpful comments, but most especially Neil Barton and Giorgio Venturi. Huge thanks also to Chris Menzel and to an anonymous referee for this volume.

[62] Given his earlier (1967, 7–9) use of the notion of "equivalent description", I conjecture that Putnam himself had in mind the (mere) *mutual interpretability* of von Neumann's function theory and **NBG**. If so, then his line of thought was rather too quick, for reasons discussed in Sect. 1 (see also Button, 2021b, §9.4, *Second*).

[63] That quote, alongside other comments from the same conference (1979b, esp. 228), was Putnam's first public avowal of (a version of) internal realism. But what *that* amounted to was never straightforward; see Button (2013, chs.8–11).

A Background on LT

The remainder of this chapter consists of appendices concerning **FLT**. Recall that **FLT** axiomatizes the Function Story of Sect. 2. Technically, it is a straightforward adaptation of the theory **LT**, which axiomatizes the Set Story. I draw on **LT** in my discussion of **FLT** and in several footnotes in the main text. So, in this appendix, I will quickly introduce **LT**.[64] I begin with the key definition (the parenthetical "Extensionality" again indicates that this definition relies upon Extensionality.)

Definition A.1 (Extensionality) For any a, let $\P a = \{x : (\exists c \in a) x \subseteq c\}$, if it exists.

Say that h is a *history*, written $Hist(h)$, iff $(\forall x \in h) x = \P(x \cap h)$.

Say that s is a *level*, written $Lev(s)$, iff $\exists h(Hist(h) \land s = \P h)$. ◊

Essentially, the levels serve as surrogates for the stages in the Set Story. We can now set up some axioms:

Extensionality $\forall a \forall b \big(\forall x(x \in a \leftrightarrow x \in b) \rightarrow a = b\big)$

Separation $\forall F \forall a(\{x \in a : F(x)\}$ exists) *with "F" a predicate-variable*

Stratification $\forall a \exists s \big(Lev(s) \land a \subseteq s\big)$

Endless $\forall s \exists t\ s \in t$

Infinity $(\exists s \neq \emptyset)(\forall q \in s)\exists r(q \in r \in s)$

Unbounded $\forall P \forall a \exists s (\forall x \in a) P(x) \in s$ *with "P" a function-variable*

Now **LT** itself is just Extensionality + Separation + Stratification. Then LT_+ is **LT** + Endless, and **ZF** is equivalent to **LT** + Infinity + Unbounded.

Using just Extensionality and Separation, we can prove that the levels are well-ordered by membership. This licenses the introduction of a powerful tool:

Definition A.2 (LT) Let ℓa be the $\in$-least level including a. So, $a \subseteq \ell a$ and $\neg(\exists s \in \ell a)(Lev(s) \land a \subseteq s)$. ◊

Note that ℓa obeys various intuitive principles, for example, that if $a \in b$ then $\ell a \in \ell b$.[65]

---

[64] For full discussion of **LT**, see Button (2021a).

[65] See Button (2021a, Lemma 3.12).

B Elementary Considerations Regarding FLT

I now turn to considering **FLT**. I stated its axioms in Sect. 3.3, but there are a few things I need to clear up.

First: I need to be a little more precise about the background logic for handling partial functions (see Sect. B.1). Second: in Sect. 3.3, I invoked the predicate, *Fev*, that is, the formal notion of a fevel (i.e. a functional-level), but I still need to define it (see Sect. B.2).

B.1 Formalizing Partial Functions

There is a "pseudo-partial" approach to partial functions, according to which, to say that f is undefined on input x is to say that $f(x) = \bot$, where $\bot$ is some "default object". I say that this is *pseudo*-partial, since when $\bot$ is counted as a value, this makes all functions total. I have no interest in this approach. I want to present a theory of genuinely partial functions. I really want to allow that f may be undefined on input x; that $\forall y\ f(x) \neq y$.

An obvious way to implement this is via a negative free logic (with no non-logical primitives). In brief, the syntax for this approach will be as follows:

- Variables (and nothing else) are primitive terms; $\tau(\sigma)$ is a term whenever both τ and σ are; nothing else is a term.
- For any term τ, we define $\tau{\downarrow}$ via $\tau = \tau$.
- x$\downarrow$ for any variable x; if α is an atomic formula containing a term τ, then $\alpha \rightarrow \tau{\downarrow}$; and $\tau(\sigma){\downarrow} \rightarrow (\tau{\downarrow} \wedge \sigma{\downarrow})$ for any terms τ and σ.

Implicitly, this is how I formulated **FLT** (and its extensions) in the main chapter. Once you get used to it, it is quite neat.

There is a downside to this approach. Whilst "$\tau(\sigma)$" intuitively stands for the application of τ to σ, no object-language symbol expresses *application*. This makes it hard to consider interpretations of theories offered with this formalism. And, of course, the main focus of this chapter is less on **FLT** itself, but more on the synonymy between $\mathsf{FLT}_{\mathsf{ZF}}$ and **ZF**, which requires considering interpretations. To remedy this, we can simply add a primitive two-place function symbol, "@", to express application. We would now regard "$\tau(\sigma)$" as an abbreviation of "$@(\tau, \sigma)$", which is a term whenever both τ and σ are. The result is a negative free logic with a primitive function-symbol.

Still, when our aim is to prove synonymy, it is easiest to work with a theory in classical logic (rather than free logic) which uses only primitive *relation* symbols (instead of *function* symbols). With that in mind, instead of expressing application via a primitive function symbol, we can express it with a primitive three-place predicate, "val". We then need an axiom stating that val is functional:

$$\textbf{FunVal}\ \forall f \forall x \forall y \forall z \big((\text{val}(f, x, y) \wedge \text{val}(f, x, z)) \rightarrow y = z \big)$$

This allows us to abandon free logic.[66] We can easily map from the functional approach to the relational approach, by (for example) treating "$f(x) = y$" as "$\text{val}(f, x, y)$", and "$f(x) \simeq g(x)$" as "$\forall y(\text{val}(f, x, y) \leftrightarrow \text{val}(g, x, y))$".

My attitude is: all these formalisms are equivalent; use whichever formalism is easiest for the purposes at hand.

B.2 Key Definitions and Axioms

My next task is to define the key predicate, *Fev*, that is, the formal notion of a fevel, which we use in **FLT**'s axiom FunStrat.

Recall from Definition 3.1 that sets$_@$ are partial identity functions. It follows from FunExt that sets$_@$ obey extensionality$_\varepsilon$.[67] In turn, this licenses us in introducing a function-theoretic analogue of ordinary set-builder notation:

Definition B.1 (FunExt) If it exists, then $\{\!\!\{x : \phi\}\!\!\}$ is the set$_@$ given as follows:

$$\text{if } \phi[^v/_x], \text{ then } \{\!\!\{x : \phi\}\!\!\}(v) = v$$
$$\text{otherwise, } \{\!\!\{x : \phi\}\!\!\}(v) \text{ is undefined}$$

I tweak this notation in obvious ways, for example, allowing $\{\!\!\{x \mathrel{\varepsilon} a : \phi\}\!\!\} = \{\!\!\{x : x \mathrel{\varepsilon} a \wedge \phi\}\!\!\}$. ◊

Using this set$_@$-builder notation, here is the define of a *fevel*:

Definition B.2 (FunExt) For any f, let $\P f = \{\!\!\{x : (\exists g \mathrel{\varepsilon} f) x \Subset g\}\!\!\}$, if it exists.[68]

Say that h is a *functional-history*, written *Fist*(h), iff $(\forall x \mathrel{\varepsilon} h) x = \P\{\!\!\{z \mathrel{\varepsilon} h : z \mathrel{\varepsilon} x\}\!\!\}$.

[66]There is a subtlety: free logics usually allow for empty domains. In abandoning free logic, we rule out the empty domain. This wrinkle is, however, insignificant. Those who find it causes issues can remedy it in either of two ways: (1) add an axiom $\exists x\ x = x$ to **FLT** (when formulated functionally); (2) when considering the interpretation $J : \textsf{LT}_+ \longrightarrow \textsf{FLT}_+$, (re)formulate **LT** in a free logic.

[67]That is, $\forall x(x \mathrel{\varepsilon} f \leftrightarrow x \mathrel{\varepsilon} g) \rightarrow f = g$, for any sets$_@$ f and g.

[68]Note that we do not initially assume that $\P f$ exists, for all f.

Say that s is a *fevel*, written $Fev(s)$, iff $\exists h(Fist(h) \wedge s = \P h)$. ◊

The utility of this definition is far from immediately obvious! But a quick glance will confirm that it arises from Definition A.1, by translating set-theoretic into set@-theoretic vocabulary.

B.3 Key Results

Of course, we need to check that Definition B.2 does what we want. Ultimately, it is vindicated by Theorem 3.2, which tells us that **FLT** and **FST** make exactly the same claims about functions. I will not prove Theorem 3.2 here,[69] but I will note that, on the way to that result, we will want to prove the fundamental theorem of fevel theory:

Theorem (3.3; FunExt, FunComp) *The fevels are well-ordered by* ε.

Proof Sketch The first step is to show that sets@ obey Separation$_\varepsilon$. So we must show that $\{\!\{x \ \varepsilon \ f : F(x)\}\!\}$ exists for any property F and any function f. To do this, fix a second-order map P such that $P(x) = x$ if $x \ \varepsilon \ f \wedge F(x)$, and $P(x)$ is undefined otherwise; now use FunComp and FunExt.

From here, the remaining steps are as for **LT** (i.e. see Button, 2021a, §3). We first show that the fevels are well-founded. This allows us to prove that any member$_\varepsilon$ of a fistory is a fevel. We can then show that a fevel is precisely the result of hitting the set@ of all earlier fevels with ¶, that is, that $Fev(f)$ iff $f = \P\{\!\{s \ \varepsilon \ f : Fev(s)\}\!\}$. Using this and induction on ε we can show that fevels are linearly ordered by ε. □

The well-ordering of the fevels also allows us to introduce a powerful notion (compare Definition A.2):

Definition B.3 (FLT) Let $\mathbb{L}f$ be the ε-least fevel including f. So $f \Subset \mathbb{L}f$ and $\neg(\exists s \ \varepsilon \ \mathbb{L}f)(Fev(f) \wedge f \Subset s)$. ◊

Together with Theorem 3.3, this licenses ε-induction. Moreover, the fevels of **FLT**, like the levels of **LT**, obey some very intuitive principles, for example, that if $g \ \varepsilon \ f$, then $\mathbb{L}g \ \varepsilon \ \mathbb{L}f$.[70]

---

[69] This is proved as for **LT**; see Button (2021a, §4).

[70] It is easy to obtain a result like Button (2021a, Lemma 3.12).

B.4 FLT and Categories

The final key result I mentioned, concerning **FLT**, is that any cumulative hierarchy of functions is a *category*. To show this, it will help to introduce some abbreviations.

Definition B.4 (FunExt) Let $\text{dom}(f) = \{\!\!\{x : \exists v\ f(x) = v\}\!\!\}$ and $\text{cod}(f) = \{\!\!\{x : \exists v\ f(v) = x\}\!\!\}$. Say that f is an *arrow* from $\text{dom}(f)$ to $\text{cod}(f)$; we may write $f : a \longrightarrow b$ to indicate that $\text{dom}(f) = a$ and $\text{cod}(f) = b$. When $f : a \longrightarrow b$ and $g : b \longrightarrow c$, the *composite* of f and g is the function $g \circ f = [\lambda x.g(f(x))] : a \longrightarrow c$. $\diamondsuit$

Lemma B.5 **(FLT)** *For any* f, g, h*:*

(1) $dom(f)$ *and* $cod(f)$ *exist*
(2) If $cod(f) = dom(g)$*, then* $g \circ f$ *exists*
(3) $f \circ a = f = b \circ f$*, when* $f : a \longrightarrow b$
(4) $h \circ (g \circ f) = (h \circ g) \circ f$*, when* $f : a \longrightarrow b, g : b \longrightarrow c, h : c \longrightarrow d.$

Proof *(1)* By Separation$_\varepsilon$ on $\mathbb{L}f$.
(2) Let $P(x)$ be given by $g(f(x))$; use FunComp on the greater of $\mathbb{L}f$ and $\mathbb{L}g$.
(3) Using (2), note that $f \circ a$ exists, since $\text{cod}(a) = a = \text{dom}(f)$ by FunExt. Now $(f \circ a)(x) \simeq f(a(x)) \simeq f((\text{dom}(f))(x)) \simeq f(x)$ for all x; so $f \circ a = f$ by FunExt. The case for b is similar.
(4) By FunExt. $\square$

This lemma immediately yields Theorem 3.4. However, note that Definition B.4 in effect identifies a function's range with its codomain. And this has certain immediate limiting effects. For example, the category we obtain from $\textsf{FLT}_+$ is *not* equipped with any of: an initial object; a terminal object; products; equalizers; or pullbacks.

The case of products is worth commenting on in a little detail. Having decided how to implement ordered pairs (see Sect. C), we can take a cue from set theory and define $f \times g := \{\!\!\{\langle x, y\rangle : x\ \varepsilon\ f \land y\ \varepsilon\ g\}\!\!\}$. So defined, $\textsf{FLT}_+$ proves that this exists for any f and g. But this is almost never a product in the *category theorist's* sense.[71] To see why, note the following.

[71]See, for example, Awodey (2010, 38–41).

Lemma B.6 (**FLT**$_+$) *Let A and B be sets$_@$ with a product diagram $A \xleftarrow{\pi_1} P \xrightarrow{\pi_2} B$. Then* $\max(|A|, |B|) = |A \times B|$,[72] *so that either $|A| < 2$ or $|B| < 2$.*

Proof By the UMP for products, for any $A \xleftarrow{q_1} Q \xrightarrow{q_2} B$, there is a unique $u : Q \longrightarrow P$ such that $q_1 = \pi_1 \circ u$ and $q_2 = \pi_2 \circ u$. And since ranges are codomains, $|P| \leq |Q|$.

I will show that $|P| = \max(|A|, |B|)$. Since ranges are codomains, for any set$_@$ Q there are arrows $q_1 : Q \longrightarrow A$ and $q_2 : Q \longrightarrow B$ iff $\max(|A|, |B|) \leq |Q|$. Since π_1 and π_2 exist, $\max(|A|, |B|) \leq |P|$. Now fix some diagram $A \xleftarrow{q_1} Q \xrightarrow{q_2} B$ with $|Q| = \max(|A|, |B|)$ and observe that $\max(|A|, |B|) \leq |P| \leq |Q| = \max(|A|, |B|)$.

I will now show that $|P| = |A \times B|$. Fix the diagram $A \xleftarrow{p_1} A \times B \xrightarrow{p_2} B$ via $p_1(\langle x, y \rangle) = x$ and $p_2(\langle x, y \rangle) = y$. Obtain $u : A \times B \longrightarrow P$ by the UMP. Suppose that $u(\langle a, b \rangle) = u(\langle a', b' \rangle)$, then $a = p_1(\langle a, b \rangle) = \pi_1(u(\langle a, b \rangle)) = \pi_1(u(\langle a', b' \rangle)) = p_1(\langle a', b' \rangle) = a'$; and similarly $b = b'$; so $\langle a, b \rangle = \langle a', b' \rangle$. Generalizing, u is an injection, so $|A \times B| \leq |P| \leq |A \times B|$. □

So, although **FLT**$_+$ merrily gives us sets$_@$ of ordered pairs, it gives us almost no categorical products.

If we want to have nice categorial things, like products and equalizers, we may want to move away from **FLT** and its extensions. In turn, this requires that we modify the Function Story (from Sect. 2). Specifically, we could tweak the Story to say that, at any stage, we find all functions whose domains and *codomains* are exhausted by the functions we found earlier. This will allow that codomains can outstrip ranges and thereby (ultimately) deliver a topos.

C Autonomous Judicial Foundations in FLT

In Sect. 4.3, I noted that we can provide "autonomous" judicial foundations within **FLT**. Essentially, we can take inspiration from **FLT**'s category-theoretic nature, to develop function-theoretic implementations of various mathematical notions (whose set-theoretic implementations are perfectly familiar). Here are some ways one might proceed.

[72] Here I write $|Z|$ for Z's cardinality, and "$\times$" is as defined a few lines ago.

Arbitrary Functions In **FLT**, we take 1-place functions as our basic currency. But it is worth considering n-place functions. Set theorists usually regard n-place functions as sets of $n+1$-tuples satisfying a functionality constraint. Function-theorists will probably prefer to curry. For example, to handle 3-place functions, we might say:

$$f(x, y, z) \text{ abbreviates } ((f(x))(y))(z)$$

Generalizing, function-theoretic foundations allow us to regard *any* object (i.e. any one-place function) immediately as a (partial) n-place function, for any n.

Arbitrary Relations Set theorists usually regard n-place relations as sets of n-tuples. Taking the above hint, function-theorists can immediately regard any object as an n-place relation, for any n. For example, to handle 3-place relations, we might say:

$$Rel_f[x, y, z] \text{ abbreviates } (\exists g \; \varepsilon \; f)g(x, y) = z$$

where the right-hand side uses currying. But we can now read "$Rel_f[x, y, z]$" as "the relation (associated with) f holds of x, y and z (in that order)".

Ordered Pairs Set theorists usually render ordered pairs via Kurotawksi's fairly arbitrary definition that $\langle a, b\rangle := \{\{a\}, \{a, b\}\}$. Function theorists might instead (fairly arbitrarily) say that $\langle a, b\rangle$ is the function which maps $a \mapsto b \mapsto b$ and is undefined otherwise. (However, ordered pairs may not play such a dominant role in function-theoretic foundations, given the earlier remarks about arbitrary functions/relations.)

Ordinals Set theorists usually adopt von Neumann's implementation of the ordinals. Function-theorists might prefer to implement ordinals via the following intuitive thought: each ordinal β is the function which maps $\alpha \mapsto \alpha \dot{-} 1$ for all $\alpha < \beta$ and which is undefined otherwise. (Here I am using $\dot{-}$ for truncated subtraction; this can of course be given a more rigorous implementation.)

These are just a few examples of mathematical concepts which can be given autonomous, function-theoretic implementations. We could continue, but I leave this for anyone who really wants to put **FLT** to work.

References

Amijee, F. (2020). Principle of sufficient reason. In *The Routledge handbook of metaphysical grounding* (pp. 63–75). Routledge.
Awodey, S. (2010). *Category theory* (2nd ed.). Oxford University Press.
Boolos, G. (1971). The iterative conception of set. *The Journal of Philosophy, 68*(8), 215–31.
Button, T. (2013). *The limits of realism*. Oxford University Press.
Button, T. (2016). Brains in vats and model theory. In S. Goldberg (Ed.), *The brain in a vat* (pp. 131–54). Cambridge University Press.
Button, T. (2021a). Level theory, part 1: Axiomatizing the bare idea of a cumulative hierarchy of sets. *Bulletin of Symbolic Logic, 27*(4), 436–60.
Button, T. (2021b). Level theory, part 2: Axiomatizing the bare idea of a potential hierarchy. *Bulletin of Symbolic Logic, 27*(4), 461–84.
Button, T. (2022a). Level theory, part 3: A boolean algebra of sets arranged in well-ordered levels. *Bulletin of Symbolic Logic, 28*(1), 1–26.
Button, T. (2022b). Mathematical internal realism. In J. Conant & S. Chakraborty (Eds.), *Engaging Putnam* (pp. 157–82). De Gruyter.
Button, T. (2023a). Symmetric relations, symmetric theories, and Pythagrapheanism. *Philosophy and Phenomenological Research, 107*(3), 583–612.
Button, T. (2023b). Wand/set theory. *arXiv:2308.06789*.
Button, T., & Walsh, S. (2018). *Philosophy and model theory*. Oxford University Press.
de Bouvère, K. (1963). Synonymous theories. In L. Henkin, J. W. Addison, & A. Tarski (Eds.), *The Theory of Models: Proceedings of the 1963 International Symposium at Berkeley*. North-Holland.
Field, H. (1982). Realism and relativism. *The Journal of Philosophy, 79*(10), 553–67.
Friedman, H. M., & Visser, A. (2014). When bi-interpretability implies synonymy. *Logic Group Preprint Series, 320*, 1–19.
Isaacson, D. (2011). The reality of mathematics and the case of set theory. In Z. Novàk & A. Simonyi (Eds.), *Truth, reference and realism*. CEU Press.
Lewis, D. (1986). *On the plurality of worlds*. Basil Blackwell.
Maddy, P. (1988a). Believing the axioms I. *Journal of Symbolic Logic, 53*(2), 481–511.
Maddy, P. (1988b). Believing the axioms II. *Journal of Symbolic Logic, 53*(3), 736–764.
Maddy, P. (1997). *Naturalism in mathematics*. Oxford University Press.
Maddy, P. (2017). Set-theoretic foundations. *Contemporary Mathematics, 690*, 289–322.
Margalit, A. (Ed.) (1979). *Meaning and use: Papers presented at the second Jerusalem philosophical encounter (1976)*. Springer.
Montague, R. M. (1957). *Contributions to the axiomatic foundations of set theory*. PhD Thesis. Berkeley.
Oliver, A., & Smiley, T. (2018). Cantorian set theory. *Bulletin of Symbolic Logic, 24*(4), 393–451.

Putnam, H. (1967). Mathematics without foundations. *Journal of Philosophy, 64*(1), 5–22.
Putnam, H. (1979a). Comments (on Kripke's "A puzzle about belief"). In Margalit (1979) (pp.284–288).
Putnam, H. (1979b). Reference and understanding, and Reply to Dummett's comment. In Margalit (1979) (pp.199–217, 226–228).
Putnam, H. (1981). *Reason, truth, and history*. Cambridge University Press.
Putnam, H. (1989). Model theory and the "factuality" of semantics. In A. George (Ed.), *Reflections on Chomsky* (pp. 213–32). Blackwell.
Putnam, H. (2014). Three blog posts: 'The modal logical interpretation and "equivalent descriptions"' (11.Dec.2014); 'Continuing' (12.Dec.2014); and 'Mathematical "existence" (13.Dec.2014). http://putnamphil.blogspot.com/2014/12/
Quine, W. V. O. (1951). Ontology and ideology. *Philosophical Studies, 2*(1), 11–15.
Robinson, R. M. (1937). The theory of classes: A modification of von Neumann's system. *Journal of Symbolic Logic, 2*(1), 29–36.
Sant'Anna, A. S. et al. (2020). Flow: The axiom of choice is independent from the partition principle. *arXiv:2010.03664*.
Scott, D. (1974). Axiomatizing set theory. In T. Jech (Ed.), *Axiomatic set theory II*. Proceedings of the Symposium in Pure Mathematics of the American Mathematical Society, July–August 1967 (pp. 207–214). American Mathematical Society.
Shapiro, S. (1997). *Philosophy of mathematics: Structure and ontology*. Oxford University Press.
Shoenfield, J. R. (1977). Axioms of set theory. In J. Barwise (Ed.), *Handbook of mathematical logic* (pp. 321–44). North-Holland.
Sider, T. (2011). *Writing the book of the world*. Oxford University Press.
von Neumann, J. (1925). Eine Axiomatisierung der Mengenlehre. *Journal für reine und angewandte Mathematik, 154,* 219–240.
von Neumann, J. (1928). Die Axiomatisierung der Mengenlehre. *Mathematische Zeitschrift, 27,* 669–752.
von Neumann, J. (1929). Über eine Widerspruchsfreiheitsfrage in der axiomatischen Mengenlehre. *Journal für reine und angewandte Mathematik, 160,* 227–41.
Wrigley, W. (2018). *Sider's Ontologese introduction instructions. Theoria, 84,* 295–308.

7

What Set Theory Could Not Be About

Toby Meadows

The practical upshot of all this is that if you stick a Babel fish in your ear you can instantly understand anything said to you in any form of language.

Set theory, in particular ZFC, is famed for its ability to provide a foundation for mathematics. But to most observers, this signature achievement is largely invisible. An analysis class may use an epsilon symbol to represent membership or some curly braces to form a set, but the way in which set theory can be argued to undergird the entirety of mathematics often remains opaque. Nonetheless, the successful performance of this foundational role is the main reason for our long-standing interest in this theory of abstract paraphernalia. Given this, it is unsurprising that so much ink has been spilled in trying to pin down the exact nature of its subject matter.[1] In this chapter, I will argue that there is an important sense in which that project is futile. In particular, I

I'd like to thank Penelope Maddy, Andrés Caicedo, Kyle Stanford, and Julia Kameryn Williams for their invaluable advice on this chapter. And thanks to Joan Bagaria and Claudio Ternullo from the Barcelona Set Theory Seminar for giving me the opportunity to present this material.

[1]For a few classic examples, see Gödel (1947); Boolos (1998); Potter (2004).

T. Meadows (✉)
Department of Logic and Philosophy of Science, University of California, Irvine, CA, USA
e-mail: meadowst@uci.edu

C. Antos et al. (eds.), *The Palgrave Companion to the Philosophy of Set Theory*, Palgrave Companions, https://doi.org/10.1007/978-3-031-62387-5_7

claim that whatever set theory is about as a foundation for mathematics, if it is about anything at all, we are never going to pin it down.

The reader may suspect that I plan to offer a formalist vision of set theory in which any serious commitment to its ontology is eschewed. This is not the case. The argument I plan to make is intended to cut across the grain of traditional debates over the ontology of mathematics. Indeed, I hope that the arguments provided will be of equal relevance to those of both platonist and formalist tendencies. In this chapter, I want to remain quiet on the matter of whether there is something that set theory is about and instead focus on what its subject matter could not be. Rather than adjudicate in existential debates over ontology, my aim is to show that any attempt at uniquely specifying the content of set theory is destined to fail. In a nutshell, I will argue for this on the basis that for any attempt to provide a plausible specification of set theory's content, there is another equally plausible account that conflicts with the first. I will then argue that, for the purposes of founding mathematics, this entails that such stories are little more than heuristic crutches: extremely useful, but not compelling.

The rest of this chapter is organized as follows. In the first section, I provide a more precise statement of the claim I aim to defend and deliver the main part of the argument. In order to do this, I'll make some effort to unpack the terms involved and remove as much ambiguity as I can. In particular, I'll provide some analysis of what set theory actually does in performing its foundational role. In the second section, I'll defend my claim using a series of examples demonstrating our inability to pin down a particular content for set theory. While the examples have a somewhat technical nature, I'll also be aiming to provide useful informal explanations of the ideas involved. The bulk of the work will reside here. In the final section, I'll discuss the metaphysical outlook on set theory assuming acceptance of my claim. In particular, I'll look at how a platonist might accept the situation.

Remark The reader will observe that this chapter has very similar goals and themes to Button's Chap. 4. Button and I developed these ideas independently as parts of larger projects that overlapped on this topic. However aside from differences of style, there is a significant difference in scope. Button provides a much more detailed analysis of function theory, which we also discuss more briefly below. However, in this chapter we aim to extend this style of argument beyond definitionally equivalent theories and, as a result, end up considering a wider variety of foundational competitors. But as the reader will see below, as the equivalence relations weaken so too could the force of the philosophical argument.

1 Refining the Claim

Here is a rendering of the strongest version of the claim that I'll defend in this chapter:

> The content of set theory – in its capacity as a foundation for mathematics – cannot be specified.

Two obvious questions emerge at this point:

(1) What is *set theory in its capacity as a foundation for mathematics*?
(2) What would it mean for the *content* of set theory to be *specified*?

These questions will occupy us until Sect. 1.3 at which point we provide the main argument.

1.1 What Is *Set Theory in Its Capacity as a Foundation for Mathematics*?

I now want to provide a description of what we shall call *set theory qua foundation*, which is to be understood as set theory in its capacity as a foundation for mathematics. My aim is to describe the basis on which set theory makes its claim to provide a foundation. I will do this by levering off Penelope Maddy's recent analysis of the goals for set theoretic foundations (Maddy, 2016, 2019). I'll then highlight what I claim is the essential mechanism that allows set theory to meet those goals. This should provide an understanding of the function of set theory qua foundation that will be sufficient for the argument that follows.

Perhaps surprisingly, the questions of how and in what way set theory provides a foundation for mathematics have rarely been tackled head on, although questions like these often lurk in the background behind debates between alternative proposals. Maddy has advanced this discussion by providing a systematic taxonomy of various foundational goals and investigated whether various systems including set theory, category theory, and homotopy type theory live up to them (Maddy, 2016, 2019). Here are a few of the most pertinent goals isolated by Maddy:

- *Metamathematical Corral* A foundation must be such that all mathematics can be embedded into it.
- *Shared Standard*: A foundation must provide a shared standard of proof for the mathematical community.

- *Generous Arena*: A foundation must provide an ontology sufficient for current and forseeable mathematics.
- *Elucidation*: Using the apparatus of the foundation, we can replace pre-theoretical notions with precise counterparts that enhance clarity and remove ambiguity.
- *Risk Assessment*: Consistency strength can be used to assess the relative riskiness of mathematical theories.

Unsurprisingly, set theory in the form of ZFC can make a plausible claim to satisfying each of these goals. However, I think a little reflection also reveals that the *Mathematical Corral* is in an important sense more fundamental than the others. The reason that ZFC can provide a *Shared Standard* is that the embeddings provided by *Metamathematical Corral* allow us to think of all mathematics as being done—so to speak—in ZFC. The reason that ZFC provides the rich ontology required for its *Generous Arena* is that the mathematics of today and beyond can be embedded in it via *Metamathematical Corral*. *Elucidation* is again facilitated by set theory's ability to absorb other areas of mathematics and clarify them in a common framework. Finally, the technical devices used to undertake *Risk Assessment* in set theory are delivered by the embeddings used in the *Metamathematical Corral*.

We shall now focus on explaining the underlying mechanism of the embeddings involved in satisfying *Metamathematical Corral*. This will provide a functional explanation of what the foundational role is and how it is upheld. To do this we turn to what is arguably the canonical example: arithmetic and the theory of the natural numbers. When we talk about the natural numbers we frequently use a language like

$$\mathcal{L}_{Ar} = \{<, 0, s, +, \times\}$$

while in set theory we use $\mathcal{L}_{\in} = \{\in\}$. To "embed" arithmetic into set theory we define a compositional translation function $t : \mathcal{L}_{Ar} \to \mathcal{L}_{\in}$ taking formulae of arithmetic and returning formulae of set theory. The idea of the translation is that it provides set theoretic definitions for the domain of arithmetic and its non-logical vocabulary.[2] So, for example, we might define the domain using a formula $\delta(x)$ which says $x \in \omega$; we might define the $<$ relation on that domain by letting $\rho(x, y)$ say that $x \in y$; we might let 0 be the empty set; and s be interpreted as the function $x \mapsto x \cup \{x\}$. $+$ and $\times$ can then be

[2]See Visser's (2004) for a comprehensive discussion of interpretation and many of the related notions used in this chapter.

defined with a recursion.[3] Thus, we have a way to take any sentence about the natural numbers and translate it into a sentence in the language of set theory. Nothing so far tells us that this translation witnesses the purported embedding required for *Metamathematical Corral*. We need a little more for that. One of the things we have is a theory of the natural numbers, say Peano Arithmetic (PA). This is a computable set of axioms in $\mathcal{L}_{Ar}$ that we believe to be true of the natural numbers. We can then show that:

Theorem 7.1 *For all sentences φ of $\mathcal{L}_{Ar}$, if $PA \vdash \varphi$, then $ZFC \vdash t(\varphi)$.*

Thus, we see that the sentences we believe to be true of the natural numbers are also provable when translated into the language of set theory.[4] There is also a more picturesque way of understanding the translation mechanism. If we are given a model M of ZFC, then the translation function t essentially describes an *internal model* N within M that is a model of PA. Given the model M, we find a domain by considering the extension of $\delta(x)$ in M; that is, $\{m \in M \mid M \models \delta(m)\}$. And we find an interpretation of the $<$ relation by considering $\{\langle m, n\rangle \in M^2 \mid M \models \rho(m, n)\}$. Thus we have a functor taking models of ZFC to models of PA, which we denote (with some overloading of the t notation) as follows:

$$t : mod(ZFC) \to mod(PA)$$

where $mod(ZFC)$ denotes the class of models of ZFC and similarly $mod(PA)$ denotes the models of PA.[5] It is in this sense that we say that mathematics can be *embedded* into set theory, although a quick glance at the underlying machinery reveals that it might be better to say *encoded* or *translated* into set theory. Given this, we then say that PA is *relatively interpreted* by ZFC. Similarly, when we say that analysis is embedded in set theory, we mean that our best theory of analysis is relatively interpreted by ZFC.

[3] A little more care needs to be taken in finding formulae to represent constants and functions since the theory needs to prove that the formula representation has—so to speak—a unique output.

[4] It's worth noting that the converse is not true. ZFC is able to prove more truths about arithmetic than PA; for example, ZFC proves that PA is consistent. There is a sense in which we come to trust the translation so much that we accept that ZFC is able to uncover truths that cannot be discovered in PA alone.

[5] Note that the translation goes from the language of PA to the language of ZFC while the functor goes in the opposite direction; as such it is known as a *contravariant* functor. See Awodey (2006) for a discussion of functors.

This discussion raises a raft of difficult questions; however, two of them are particularly pertinent to our current problem:

(1) Is there anything special about the t interpretation?
(2) Is there anything special about $\mathcal{L}_{Ar}$?

With regard to t, the first thing to note is that there are—of course—many different interpretations that would also do the same job. Indeed, the Benacerraf article—whose title this one riffs upon—takes this fact as its main case study (Benacerraf, 1965). More specifically, Benacerraf considers two interpretations. The first one interprets the domain of the natural numbers as von Neumann's ω. This is what we did above. However, Zermelo offered an alternative where the domain of the natural numbers is the smallest set containing $\emptyset$ and closed under the function $n \mapsto \{n\}$. From the perspective of ZFC, these interpretations define two structures $\mathbb{N}_{vN}$ and $\mathbb{N}_Z$ which are isomorphic. Given this and the fact that neither structure has the "actual" natural numbers in its domain (whatever that would mean), it seems clear that for the project of providing a foundation in which arithmetic can be accommodated, there is no pressing need to choose one or the other. From a foundational point of view, we just need to know that there is an interpretation, not that there is a particularly good one.[6]

Regarding $\mathcal{L}_{Ar}$, there is a little niggle that we shall exploit and generalize further in this chapter. I introduced the language of arithmetic as $\mathcal{L}_{Ar} = \{<, 0, s, +, \times\}$; however, it's clear that I could also have used the language $\mathcal{L}^*_{Ar} = \{0, 1, +, \times\}$ and described a theory PA^* for this language. It's obvious that this makes no substantive difference to our ability to do arithmetic or to set theory's claim to have provided a foundation for it. If a proponent of PA were to debate with a proponent of PA^* over which was the correct theory of arithmetic, it would be fair to say that their debate was merely verbal. We can make this idea more precise by observing that there are further interpretations

$$u : \mathcal{L}^*_{Ar} \to \mathcal{L}_{Ar} \text{ and } w : \mathcal{L}_{Ar} \to \mathcal{L}^*_{Ar}$$

such that $u : mod(PA) \to mod(PA^*)$ and $w : mod(PA^*) \to mod(PA)$. Thus, each theory can—so to speak—define an internal model of the other.[7] However, we can obtain more: u and v can be defined in such a way that

(1) $w \circ u(M) = M$ for all models M of PA; and

[6]This of course is the launching point into the rich literature on mathematical structuralism. A good introduction can be found in Burgess (2015).

[7]In other words, PA and PA^* are mutually interpretable.

(2) $u \circ w(N) = N$ for all models N of PA^*.

Informally speaking, this tells us that if we translate $\mathcal{L}_{Ar}$ into $\mathcal{L}^*_{Ar}$ and back again we end up exactly where we started, and similarly if we start in $\mathcal{L}^*_{Ar}$.[8] We might say that these seemingly distinct theories are in fact just two different ways of describing exactly the same content. Theories with this close relationship are known as *definitionally equivalent* or *synonymous*.[9] As with the alternative interpretations, we see again that from a foundational point of view, there is nothing special about $\mathcal{L}_{Ar}$ as a language for talking about arithmetic.[10] There are many other languages that will also work fine. The important thing is that we interpret one of them; for then the rest follow.

Putting this together we see that the foundational role of set theory is demonstrated by its ability to interpret different fields of mathematics (like arithmetic) into the language of set theory in such a way that truth is preserved and trivial differences about the ways we talk about mathematics can be ignored. We have a generous arena within which the diverse realms of mathematical activity can be united. This is the heart of the foundational project.

This provides an overview of what I believe is most important in a foundation for mathematics and it is with regard to this that we shall be most concerned in what follows. However, other thinkers place greater value in the conceptual and epistemic roles that a foundation might perform.[11] For example, one might ask whether one foundational framework is conceptually or epistemically prior to another. Perhaps one needs mastery of the concepts from one foundation in order to grasp another.[12] While I have less to say on these matters, I believe the argument below still works even if we incorporate conceptual and epistemic roles into our foundational desiderata. More specifically, I claim that each of the foundational frameworks considered below are on an equal footing conceptually and epistemically.

---

[8]Note that the translations don't take sentences back to the same sentences but rather sentences which are equivalent according to the theory.

[9]See Visser (2006) for a comprehensive discussion of definitional equivalence and interpretation.

[10]This is not to rule out the possibility that in specific applications one language might be more useful than another. See Baldwin (2018) for some excellent examples of this phenomenon.

[11]I thank an anonymous reviewer for highlighting the importance of this issue.

[12]For a more thorough discussion of these issues see Linnebo and Pettigrew (2011).

1.2 What Would It Mean for the *Content* of Set Theory to Be *Specified*?

First consider a more general question: how would one specify the content of an arbitrary theory? Let's consider what a positive example might look like. Suppose that T is a theory that, like set theory, can be computably axiomatized via first-order logic in a finite language $\mathcal{L}$. Moreover and perhaps in contrast with set theory, suppose the language of T is intended to be understood as representing ordinary objects and relations that hold between them. I don't want to attempt a precise description of what an ordinary object or relation is, however, this will not matter much below. Given such a theory, we suppose that the names and predicates of $\mathcal{L}$ have well-known counterparts in our natural language and these give us the *intended interpretation* of $\mathcal{L}$.[13] I will then say that the content of T can be specified through this mapping between the elements of the formal language $\mathcal{L}$ and their natural language counterparts. We might call this a *naïve linguistic analysis*. The most obvious example of this kind of analysis is found in the effort to provide a formal semantics for natural language.[14]

In a nutshell, a content specification is an interpretation of the non-logical vocabulary of a formal language into the informal language that we speak. It describes a link between the formal and the informal and as such, the content specification is something over and beyond the theory itself.[15] On this understanding, the content of a theory would then be comprised by the semantic items corresponding to the natural language counterparts of the formal language.[16] I should note that, in general, scientific theories that make use of mathematics are unlikely to be susceptible to a full specification of their content.[17] The reasons for this will emerge below, but in many cases, a satisfying partial specification of their content is still possible.

---

[13]Indeed, one could be tempted to say that a content specification just is an intended interpretation. I have opted not to do this since I think the latter term has a number of natural disambiguations that I want to avoid.

[14]A classic example of this is Montague semantics; see Janssen and Zimmermann (2021). Note, however, that the analogy is not perfect in that semantics projects are not focused on providing axiomatic theories to describe their subject matter. Moreover, given the frequently everyday subject matter of semantics, this kind of account is not generally used as a lever to give purchase ontological claims.

[15]Indeed, one might go on to formalize the content specification itself in a formal metalanguage using the semantic tools provided by model theory. The ensuing regress is nicely described and investigated in Lewis (1984).

[16]I thank an anonymous referee for helping me be more precise about this.

[17]For a somewhat artificial example, we might consider Montague's axiomatization of Newtonian mechanics (Montague, 1974). More sophisticated work in this area has been done by Andréka et al. (2006).

So how might one attempt to specify—even partially—the content of set theory? Following the template above, we would like to find natural language counterparts to the non-logical vocabulary of set theory. Thus, we are interested in counterparts for the predicate Set and the relation symbol $\in$.[18] We then have obvious choices for counterparts: Set is to be interpreted by those things that are *collections*; and $\in$ is the relation that holds between some object and collections of which it is a *member*.[19] There is, of course, an extensive literature taking off from this cue. Moreover, it is clear that Cantor's early development of set theory levers off such intuitions. Contemporary work in this area tends to see set theory as invested in a project of analyzing the pre-theoretical concepts of membership and collection. Great examples in this area include (Boolos, 1971; Potter, 2004; Linnebo, 2010). Such projects take as a starting assumption that the content of set theory is somehow specified by this analysis. They take it that the natural language counterparts give us the intended interpretation of the language of set theory. In contrast, I will argue that the content set theory *qua foundation* cannot be specified by such a naïve linguistic analysis.

1.3 The Argument

We are now in a position to lay out the argument for our main claim. The following premise provides the main engine behind the argument and will be defended in the next section.

> There are many equally plausible content specifications for theories equivalent to set theory qua foundation.

We'll defend this claim by providing a series of examples of theories that each delivers as good a foundation for mathematics as set theory. Indeed, each of them will be—in some salient sense—equivalent to set theory. Moreover, each theory will come along with a plausible attempt at content specification that

[18] It is conventional to articulate set theory in the language with just the membership relation, $\in$, however, this only works when we are only quantifying over sets. Thus, if we wish to incorporate set theory into natural language it is convenient to add an additional predicate to delimit the range of the quantifiers. This is why we introduce the predicate Set.

[19] There is a small issue here in that set theory as used in the foundations of mathematics only considers objects that are sets. This means that there is only one object that has no members: the empty set. However, in ordinary parlance we often consider sets of ordinary objects, none of which have members. This can be handled by moving to a theory like ZFA (ZF with atoms) from which an inner model of the pure sets is easily recovered. See Jech (2008) for details on ZFA.

is distinct from the orthodox approach described above. The key point about these alternatives is that they will be equivalent in a sufficiently strong sense for us to think that *qua foundation* they are the same theory. To establish this, I'll demonstrate that each of these theories is capable of fulfilling the fundamental interpretational requirements discussed above. More surprisingly, each of these theories is so closely related that any one of them is able to define any of the others.

Supposing the reader agrees that these theories are equivalent, we then claim that no content specification can be given for set theory qua foundation that is strictly superior to that provided by a diverse collection of alternative specifications. As the reader will see, the naïve linguistic analyses provided below are, like the collection and membership story, intuitively appealing and heuristically useful. Nonetheless, in the midst of such diversity none of them can lay any serious claim to being definitive. As such, I contend that a naïve linguistic analysis is incapable of definitively specifying the content of set theory qua foundation. In the absence of any other means of specifying set theory's content, I propose we take it that set theory's content qua foundation simply cannot be specified.

The final claim above is a noticeable strengthening of the previous conclusions and is intended as a kind of challenge: if the naïve analysis doesn't work, let's find something that does. Until such a proposal emerges, I think the strong conclusion should be our default position. Note that a successful proposal would need to do something quite novel as variations on the naïve linguistic analysis provide the favored approach in the ontology of set theory. I should finally remark on an upper bound on the strength of the claims that are being made here. While I am claiming that the content of set theory qua foundation cannot be specified using the naïve linguistic analysis, I am *not* arguing that there is no other understanding of content which could be said to correspond to set theory. Indeed, the argument below can be construed as uncovering what such an account would need to do.

2 Defending the Claim

Our goal now is to demonstrate that set theory qua foundation could be reasonably construed to have a number of plausible and yet distinct content specifications. In a perfect world, just one new example would be sufficient. However, I think the case is more persuasively made by also observing how deep this phenomenon runs. As with our discussion of alternatives in arithmetic interpretation in Sect. 1.1, we will now consider examples in which both interpretation and language are varied. Thus, we might see ourselves as generalizing Benacerraf's problem for the natural numbers by applying the analogous considerations to set theory itself.

2.1 An Alternative Interpretation

I will now introduce an alternative content specification for the standard axiomatization of the set theory, ZFC. Recall that when we considered alternative interpretations of arithmetic, we focused on von Neuman and Zermelo's interpretations of the natural numbers. From there, we were able to see that the resulting mathematical structures are isomorphic. The isomorphism was given using set theory by comparing the two structures. We'd like to do something similar here, but given that we are now considering set theory itself, this avenue of comparison is no longer available. As such, we take a modified version of this course. We've noted above that set theory is often understood to provide a theory of membership and collection. This idea is frequently justified by recourse to the cumulative way in which the universe of sets emerges. Starting with the empty set, we take all its subsets, then we take the subsets of that and so on. It can then be seen that ZFC is true in such a structure and it is often claimed that the heuristic clarity of this construction provides intrinsic evidence that ZFC is true (Boolos, 1971; Potter, 2004). This kind of story appears to lean heavily on a deep connection between set theory and its purported analysis of the membership relation. Our goal in this section is to subvert this idea. While we accept that the story of the cumulative hierarchy is good and useful, we do not accept that it lends much evidence toward the truth or even consistency of ZFC. To enact this subversion, we offer an alternative heuristic which I claim is as good and useful as the cumulative hierarchy is for the purposes of understanding ZFC. For this we turn to graph theory.

To describe the structure, we make use the theory GBN, which conservatively extends ZFC with a layer of classes and a predicative comprehension axiom.[20] Let $\mathcal{G} = \langle G, A\rangle$ be a directed graph coded as a class. We say that $\mathcal{G}$ is:

- *well-founded* if it has no infinitely descending paths;
- *extensional* if no two vertices have arrows pointing into them from exactly the same vertices; and

[20] See Chapter 1.12 of Kunen (2006) for a proper description of GBN.

- *set-like* if for every vertex v there is a set containing the vertices with arrows pointing into x.

We say that graph $\mathcal{G}_0$ can be *initially embedded* in $\mathcal{G}_1$, if there is an embedding $f : \mathcal{G}_0 \rightarrow \mathcal{G}_1$ such that whenever we have some $b \in G_1$ with $b \in f(a)$ for some $a \in G_0$, b is already in the range of f. Informally speaking, the range of the embedding is an initial segment of $\mathcal{G}_1$. This initial segment cannot contain, so to speak, any holes and thus G_1 only exceeds the range of f by adding elements beyond the end of that range. Given some property expressed by $\varphi(\cdot)$ in the language of set theory, we say that a graph $\mathcal{G}$ is *end-maximal in* $\varphi(\cdot)$ if $\varphi(\mathcal{G})$ holds and $\mathcal{G}$ cannot be initially embedded into any another graph satisfying $\varphi(\cdot)$. The idea here is that such a graph is as large as it can be and still satisfy the property in question. We then say that $\mathcal{G}$ is a *foundational model* if $\mathcal{G}$ is end-maximal in the properties of being well-founded, extensional, and set-like. Such a structure seems sufficiently rich to be able to provide a foundation for mathematics. Indeed, it is easy to see that if we take the Mostowski collapse of a foundational model, we end up with the universe V of sets with the standard membership relation upon it. Thus, we already know exactly how to axiomatize it: we just use ZFC.

So what does this tell us? It gives us a natural way to describe structures satisfying ZFC while making no mention of membership or collection. As such, it provides us with a way to do set theory without needing to talk about sets. This alternative way of motivating set theory places no obstacles in the way of using ZFC to interpret other theories. Thus, the foundational role of ZFC is preserved. However, from this point of view—if we were pressed to divulge the content of the theory—it would be most natural to say that it was about vertices and arrows organized in a fashion to ensure every possible pattern is represented while avoiding redundancy and enforcing an inductive order.[21] As such, we have an alternative specification of the content of set theory, which has no impact on set theory's ability to perform its foundational role.

There are—I think—a few natural concerns one might have. I'll discuss two of them. First, one might worry that the third property of a foundational model, set-likeness, is a property that depends on notions of membership or collection in order to be described.[22] This dependency was not present

[21] If the reader is concerned that this approach is merely the concoction of the author, many examples of this graph theoretic understanding of set theory can be found in pedagogical contexts including (Kunen, 2009) and research contexts including (Enayat et al., 2011).

[22] One might consider this as a failure of conceptual autonomy as described in Linnebo and Pettigrew (2011).

in the rendering of extensionality or well-foundedness. Set-likeness, however, required for any vertex v the existence of a *set* of vertices with arrows pointing into v. As such, we seem to be leaning on the notion of set to define set-likeness. I think this is misleading. In order to provide a precise description of a foundational model, I made use of an already existing foundational theory, GBN. We used this theory because it is well-understood and because it allows us to naturally articulate the maximality notion we required. However, given that we have shown that ZFC can be understood as a theory of graphs, we can also make an analogous move for GBN. Thus, we can construe it as a two-sorted expansion of ZFC in which there are two kinds of vertex: *small* vertices and *large* vertices, playing the analogous roles of sets and classes, respectively. We can then say that a relation R is *locally small* if for any small vertex a there is some small vertex b *representing the trace of a in R*; that is, b is such that for all small vertices x, xRa iff there is an arrow from x to b. I claim that this re-imagining of set-likeness from the graph theoretic perspective reveals that set-likeness does not rely on membership and collection, but rather leans—if anywhere—on notions of size and representability.

The second concern regards the banality of the proposal. I think that a natural and contrasting response to this example is that it's so obvious as to be all but pointless. Perhaps we have already taken the lessons of Benacerraf seriously and developed our own structuralist approach to mathematics. As such, we might uphold the mantra that we only care about mathematical structures—as they say—up to isomorphism. Given this, it seems difficult to care too much about the obvious fact that there are alternative interpretations of the language of set theory that are— in some sense—isomorphic to each other. I think this is a reasonable thing to think and I mostly agree. One caveat I'd place on making this move is that the user of it should also understand that they are also accepting the claim that the content of set theory qua foundation cannot be uniquely specified. Moreover, this kind of generalized structuralism will not be sufficient to deal with all of the examples that follow.

2.2 Some Alternative Languages

As in our discussion about interpreting arithmetic, we now move to discuss changes of language. This presents a significant step beyond a basic structuralist position in that we are going beyond the mantra that *isomorphism is identity* by claiming that theories can be equivalent even when they are articulated in different languages.[23] We will provide three different alternative attempts at specifying the content of set theory qua foundation, each of which we claim is equivalent qua foundation to any of the others. They will progress from what I believe is an uncontroversial example of genuine theoretical equivalence through to a weaker but still plausible form of equivalence. The first will be a variation on ZFC using functions. The second example will be based on Lawvere's elementary theory of the category of sets. The third example draws on the generic multiverse. I hope that most readers will come along for the entirety of the ride, but if not, I think it will still be instructive to identify the point at which one gets off. We'll also provide some discussion about the philosophical positions that are compatible with these notions of theoretical equivalence. For a detailed discussion of similar issues in philosophy of science see Halvorson (2019).

2.2.1 Function Theory

For our first alternative, we consider a foundation for mathematics that is formulated using functions as its basic elements rather than sets. This idea has a long history, going back to von Neumann's axiomatization of GBN (von Neumann, 1967).[24] Button's chapter presents another axiomatization of this idea with his theory FLT and its extensions. Here, we will present a theory which is closer to von Neumann's original axiomatization, which I call vNC.[25]

---

[23]More sophisticated versions of structuralism have offered coarser equivalence relations to identify different instances of a structure. For example, in Resnik (1997), Wilson (1981), and Shapiro (1997) a relation similar to definitional equivalence (discussed further below) is proposed in order to address issues of language variation as in our arithmetic example from Sect. 1.1. However, the structuralist literature seems to have largely left this issue aside and a definitive account of identity conditions on structures is yet to appear. Such an equivalence relation may help with the example in Sect. 2.2.1, but it will not work for the examples of Sects. 2.2.2 and 2.2.2. Given this, we might hope that the sequence examples from this section would be useful for developing a better understanding of the appropriate identity conditions between structures. Moreover, a successful story of this kind might provide an alternative conception of content that would meet the challenge posed at the end of Sect. 1.3.

[24]Indeed I should remark that the functional approach has been discussed in recent debates around category theoretic foundations (Hellman, 2003). However, the approach below does not immediately give a category since application does not compose. Nonetheless, a category can easily be defined, as it can in ZFC, by finding appropriate objects to represent domains and codomains.

[25]By coincidence, Button and I were working on very similar ideas at the time these chapters were written. As such, the reader may want to gloss more quickly here. While it may appear that we now inventing the wheel for a second time, we have opted to retain the brief exposition in this chapter for a few simple reasons. First, we believe that function theoretic approaches to foundations are of independent logical interest, and consequently, having two alternative axiomatizations in this volume is a boon rather than a burden. Second, the motivating ideas behind the respective axiomatizations are quite different. Button's axiomatization begins from a stage-based approach originating from work made famous by Scott,

The main variation is the development of a theory that only talks about small objects and avoids large ones akin to proper classes. We provide an overview of the theory below and then discuss its significance for our claim. We work in the language $\mathcal{L}_f = \{@\}$ where @ is a three-place relation symbol whose intended meaning is function application.[26] $@(f, x, y)$ is thus intended to mean the result of applying the function f to x is y. We'll use lower case Roman letters for variables, but note that everything in this system should be thought of as a function. Our first axioms ensure that we have functions that are extensional.

(Functionality) $@(f, x, y_0) \wedge @(f, x, y_1) \to y_0 = y_1$.
(Extensionality) $\forall x \forall y(@(f, x, y) \leftrightarrow @(g, x, y)) \to f = g$.

We then introduce a couple of helpful abbreviations that make our exposition feel a little more natural:

- $f(x) = y$ means $@(f, x, y)$;[27] and
- $f(x) \downarrow$ means that $\exists y\ @(f, x, y)$.

We then allow that there is a function.

(Existence) $\exists f\ f = f$.[28]

We defer the rest of the axioms until the Appendix; however, we remark that the basic idea is quite simple: we get the theory to work by adding counterparts to the axioms of ZFC that are intended to work naturally in the functional context. Let us call the resultant system vNC. It is not difficult to see that this can be done and indeed that some axioms, like Replacement, have more natural counterparts in a function theory where it is expressed as a composition axiom. By design, vNC is very similar to ZFC, and as such, it has no problem

Boolos, and others (Boolos, 1971; Potter, 2004). Button's trilogy is a tour de force on the subject (Button, 2021a,b, 2022). Besides providing a pleasing analysis of the cumulative aspects of set theory, Button's approach should be generalizable to any theory amenable to axiomatization through stages. By contrast, the axiomatization presented here (in the Appendix) is intended to be a close copy of ZFC based on von Neumann's motivations. It is hoped that this kind of axiomatization will be easier to weaken in order to accommodate the kinds of weak set theories used in fine structure theory and partial combinatory algebras: see Chapter 9 of Foreman and Kanamori (2009) and Feferman (2006). Finally, our little discussion below emphasizes the semantic, functorial approach to interpretation while Button's is more syntactic. The reader's preferences may vary, but it seems useful to have both perspectives available.

[26] Partial functions are tedious to accommodate in ordinary first-order logic. This is a relatively simple way of implementing them in that context.

[27] This is quite a serious abuse of language, but it's also very convenient. Strictly speaking f is not a function symbol of $\mathcal{L}_f$, so $f(x)$ is not a term. This also means that the $=$ is not the real identity symbol, but rather we just have an abbreviation.

[28] Strictly, this is unnecessary since $\exists x\ x = x$ is a theorem of first-order logic, however, it is also often included; for example, see Kunen (2006).

in providing a foundation for mathematics. However, as with our investigation of arithmetic theories, we can say something stronger.

Proposition 7.2 *ZFC and vNC are bi-interpretable. That is, there are interpretations $z : mod(ZFC) \rightarrow mod(vNC)$ and $v : mod(vNC) \rightarrow mod(ZFC)$ such that:*[29]

(1) $v \circ z(M) \cong M$ *for* $M \models ZFC$; *and*
(2) $z \circ v(N) \cong N$ *for* $N \models vNC$.

Moreover, the relevant isomorphisms are uniformly definable over all models of ZFC and vNC, respectively.

I'll just provide a sketch of this because I think—as is often the case—the interpretations linking the theories illuminate our understanding of them.

Proof (Sketch) The interpretation z must define an internal model of vNC inside a model of ZFC. We use ZFC to define the domain by transfinite recursion as follows:

$$\begin{aligned} D(0) &= \emptyset \\ D(\alpha+1) &= \{g \subseteq D(\alpha) \times D(\alpha) \mid g \text{ is a function}\} \\ D(\lambda) &= \bigcup_{\alpha \in \lambda} D(\alpha) \text{ for limit } \lambda. \end{aligned}$$

We then define function application by translating $@(f, x, y)$ as $f(x) = y$ (i.e., $\langle x, y\rangle \in f$). In the other direction, v defines an internal model of ZFC inside a model of vNC. We let the domain be those identity functions (i.e., those f such that $dom(f) = f$) whose fields consist of identity functions whose fields consist of identify functions and so on.[30] Then to define the membership relation we translate $f \in g$ as $g(f) \downarrow$. We leave the remaining calculations verifying the existence and definability of the isomorphisms to the reader.

□

[29] See Visser (2004) for a thorough account of bi-interpretability.

[30] More specifically, we define an analogue to transitivity in set theory by saying that a function g is *transitive* if for all f, if $g(f) \downarrow$, then $f \sqsubseteq g$. We then let f be in the domain of the vNC interpretation if there is a transitive, identity function g such that $f \sqsubseteq g$.

In fact, by results of Visser and Friedman (2014), we can go further and see that ZFC and vNC are definitionally equivalent.[31] This gives us good reason to think that ZFC and vNC are—at the very least—equivalent for the purposes of providing a foundation of mathematics, as any of the foundational roles described above can be equally well performed by either vNC or ZFC. To see this, return to the example of arithmetic and its embedding in ZFC. By composing the translations $t : \mathcal{L}_{Ar} \rightarrow \mathcal{L}_{\in}$ and $v : \mathcal{L}_{\in} \rightarrow \mathcal{L}_{fun}$ it is easily seen that for all $\varphi \in \mathcal{L}_{Ar}$

$$PA \vdash \varphi \Rightarrow vNC \vdash v \circ t(\varphi).$$

In other words, we can use the bi-interpretation to piggy-back on our original translation from arithmetic into set theory. It is then clear that the same trick will work for any other theory which can be embedded into set theory. Nonetheless, it seems obvious that function theory is about functions and not sets, and as such, we have another candidate specification of content for set theory qua foundation that makes no mention of collection or membership. Moreover, I think it is fair to say that vNC requires no prior conceptual mastery of ZFC or sets in order to understand it. Of course, something like sets are present in vNC just as something like functions are present in ZFC. Perhaps one might argue that sets are conceptually prior to functions, however, the prospects for a successful argument of this kind seem bleak.

The objection might be raised that there is something unnatural or ad hoc in this kind of piggy-backing. I'll respond to this in two ways. First, a quick perusal of vNC reveals that more natural translations will be easy to develop. We should expect that there is a more direct and natural way of embedding arithmetic into vNC than composing two interpretations. Indeed, for mathematical objects that make heavy use of function spaces, it is likely that vNC would provide more natural translations than ZFC. Just because these translations are currently unknown should not be a mark against vNC's ability to perform the foundational role. Second, it is important to note that the kind of equivalence offered here, definitional equivalence, is taken for granted in the actual practice of set theorists. To explain, no working set theorist actually does set theory in the austere language $\mathcal{L}_{\in}$ consisting of a single two-place relation symbol. This would be akin to writing computer programs in machine code. Rather, we expand the language of set theory with various useful relations and

[31]Note Button refers to definitional equivalence as "synonymy." Both terms are common in the relative interpretability literature.

terms; for example, $\subseteq$ for the subset relation and $\mathcal{P}$ for the powerset function. The resultant language is no longer $\mathcal{L}_\in$ and—in a very weak sense—we are using a different theory. Of course, we don't worry about this since we know that all of these relations and terms can—in principle—be eliminated using their definitions. The key point then is that the result above shows us that ZFC and vNC are related in exactly the same way: they define each other.[32]

2.2.2 $ETCS^+$

For our next example, we turn to category theory. Again, we highlight an equivalence that we claim entails that for all foundational intents and purposes, the proposed theory is as good as ZFC and yet suggests a different specification of its content. Again, our evidence will be based on an equivalence between ZFC and the proposed theory; however, this time the equivalence will be a little weaker than that seen in the previous example. First we recall that the *Elementary Theory of the Category of Sets* ($ETCS$) can be axiomatized by the following three axioms (Lawvere, 1964):

(1) the universe is a well-pointed topos;
(2) there is a natural number object; and
(3) the axiom of choice holds (every epic splits).

This theory is known to be weaker than ZFC; however, we may add the following category theoretic rendering of the Axiom of Collection:[33]

If $F(m, C)$ is a formula in the language of category theory, $\mathcal{L}_{Cat}$, then

$$\forall A \exists B \forall m : A \rightarrowtail 2(\exists C\ F(m, C) \rightarrow \exists C < B\ F(m, C)).$$

It turns out that these theories are very close to each other:

Theorem 7.3 (Mitchell, 1972; Osius, 1974) *There are interpretations* $sets : \mathcal{L}_{cat} \rightarrow \mathcal{L}_\in$ *and* $cats : \mathcal{L}_\in \rightarrow \mathcal{L}_{Cats}$ *such that:*

---

[32]One might worry that the definitions used in set theory are simpler than those used to move between ZFC and vNC. However, set theory has many commonly used definitions which are much more complicated—from a heuristic point of view. For example, one might consider the definition of a mouse from inner model theory (Steel, 1995).

[33]See Osius (1974) for an explanation of this notation.

(1) $cats \circ sets(M) \cong M$ *for* $M \models ZFC$; *and*
(2) $sets \circ cats(N) \simeq N$ *for* $N \models ETCS^+$.

where $M \simeq N$ *if* M *and* N *are categorically equivalent.*[34]

Because of the use of categorical equivalence instead of isomorphism in (2) above, we have a slight weakening of bi-interpretability. I'll try to informally explain what is happening here without going into the full details of the interpretations involved. This is important as it helps us to understand why the relationship between ZFC and $ETCS^+$ is weaker than that between ZFC and vNC, but still a compellingly close one. Before we do this, first we note that we now have another example of a theory that pulls the same foundational weight but which appears to be about objects and arrows, not sets.

The functor $sets$ takes a model M of ZFC and defines an internal model of $ETCS^+$ by letting the object domain be M and the arrow domain be M's functions; composition is then defined in the obvious way. This is just the standard definition of the category of sets.[35] The other direction is more complex, so we just sketch the outline. The $cats$ functor takes a model N of $ETCS^+$ and defines a domain consisting of equivalence classes of arrows that are designed to represent sets and their transitive closures.[36] The membership relation is then defined on these equivalence classes.[37] Now consider the back-and-forth translation for a model M of ZFC. First we go to M's version of the category of sets, $sets(M)$. Then we translate again to $cats \circ sets(M)$ and we end up with a model of ZFC whose sets are equivalence classes of sets from M. It is then not hard to see that there is an isomorphism between M and $cats \circ sets(M)$ which is definable in M. This is how (1) from Theorem 7.3 works. From the other side, consider a model N of $ETCS^+$. The $cats$ functor takes us to a model of ZFC whose objects are equivalence classes of arrows from N. We then translate again to obtain $sets \circ cats(N)$ whose objects and arrows are both equivalence classes of arrows from N. However, it is not generally the case that $sets \circ cats(N)$ is isomorphic to N. In essence, this is caused by the quotient taken to obtain the equivalence classes of arrows in forming $cats(N)$. In category theory, the empty set is naturally represented by what is known as an initial object.[38] However, the axioms of $ETCS^+$ make no

[34]See Section 7.8 of Awodey (2006) for a definition and discussion of categorical equivalence.
[35]See, for example, Section 1.2 in Awodey (2006).
[36]At this level of detail, there is little difference between the approach of Mitchell (1972) and Osius (1974).
[37]It should be noted that these classes are not represented by any particular thing—set or arrow—in N.
[38]This is an object that has exactly one arrow between it and any other object.

restriction on how many initial objects there are, while ZFC demands there is only one empty set. Thus, the $cats$ interpretation puts all the initial objects in one equivalence class. Once this is done, those extra initial objects cannot be recovered by further translation: the information is lost. But what exactly is this information? Each of the initial objects are isomorphic to each other. Similarly, each equivalence class of the translation contains those objects that are isomorphic to each other. As such, the information lost is information that plays no role in foundations. No isomorphism types are gained or lost in the move between N and $sets \circ cats(N)$.[39] Moreover, category theorists call this relationship *categorical equivalence* and they take it as the natural standard of equivalence between mathematical objects represented in category theory.[40]

One might also be concerned that this example is attempting to make a decisive intervention into the well-worn and often fraught debate between set theory and category theory as to which is the superior foundation of mathematics.[41] One might think this on the basis that we have established that a strong form of category theory, $ETCS^+$, is—in a significant sense—equivalent to our preferred set theory, ZFC. I think such a conclusion is premature.[42] Category theory's claim to a place—perhaps *the* place—in the foundations of mathematics arguably rests on a different basis to that of set theory. While set theory seeks to provide a deep ontology in which all of mathematics can be housed, category theoretic practice is generally not focused on these questions. Rather, it tends to seek out common patterns in diversity and then bring them all together. While set theory also brings disparate fields together, it often loses the essence of the pattern along the way.[43] We have largely ignored such issues in this chapter since we are focused on how set theory, not category theory, performs its foundational role. As such, we are not claiming to have made any deep intervention in this debate, but rather to

[39] For a powerful argument regarding the importance of representing isomorphism types in set theoretic foundations, see Maddy (1997).

[40] See Section 7.9 in Awodey (2006) for a helpful discussion.

[41] Some recent moves in this debate include Hellman (2003); Awodey (2006); Linnebo and Pettigrew (2011).

[42] For a discussion of similar equivalences from a more category theoretic perspective see Awodey (2011).

[43] This brief remark is intended to gently push back on Linnebo and Pettigrew's project whereby category theory's foundational success is mostly measured against the existing successes of set theory (Linnebo & Pettigrew, 2011). I am suggesting that to construe category theory in this way misses something important about category theoretic research. Some of this missing content is discussed in Awodey (2004). Moreover, Maddy's recent discussion of set theory foundations leaves some room for this under the banners of Metaphysical Insight and Essential Guidance. While Maddy is unwilling to admit these goals into her pantheon they are, nonetheless, recognized.

have merely observed that $ETCS^+$ is—via the results above—a species of set theory qua foundation.

Finally, one might also worry that although $ETCS^+$, while an independent theory, is somehow dependent on the concepts used in the orthodox interpretation of set theory as being about membership and collection. Such a worry is discussed by Linnebo and Pettigrew (2011), when they say of $ETCS$ that, "whenever we come to explain these axioms to those unfamiliar with them, we inevitably appeal to the membership relation, the subset relation, the notion of ordered pair, the notion of a function as a set of ordered pairs, and so on."[44] If this were the case, then the basis of our claim that $ETCS^+$ comes with an independent content specification would be significantly weakened. Linnebo and Pettigrew offer some helpful responses. In particular, they recall Lawvere and Rosebrugh (2003)'s textbook, *Sets for Mathematics*, which gives a comprehensive introduction to set theory from a categorical perspective.[45] Beyond this, we also see that there is something about the list of concepts used in the objection. While one might be tempted to argue that the membership relation is naturally represented by the $\in$ relation, it is obvious that ordered pairs and thus function are merely coded by sets. It's important that set theory can do this, but as we saw in the previous section there's nothing stopping us from starting with a theory of functions and using functions to represent sets. The idea that one conceptual framework is then more fundamental than another seems very odd to me. I am attempting to push back on this with the current series of intuitive examples. But regardless of whether I succeed, I think a greater burden of proof, than is usually recognized, lies with the defender of these hierarchies of fundamentality.

2.2.3 MV

Our final example comes from recent work in contemporary set theory. The generic multiverse has been proposed as a way of understanding the practice of set theorists in the face of the large cardinal program's inability to transcend the undecidability of fundamental problems in set theory (Woodin, 2012; Steel, 2014). One might think of the large cardinal program as set theory's best attempt to provide the interpretationally strongest mathematical foundation possible (Maddy, 1997). It has had significant success in the realm of analysis;

[44] This is part of an argument that $ETCS$ lacks conceptual autonomy attributed to Dan Isaacson.

[45] Moreover, a brief perusal of the website *nLab*, reveals that there is a generation of younger mathematicians far more familiar with categorical approaches than ZFC.

however, the continuum hypothesis cannot be decided by large cardinals since forcing leaves large cardinals in tact while being able to change the cardinality of the reals to be essentially anything not ruled out by ZFC.[46] The generic multiverse proposal addresses this issue by taking this incompleteness seriously and thus taking the generic extensions of a particular universe as genuine alternatives.[47]

John Steel provides an axiomatization MV of the multiverse idea (Steel, 2014). We work in a language $\mathcal{L}_{MV}$ with the $\in$-relation and two sorts; one sort for sets; the other sort for worlds (aka universes).[48] Then we have the following axioms:[49]

(1) Every world W is transitive proper class satisfying the axioms of ZFC.
(2) If W is a world and W is a generic extension of U, then U is a world.
(3) If W is a world and $\mathbb{P}$ is a poset in W, then there is a world U where U is a $\mathbb{P}$-generic extension of W.
(4) If W_0 and W_1 are worlds, then W_0 and W_1 can both be generically extended to some world U.

There is a strong sense in which—despite the alien appearance—the practice of set theorists can be construed as taking place in MV. In essence, we just make use of the third axiom whenever we encounter a forcing argument. If anything, it makes life a little simpler. Moreover, this theory can be seen as answering an old problem posed by Dana Scott regarding the plurality of set theories opened up by the advent of forcing:

> I see that there are any number of contradictory set theories, all extending the Zermelo–Fraenkel axioms: but the models are all just models of the first-order axioms, and first-order logic is weak. I still feel that it ought to be possible to have strong axioms, which would generate these types of models as submodels of the universe, but where the universe can be thought of as something absolute. Perhaps we would be pushed in the end to say that all sets are countable (and that the continuum is not even a set) when at last all cardinals are absolutely destroyed. But really pleasant axioms have not been produced by me or anyone else, and the suggestion remains speculation. A new idea (or point of view) is

[46]Here we mean small set forcing, that is, forcing posets $\mathbb{P}$ with cardinality less than the large cardinal in question. See the section on Easton forcing in Chapter VIII in Kunen (2006) for the variety of consistent and different values of the continuum's cardinality. For discussion of resilience of large cardinals to forcing see Chapter 10 in Kanamori (2003).

[47]See Maddy and Meadows (2020) for a thoroughgoing discussion of the generic multiverse and its motivations.

[48]Strictly, we also demand that sets can be members of sets and sets can be members of worlds, but worlds cannot be members of sets or worlds.

[49]The full versions of the axioms can be found in Maddy and Meadows (2020).

needed, and in the meantime all we can do is to study the great variety of models. (page xv in Bell, 2005)

Steel's theory can be seen as providing the key new idea in axiomatizing the plurality of set theoretic worlds given by forcing: retaining parameters to quantify over them.[50] In the background of MV, we end up with a universe of sets where everything is indeed countable, but not relative to every world.[51]

We might then wonder if this system is consistent. In fact, we can do better. To illustrate this we make some minor modifications to our theories. Let ZFC^* be ZFC plus the following assumptions: the universe is not a proper generic extension of any inner model, and there is a definable well-ordering of the universe.[52] Let MV^* then be MV plus the assumption that there is some world W that is not a proper generic extension of some other world and has a definable well-ordering of its domain. Then we see that:[53]

Theorem 7.4 (Meadows, 2021) *ZFC^* and MV^* are sententially equivalent;[54] there are translations $uni : \mathcal{L}_{MV} \to \mathcal{L}_{\in}$ and $mv : \mathcal{L}_{\in} \to \mathcal{L}_{MV}$ such that:*

(1) $mv \circ uni(M) \equiv M$ *for* $M \models ZFC^*$*; and*
(2) $uni \circ mv(N) \equiv N$ *for* $N \models MV^*$.

Thus, if we start with a model M of ZFC^* and use the translation to obtain a model of MV^* and then translate again to obtain another model N of ZFC^*, we see that M and N are elementary equivalent, that is, they instantiate exactly the same theory.[55] This is not as strong as definitional equivalence or bi-interpretability: we don't end up with the same or an isomorphic model. But what difference does that make from a foundational perspective? It is easy to see that MV^* is able to use the mv translation to piggy-back on

---

[50]The other key ingredient is a theorem of Laver and Woodin that allows us to formalize Axiom 2 above.

[51]The idea here is that while each world W can be thought of a universe of sets, MV also allows us to quantify over all of the sets regardless of which world they are in.

[52]So in common set theoretic notation, we are assuming $GA + V = HOD$.

[53]Note that ordinary ZFC and MV are indeed equiconsistent.

[54]See Visser (2004) for a definition of sentential equivalence.

[55]One might be concerned that the extra assumptions are ad hoc. In fact, a syntactic version of sentential equivalence can be established between: ZFC plus there is an extendible cardinal and the universe is not a proper generic extension of any inner model; and MV plus there is some world with an extendible cardinal (Maddy & Meadows, 2020). It has been shown that an extendible cardinal implies that there is a world that is not a proper generic extension of any world (Usuba, 2017). We should, however, mention that the translation from $\mathcal{L}_{MV}$ into $\mathcal{L}_{\in}$ is not strictly an interpretation since is not compositional in that it involves the forcing relation. It is nonetheless computable.

any of the translations that have been procured for ZFC^*. Moreover, it can be seen that MV^* does not prove anything that couldn't already be proven in ZFC^*.[56] And better, any strengthening of one system can be uniformly matched using the interpretations in a way that preserves the relation of sentential equivalence.

Assuming we accept that MV can perform the foundational role as well as ZFC and that they are in a sense the same theory, we see that we have another theory whose linguistic presentation suggests that set theory has content distinct from that of ZFC. This again is further evidence that the set theory's content cannot be uniquely specified with standard linguistic techniques. However, unlike the previous two cases, we do see that sets make an appearance in the multiverse language. There are a couple of things to say about this. First, in the presence of our previous examples, we still have sufficient evidence for our claim that set theory is not really about membership and collection. Second, the sets described in the theory of the generic multiverse are very unlikely to be acceptable to those who think that set theory provides some kind of conceptual analysis of membership. This is because the generic multiverse provides an axiomatization in which the domain of sets is in flux. For a particularly stark example of this, note that with a generic multiverse any set whatsoever is countable in some universe.[57]

In contrast to the previous examples, we might also worry that mastery of the concepts employed in ZFC is required before one can understand MV. This is more plausible here than in the previous cases since every world in the multiverse must already satisfy ZFC. I think we just have to take this one on the chin, however, there is a little more we might say in favor of the conceptual significance (rather than priority) of MV. While it is plausible that one needs to know ZFC prior to learning MV, once MV is understood along with the result above, a radical shift in conceptual perspective is delivered. We have a theory that is, in a very precise sense, equivalent to ZFC that upends the orthodox idea that the membership relation is univocal. This does not establish an alternative story of conceptual priority; rather we have a kind of conceptual disruption in that the new framework puts pressure on intuitive conceptions about set theory and how it should be understood.

[56]This is because sentential equivalence implies mutual faithful interpretability.

[57]It is of course possible that one might take the generic multiverse as a way of analyzing membership as semantically or epistemically under-determined. However, this kind of pluralistic analysis sits in stark contrast with more orthodox analyses that focus on a fixed idea of *the* universe of sets. As such, the pluralists and the orthodoxy would already be in a debate regarding set theory's content.

2.3 Summary

Putting all this together, we have now provided four alternative approaches to set theory, each of which is capable of fulfilling the foundational role and each of which came with a seemingly plausible and distinct naïve linguistic analysis. With our first example, we provided a means of motivating the axioms of ZFC that is graph theoretic in spirit and thus unconcerned with the concepts of membership and collection. Generalizing from our arithmetic example, we thought of this approach as providing an interpretation that was—loosely speaking—isomorphic to the orthodox one. At this point, part of our burden has been discharged: we have shown that set theory can be described in a way that is independent of the notions of membership and collection. However, we might still wonder if this problem is not so deep. Perhaps, set theory really is about either membership or graphs but we just don't know which. This might not be the cause for much worry. This then motivated our second batch of examples. Rather than merely altering the background story we use to motivate the axioms of set theory, we modified the underlying linguistic basis through which set theory qua foundation is articulated. We outlined three different languages in which theories *equivalent* to ZFC could be developed. These variations go beyond merely changing the motivating story. We aren't merely changing the informal concepts that link to the formal language of set theory, we are changing the linguistic framework itself.

Given these equivalences, we make two claims: that each of these theories can perform the foundational role of set theory and that each of these theories has a plausible claim to being a merely notational variant of our conventional approach to set theory. Nonetheless, the obvious ways to attempt to specify the content of these accounts are not entwined with notions of collection and membership. Indeed, if the linguistic resources of these theories were to be taken as a guide to their content, then they would seem to be about three different topics. Thus, given that we take the equivalences seriously, we conclude that we cannot take the linguistic basis of set theory (or any of the other theories) as a guide for the specification of its content. The content of set theory simply cannot be specified in this way.[58]

[58] I should note that this position sits in stark contrast to many contemporary projects in analytic metaphysics. For example, Sider (2011) aims to uncover or at least posit the existence of the fundamental language of the world. As such, he makes some effort to cast doubt on the interpretational techniques deployed in this chapter. Deep engagement with this literature is outside the scope of this chapter, however, it would be remiss not to note that a perhaps deeper philosophical debate lurks under the surface of this article.

This is our argument. However, it seems fairly likely that some readers will not have completed the journey to acceptance of each of the examples provided as legitimately equivalent ways of articulating set theory qua foundation. I claim that this should not affect whether the reader accepts the conclusion of the main argument of this chapter. This warrants some explanation. While I think there can be no problem with the graph theoretic example—since it's just ZFC again—the relationship between vNC and ZFC may give some pause. This would be misguided. The relationship between vNC and ZFC, definitional equivalence, is not only extremely close, it is based on a technique which is integral to the practice of set theory: defining new relations and terms. As such, the problem must be with one of the latter two examples: $ETCS^+$ or MV. With these cases, some room for controversy is opened. We'll discuss this below; however, we note that the main argument still goes through even if we only accept the equivalence of vNC and ZFC. We just have less evidence.

The last two examples are interesting in that they illustrate natural weakenings of definitional equivalence that have plausible and interesting overlaps with common positions in philosophy of mathematics. As noted above, $ETCS^+$ is not definitionally equivalent to ZFC. Rather the back and forth interpretations take us: from a model of ZFC to a model that is isomorphic to the first and from a model of $ETCS^+$ to a model that is categorically equivalent to the first. So we don't end up back where we started, but somewhere very close by. Is it close enough? I think this is dependent on one's antecedent philosophy of mathematics. In particular, if one has a sympathy toward structuralist philosophies of mathematics whose slogan is roughly, "isomorphism is identity," then I think both isomorphism and categorical equivalence should be accepted as getting back close enough to where we started for the purposes of providing a foundation for mathematics. Moreover, given that there is ample evidence that practicing mathematics implicitly accepts a structuralist attitude, there is not a lot of room to push back against ZFC's equivalence with $ETCS^+$. The relationship between ZFC^* and MV^* is weaker again. The back and forth translations take us from a model of ZFC^* (respectively MV^*) to a model that is elementary equivalent to the first. Again, we don't end up exactly where we started. It is close by but it is not as close as an isomorphism. Indeed, the internal model obtained (in either case) will not be well-founded from the perspective of the original model. So we end up with a model that satisfies the same complete theory, but has lost arguably important structural features of the original model. As such, this last stop on the ride may generate a little pushback and some will reasonably want to get off at this point. From a syntactic point of view, however, instantiation of the same complete theory is exactly what one would want. Thus, for those who lean toward the formalist[59] end of the mathematical ontology spectrum, the equivalence of ZFC^* and MV^* seems obvious.

3 Reviewing Some Consequences

In this final section, I would like to briefly review some of the upshot of my claim. I'll consider three issues. First, it might be thought that if we accept that the content of ZFC cannot be uniquely specified, then perhaps we ought to abandon ZFC and perhaps take up one of the alternatives discussed above. This would be misguided. Just because there are many different languages and ways of doing set theory does not mean that it's not extremely valuable to have a canonical way of doing this. The point we are making here is that the claim to canonicity is one of convention and nothing deeper than that. However, without an agreed convention it would be almost impossible—from a practical point of view—to engage in set theoretic research. For a contrasting example, one might say that transitive sets are the canonical representatives of well-founded, extensional relations. This claim is also contentious and places pressure on us to clarify exactly what "canonical" means. I don't propose to attempt that here; however, there is something more we can say to differentiate these examples. With regard to ZFC and its alternative content specifications, I have argued that each of these analyses is as good as any of the others. But in the case of representing well-founded, extensional relations, transitive sets provide a very natural proxy with no obvious competitors. As such, one might argue that in this case there is more to the canonicity than mere convention. I claim this is not the case in the search for a content specification for set theory.

Second, I want to take some care to say that the claim made in this chapter may not transfer to category theory; for example, it may be the case that category theory understood in its foundational role is—in some sense—concerned with objects and arrows. In particular, while analogous equivalent alternatives could likely be developed for category theory, the force of those examples could be weaker. This is because—unlike set theory—category theory generally attempts to provide representations of mathematical objects that are canonical in a sense that is deeper than mere convention.[60] As

[59]Similar remarks apply to Maddy's arealist despite the distinctions between those positions (Maddy, 2011).

[60]For some contrasting discussion of these issues, see Awodey (2004) and Maddy (2016). My opinion is that representing and building mathematical objects in category theory is more difficult than in set theory.

to the challenging questions of whether this project is worthwhile or whether category succeeds in achieving it, we leave that aside.

For the third issue, recall that in the previous section I argued each member of a sequence of weakening equivalence relations between theories should be accepted as showing those theories are, at the least, equivalent qua foundation. I also accepted that some readers would be less likely to accept sentential equivalence than definitional equivalence or bi-interpretability. In particular, those of a platonist disposition may have good reason to think that going forth and back via interpretations and ending up at a merely elementary equivalent structure is a situation in which important information has been lost. For our purposes, a *platonist* will be someone who believes that mathematical objects exist and that their existence is independent of the people who talk about them. Much more could be said, but a broad brushstroke will suffice for our purposes since I aim to be as inclusive as possible. Now from a platonist's perspective, one might worry that the argument of this chapter would force us into some form of formalism or anti-realism. We might think that if the content of set theory cannot be specified, then perhaps set theory is just so much more scratching of symbols onto the world with nothing substantial answering to them. Regardless of whether the reader accepts the antecedent of this conditional, I claim that the conditional itself is false. I'll argue for this by examining what is left over after one accepts my main argument. Recall that we've argued that the content of set theory cannot be specified since there are alternative interpretations and languages that are—for foundational intents and purposes—the same. The upshot of this is that we cannot simply take these different ways of doing set theory at their word. Just because set theoretic investigation began with the metaphors of membership and collection and is articulated in a language with a relation symbol that riffs on that metaphor, this does not mean that we should think that set theory is about membership any more than we think it is about graphs, functions, objects, arrows, or worlds. This pill will be too difficult to swallow for some, but even when we do accept it, it does not entail that there is nothing out there that set theory is about. It just means that we cannot properly specify it. Whatever set theory is about, it cannot be pinned down by language and logic alone. In the absence of anything concrete to anchor our talk, there will always be another way to talk about it; another way that is equally good for foundational purposes and, when we look a little deeper, arguably the same. To put it bluntly, while some

However, very frequently the extra challenges of making the representation in category theory reveal deep insights and connections with other areas of mathematics.

philosophers aim to carve nature at its joints, I'm saying that in the case of set theory, nature is a vegetable.

3.1 Conclusion

We've argued that the content of set theory qua foundation cannot be specified. Consequently there is no serious connection between ZFC with the concepts of membership and collection They are rather metaphors that allow us to better engage in the practice of set theory. Their heuristic value is transparent and should not be underrated, but nor should they be overrated. At bottom, the value of these metaphors is simply not sufficient evidence that some abstracts correspond in any deep or interesting way to the contingent means by which we talk about them. We showed that if we are interested in performing the foundational role that set theory performs, then there is nothing stopping us from thinking that our theory is about any number of alternative ontologies. In particular, we have seen that the fundamental instrument of foundations, interpretation, can be deployed upon set theory itself to reveal that it is pointless to attempt to definitively specify its content.

Appendix

We continue the exposition of function theory vNC. First we work toward defining the empty function. Toward this goal we introduce our version of the Separation Schema, which we call Restriction.

(Restriction) For arbitrary formulae $\varphi(x, \bar{y}) \in \mathcal{L}_f$ and arbitrary f there is some g such that for all x, z

$$g(x) = z \Leftrightarrow f(x) = z \wedge \varphi(x, \bar{y}).$$

This tells us that we may take the restriction of any function to a definable part of its domain. For any such $\varphi(x, \bar{y})$ and f, it is clear by Extensionality that there is exactly one such g so we may denote this with a term

$$f \restriction \varphi(\cdot, \bar{y}).$$

We then observe that given some f as given by Existence,

$$f \restriction (\cdot \neq \cdot)$$

is the function with the empty domain and that it is unique by Extensionality. We denote this function as $\emptyset$. We now introduce two axioms that have no natural counterpart in set theory:

(Domain) $\exists g \forall x \forall y (g(x) = y \leftrightarrow y = x \wedge f(x) \downarrow)$.
(Range) $\exists h \forall x \forall y (h(x) = y \leftrightarrow y = x \wedge \exists z\, f(z) = x)$.

In other words, for any function f there is another function with the same domain that is the identity on that domain and a function that is the identity on f's range. Given such a g is unique we denote it by $dom(f)$; and given such an h we denote it by $ran(f)$. We now introduce useful axioms for building new functions:

(Pairing) $x_0 \neq x_1 \rightarrow \exists g\ (g(x_0) = y_0 \wedge g(x_1) = y_1)$.

This tells us that given $x_0 \neq x_1$, y_0, and y_1 we may form a function with a pair of inputs and outputs. Toward a counterpart to the Union axiom, let us say that a function f has a *coherent domain* if for all g_0, g_1 such that $f(g_0) \downarrow$ and $f(g_1) \downarrow$ if $g_0(x) = y_0$ and $g_1(x) = y_1$, then $y_0 = y_1$. Let us write $h \sqsubseteq f$ to mean that f *extends* h as a function, that is, for all x, y if $h(x) = y$, then $f(x) = y$. The following version of Union allows us to *paste* together functions which form a coherent domain.

(Pasting) If f has a coherent domain, then there is some g such that whenever $f(h) \downarrow$, $h \sqsubseteq g$.

We may then define something analogous to powerset for functions. First let us say that h is a *partial function within* f if $dom(h) \sqsubseteq dom(f)$ and $ran(h) \sqsubseteq ran(f)$.

(Power) There is some g such that for all h, $g(h) = h$ iff h is partial within f.

In other words, for any function f there is a function g whose domain consists of all the partial functions from subsets of the domain of f to subsets of its range. Using Extensionality, we may then define a term $\mathcal{P}$ such that

$$\mathcal{P}(f)(h) = h \Leftrightarrow h \text{ is partial within } f.$$

The following is an inductive principle that plays a similar role to Foundation. Let $field(f)$ be the result of pasting together the functions $dom(f)$ and $ran(f)$.[61]

(Induction) If $\forall f(\forall g(field(f)(g) = g \rightarrow \varphi(g)) \rightarrow \varphi(f))$, then $\forall f \varphi(f)$ where φ is any formula of $\mathcal{L}_f$.[62]

Replacement has a particularly natural counterpart, which we call Composition.

(Composition) If $\forall x \exists! y \varphi(x, y, \bar{z})$, then there is some g such that

$$g(x) = y \Leftrightarrow f(x) = w \wedge \varphi(w, z, \bar{y}).$$

In other words, we can always compose a class function with an ordinary function to obtain another function. Given such an f it is clear that we get a unique such g. Let us denote as $\varphi(\cdot, \cdot, \bar{y}) \circ f$. For infinity, we use the following which will also give us our version of the naturals.

(Infinity) $\exists f(f(\emptyset) \downarrow \wedge \forall g(f(g) \downarrow \rightarrow \exists h(f(h) \downarrow \wedge h(g) \downarrow))$.

Observe that for all f we can show that there is a $\sqsubseteq$-smallest function g such that $g(f) = f$ and for all h, $g(h) = f(h)$ if $f(h) \downarrow$. We call this the *successor* of f and denote it $s(f)$. It can then be shown that there is a $\sqsubseteq$-smallest function ω such that $\omega(\emptyset) = \emptyset$ and for all n such that $\omega(n) = n$ then $\omega(s(n)) = s(n)$. Finally, we provide our version of the axiom of choice.

(Choice) $\exists g(dom(g) = ran(f) \wedge \forall x \forall y(g(x) = y \rightarrow f(y) = x)$.

Finally we observe—with thanks to Tim Button—that these axiomatizations yield the same theory.

Proposition 7.5 *vNC and $FLTC$ are the same theory.*[63]

Proof It can be seen that the same interpretations are used by Button and Meadows to witness, respectively, that vNC and $FLTC$ are bi-interpretable

[61] More specifically, we using Pairing and Restriction to obtain a function g such that $g(dom(f)) = dom(f)$ and $g(ran(f)) = ran(f)$. Then we see that g has a coherent domain so we may use Pasting and Restriction to obtain a function h such that: $dom(f) \sqsubseteq h$; $ran(f) \sqsubseteq h$ and if $h(x) \downarrow$, then either $dom(f)(x) \downarrow$ or $ran(f)(x) \downarrow$.

[62] We shall take it that parameters are allowed in φ, but we shall avoid notating them here in the interests of visual clarity in what is a quite *busy* formulation.

[63] We note that Button does not use choice, so we have added it to the first-order version of FLT to obtain $FLTC$. The result also holds if we remove choice from vNC and compare it to FLT.

with ZFC. Since these interpretations witness bi-interpretability, they also witness mutual faithful interpretability. This means that for all $\varphi \in \mathcal{L}_f$ we have:

$$\begin{aligned} vNC \vdash \varphi &\Leftrightarrow ZFC \vdash z(\varphi) \\ &\Leftrightarrow FLTC \vdash \varphi \end{aligned}$$

as required. □

References

Andréka, H., Madarász, J. X., & Németi, I. (2006). *Logical axiomatizations of space-time. Samples from the literature* (pp. 155–185). Springer.
Awodey, S. (2004). An answer to Hellman's question: 'does category theory provide a framework for mathematical structuralism?'. *Philosophia Mathematica, 12*(1), 54–64.
Awodey, S. (2006). *Category theory.* Clarendon Press.
Awodey, S. (2011). *From sets to types, to categories, to sets* (pp. 113–125). Springer Netherlands.
Baldwin, J. T. (2018). *Model theory and the philosophy of mathematical practice: Formalization without foundationalism.* Cambridge University Press.
Bell, J. L. (2005). *Set theory: Boolean-valued models and independence proofs.* Clarendon Press.
Benacerraf, P. (1965). What numbers could not be. *The Philosophical Review, 74*(1), 47–73.
Boolos, G. (1971). The iterative conception of set. *Journal of Philosophy, 68*(8), 215–231.
Boolos, G. (1998). Must we believe in set theory? In *Logic, logic, and logic* (pp. 120–132). Harvard University Press.
Burgess, J. P. (2015). *Rigor and structure.* Oxford University Press.
Button, T. (2021a). Level theory, part 1: Axiomatizing the bare idea of a cumulative hierarchy of sets. *Bulletin of Symbolic Logic, 27*(4), 436–460.
Button, T. (2021b). Level theory, part 2: Axiomatizing the bare idea of a potential hierarchy. *Bulletin of Symbolic Logic, 27*(4), 461–484.
Button, T. (2022). Level theory, part 3: A boolean algebra of sets arranged in well-ordered levels. *Bulletin of Symbolic Logic, 28*(1), 1–26.
Enayat, A., Schmerl, J. H., & Visser, A. (2011). *ω-models of finite set theory.* Lecture Notes in Logic (pp. 43–65). Cambridge University Press.

Feferman, S. (2006). A language and axioms for explicit mathematics. In *Algebra and Logic: Papers from the 1974 Summer Research Institute of the Australian Mathematical Society, Monash University* (pp. 87–139). Springer.

Foreman, M., & Kanamori, A. (2009). *Handbook of set theory*. Springer.

Gödel, K. (1947). What is Cantor's continuum problem? In S. Feferman, J. Dawson, & S. Kleene (Eds.), *Kurt Gödel: Collected works Vol. II* (pp. 176–187). Oxford University Press.

Halvorson, H. (2019). *The logic in philosophy of science*. Cambridge University Press.

Hellman, G. (2003). Does category theory provide a framework for mathematical structuralism? *Philosophia Mathematica, 11*(2), 129–157.

Janssen, T. M. V., & Zimmermann, T. E. (2021). Montague semantics. In E. N. Zalta (Ed.), *The stanford encyclopedia of philosophy*. Metaphysics Research Lab, Stanford University, Summer 2021 Edition.

Jech, T. J. (2008). *The axiom of choice*. Dover Books on Mathematics Series. Dover Publications.

Kanamori, A. (2003). *The higher infinite: Large cardinals in set theory from their beginnings*. Springer.

Kunen, K. (2006). *Set theory: An introduction to independence proofs*. Elsevier.

Kunen, K. (2009). *The foundations of mathematics*. Mathematical Logic and Foundation. College Publications.

Lawvere, F. W. (1964). An elementary theory of the category of sets. *Proceedings of the National Academy of Sciences, 52*(6), 1506–1511.

Lawvere, F.W., &. Rosebrugh, R. (2003). *Sets for mathematics*. Cambridge University Press.

Lewis, D. (1984). Putnam's paradox. *Australasian Journal of Philosophy, 62*(3), 221–236.

Linnebo, Ø. (2010). Pluralities and sets. *Journal of Philosophy, 107*(3), 144–164.

Linnebo, Ø., & Pettigrew, R. (2011). Category theory as an autonomous foundation. *Philosophia Mathematica, 19*(3), 227–254.

Maddy, P. (1997). *Naturalism in mathematics*. Oxford Scholarship Online. Philosophy Module. Clarendon Press.

Maddy, P. (2011). *Defending the axioms: On the philosophical foundations of set theory*. Oxford University Press.

Maddy, P. (2016). *Set-theoretic foundations*. American Mathematical Society.

Maddy, P. (2019). What do we want a foundation to do? In S. Centrone, D. Kant, & D. Sarikaya (Eds.), *Reflections on the foundations of mathematics: Univalent foundations, set theory and general thoughts* (pp. 293–311). Springer.

Maddy, P., & Meadows, T. (2020). A reconstruction of steel's multiverse project. *Bulletin of Symbolic Logic, 26*(2), 118–169.

Meadows, T. (2021). Two arguments against the generic multiverse. *Review of Symbolic Logic, 14*(2), 347–379.

Mitchell, W. (1972). Boolean topoi and the theory of sets. *Journal of Pure and Applied Algebra, 2*(3), 261–274.

Montague, R. (1974). *Formal philosophy: Selected papers of richard montague*. Yale University Press.
Osius, G. (1974). Categorical set theory: A characterization of the category of sets. *Journal of Pure and Applied Algebra, 4*(1), 79–119.
Potter, M. D. (2004). *Set theory and its philosophy: A critical introduction*. Oxford University Press.
Resnik, M. D. (1997). *Mathematics as a science of patterns*. Oxford University Press.
Shapiro, S. (1997). *Philosophy of mathematics: Structure and ontology*. Oxford.
Sider, T. (2011). *Writing the book of the world*. Oxford University Press.
Steel, J. R. (1995). Projectively well-ordered inner models. *Annals of Pure and Applied Logic, 74*(1), 77–104.
Steel, J. R. (2014). Gödel's program. In J. Kennedy (Ed.), *Interpreting Gödel: Critical essays*. Cambridge University Press.
Usuba, T. (2017). The downward directed grounds hypothesis and very large cardinals. *ArXiv e-prints*.
Visser, A. (2004). Categories of theories and interpretations. In *Utrecht Logic Group Preprint Series, 228*.
Visser, A. (2006). Categories of theories and interpretations. . In A. Enayat, I. Kalantari, & M. Moniri (Eds.), *Logic in Tehran. Proceedings of the workshop and conference on Logic, Algebra and Arithmetic, held October 18–22, 2003* (Vol. 26, pp. 284–341).ASL.
Visser, A., & Friedman, H. M. (2014). When bi-interpretability implies synonymy. *Logic Group preprint series, 320*, 1–19.
von Neumann, J. (1967). An axiomatization of set theory. In J. van Heijenoort (Ed.), *From Frege to Gödel: A Source Book in Mathematical Logic, 1879–1931* (pp. 393–413). Harvard University Press.
Wilson, M. (1981). The double standard in ontology. *Philosophical Studies, 39*(4), 409–427.
Woodin, W. H. (2012). *The continuum hypothesis, the generic multiverse of sets, and the Ω conjecture*. Cambridge University Press.

8

Hamkins' Analogy Between Set Theory and Geometry: Pluralism by Leveling Up?

Sharon Berry

1 Introduction

In work like Hamkins (2012), set theorist Joel David Hamkins develops an influential multiverse approach to set theory. On this view, there are many different hierarchies of sets, and there's no fact of the matter about whether certain set-theoretic statements are true, beyond the fact that they are true of some hierarchies of sets within the multiverse and false in others. On this view (contrary to more traditional forms of set-theoretic Platonism) there's no full-width intended hierarchy of sets, which contains all possible subsets of the sets it contains—or even all subsets of its copy of the natural numbers. Rather, for *every* set-theoretic universe V in the multiverse, there's a forcing extension—a 'fatter' universe of sets that includes all sets in V but also a certain missing subset of some infinite set already included in V. Hamkins gives two main motivations for this view in Hamkins (2012): an appeal to the phenomenology of forcing arguments and an analogy between set theory and geometry.

In Sect. 2 of this chapter, I'll review some basic facts about Hamkins' multiverse project and then note that Hamkins develops the long-familiar pluralist analogy between set theory and geometry in an elaborate and somewhat unusual way. Specifically, he seems to advocate what I'll call a 'leveling up' (rather than 'leveling down') pluralism about both geometry and set

S. Berry (✉)
Indiana University Bloomington, Bloomington, IN, USA
e-mail: seberry@invariant.org

C. Antos et al. (eds.), *The Palgrave Companion to the Philosophy of Set Theory*,
Palgrave Companions, https://doi.org/10.1007/978-3-031-62387-5_8

theory. This position is both interesting in its own right and seemingly argumentatively useful to Hamkins. For, it promises to let him motivate choosing multiverse theory in particular (which appears to provide a kind of leveling up) over other forms of set-theoretic pluralism—not just adopting some form of set-theoretic pluralism.

In Sect. 3 I propose a way of filling in Hamkins' suggestive sketch of *geometrical* pluralism via leveling up. In Sect. 4, I'll question whether Hamkins' multiverse theory actually provides leveling up pluralism about set theory. And in Sect. 4 I'll sketch a counterpossible-counterfactual variant on Hamkins' multiverse theory which faces various challenges, but does clearly qualify as leveling up pluralism about set theory.

I don't claim the latter view is correct (I personally favor a far more truth value realist position Berry, 2022). But I'll suggest that it has some appeal and illuminatingly contrasts with Hamkins' multiverse theory in a few ways.

2 Background

2.1 Hamkins' Platonist Multiverse

Let us begin with some background about Hamkins' proposal.

Hamkins' multiverse view is, as he says, a form of Platonism (Hamkins, 2012). It combines ontological realism (the view that mathematical objects like sets exist) with significant truth value anti-realism (the view that there isn't a single right answer to many questions in the language of set theory).[1] Hamkins' approach differs from both formalism and conventional Platonist approaches to set theory, in taking there to be many differently structured universes of sets, that are all on equal footing[2]—rather than a single intended set-theoretic universe.

Most centrally, Hamkins claims that every set-theoretic universe V has a forcing extension $V[G]$ which adds some[3]) 'missing subsets' to a set in V.

[1]For example, Hamkins writes, "In this article, I shall argue for…the multiverse view, which holds that there are diverse distinct concepts of set, each instantiated in a corresponding set-theoretic universe, which exhibit diverse set-theoretic truths. Each such universe exists independently in the same Platonic sense that proponents of the universe view regard their universe to exist." (Hamkins, 2012).

[2]That is, all these variant universes are on equal metaphysical footing, the mathematical study of them counts as equally set theoretic, etc. This is not to say that all universes in the multiverse are equally interesting, beautiful, or useful to study.

[3]More specifically, Hamkins accepts the "Forcing Extension Principle" that "For any universe V and any forcing notion P in V, there is a forcing extension $V[G]$, where $G \subseteq P$ is V-generic." (Hamkins, 2012).

This contrasts with more conventional (single universe) forms of Platonist set theory. Such conventional single universe Platonists take the intended hierarchy of sets to already contain all possible subsets of sets it contains, so it cannot be thus expanded. Accordingly, they regard forcing arguments as telling us about how *countable models* of the ZFC axioms (within the true universe of sets) could be extended—or about boolean valued models[4]—not how the total universe we are currently working in could be extended.

In addition to this claim about forcing extensions, Hamkins advocates some even bolder closure conditions on the multiverse in Hamkins (2012) like the following. Every universe looks countable from the point of view of some larger universe (Countability Principle). And every universe's copy of the ordinals looks ill founded from the point of view of some larger universe (Well-Foundedness Mirage).

Hamkins motivates his claim that all universes have forcing extensions (and thus indirectly his multiverse proposal) in three ways in Hamkins (2012). First, he appeals to the phenomenology of making forcing arguments. He reports that his experience of making such arguments suggests contact with a genuine universe of sets, properly extending whatever universe he is working in, when making these arguments.[5] Second, he argues that standard ways of reporting theorems proved by forcing arguments have changed, so as to be more susceptible to a multiversist reading.

But third (and most importantly for this chapter), Hamkins appeals to a claimed "very strong analogy" (Hamkins, 2012) between historical changes in attitudes toward geometry and the revision in understandings of set theory which he is advocating.

2.2 Hamkins' Analogy

So let's now turn to that analogy. Using a comparison with geometry to motivate and explain set-theoretic pluralism/truth value anti-realism is very common. Set-theoretic pluralists often say things like, 'there's no general right answer to set-theoretic questions like the continuum hypothesis, just like there's no general fact about whether the parallel postulate is true'.[6] However, in this section, I'll note that Hamkins fleshes this familiar comparison out

[4]Here I mean models of set theory that don't assign determinate binary facts about elementhood.
[5]By 'working in' I mean quantifying over and using 'set' and 'element' to mean set and element within.
[6]See, for example, Clarke-Doane (2020).

in a somewhat eccentric way, which is both interesting in its own right and potentially useful to his argument.

Hamkins writes that there's "a very strong analogy between the multiverse view in set theory and the most commonly held views about the nature of geometry" (Hamkins, 2012). He supports this claim by describing three stages of progress in attitudes to geometry which (he proposes) we should emulate in the case of set theory, as follows.

In the first stage, mathematicians take there to be a unique background geometrical universe, which fixes the intended interpretation of 'point' and 'line' in all contexts[7]

In the second stage, mathematicians embrace a limited geometrical pluralism. They accept that studying variants on Euclid's axioms can have mathematical interest and be part of mathematics. However, they regard these variant axiom systems as having a secondary status, merely being true on certain "toy models" (Hamkins, 2012)/unintended interpretations, like ones which interpret 'line' as meaning great circle on the surface of a sphere in Euclidean space. And they deny theorems proved in these axiom systems have *genuine geometrical content*. Instead, they regard such theorems as a legitimate part of mathematics, but merely showing something about provability and models (as work in any syntactically consistent axiom system would).

Finally, in the third stage, mathematicians accept full geometrical pluralism, which regards work in various geometrical axiom systems as equally legitimate and fully "geometrically meaningful" (Hamkins, 2012). Hamkins describes this change as arising from a process where "geometers gained experience in the alternative geometries, developing intuitions about what it is like to live in them" (Hamkins, 2012).

I think this picture of the transition from traditional to pluralist geometry is interesting in two ways.

First, according to this picture, fully adopting geometrical pluralism requires coming to see theorems in variant axiom systems as having *genuine geometrical content* (as not just providing mathematical knowledge of some kind). Accordingly, we face the question: What does it mean to regard a theorem as having genuinely geometrical (or set-theoretic) character? In a way, this question is naturally motivated by disputes between multiverse theory and traditional single universe Platonist set theory (which Hamkins compares to the disagreement between stage two and stage three attitudes to geometry).

[7] Hamkins writes, "For two thousand years, mathematicians studied geometry, proving theorems about and making constructions in what seemed to be the unique background geometrical universe." (Hamkins, 2012).

For note that the traditional single universe Platonist will heartily agree that studying variant set-theoretic systems is a legitimate part of *mathematics*, and indeed reveals truths about hierarchies of sets (extending countable models of ZFC). So their real disagreement with the multiverse theorists seems to concern whether such mathematical work has a further feature—which we might call expressing genuinely *set-theoretic* content.

Second, Hamkins' description of a path to geometrical pluralism suggests what I'll call a leveling up (as opposed to leveling down) understanding of geometrical pluralism, in the following sense.

According to what I'll call *leveling down pluralism* about geometry, we can go from traditional views to a suitable form of geometrical pluralism by (so to speak) mere subtraction. We come to accept variant axiom systems as 'fully geometrical.' But doing this is purely a matter of rejecting/debunking traditional claims that attributed previously privileged axioms systems a (geometrically relevant) special status. On this view, going pluralist doesn't require any substantive change to our stage two (pre-pluralist) understanding of work in *non-Euclidean axiom systems*.[8] It just requires renouncing (or re-classifying as irrelevant to mathematics) traditional claims about ways Euclidean geometry was uniquely correct or intended (e.g., rejecting assumptions that Euclid's axioms expressed a priori metaphysically necessary truths on the intended physical interpretations of 'point' and 'line'). We cease to regard alternative geometries as telling us about *mere* toy models, by renouncing the distinction between intended and toy models.

In contrast, Hamkins' description of a three stage path to geometrical pluralism suggests a different species of geometrical pluralism, which I'll call pluralism by leveling up. Becoming a geometrical pluralist in this sense will still involve some debunking and rejection of assumptions about traditionally favored axiom systems (at the very least, we renounce anti-pluralist claims that these systems are special). But it also requires somehow enriching or changing your attitude to traditionally disfavored geometrical axiom systems, by coming to see work in these variant systems as telling you about something different from mere provability and toy models. For note that Hamkins describes a transition to full geometrical pluralism arising from a process of "geometers gain[ing] experience in the alternative geometries, developing intuitions about what it is like to live in them" (Hamkins, 2012). And he seemingly continues

[8]Obviously, it requires the Cambridge change-like transition of ceasing to take these axioms to exist alongside other axioms with a favored status like being true on the uniquely intended interpretation.

to accept and appeal to an important distinction between mere toy models and something else ('full grown' models) after going pluralist.[9]

In the next section, I'll suggest a way of fleshing out this (admittedly somewhat cryptic) vision of leveling up geometrical pluralism. I think this view is interesting and worth exploring (even aside from any connection to Hamkins) as a counterpoint to more common leveling down pluralism about geometry. However, I don't claim leveling up pluralism is overall preferable. Prima facie, I think both styles have some appeal.

For example, leveling down pluralism has a kind of logical positivist adjacent Gordian knot cutting appeal—promising to help naturally eliminate many prima facie vexing philosophical questions. For example, if we entirely reject the distinction between intended and unintended interpretations of 'point' and 'line,' then we don't need an account of the meaning of these bits of geometrical vocabulary. Also, we can (but need not) give a simple picture of the relationship between physics and geometry, by saying the following. Scientific theories are free to appeal to mathematical facts by incorporating bridge laws. But no particular principles connecting geometrical terms to physical facts or empirical expectations have special a priori or quasi-analytic status.

On the other hand, this very simplicity of leveling down pluralism will strike some as more implausibly Procrustean than attractively Gordian knot cutting in a few ways. For example (by rejecting all a priori preferred physical interpretations of geometrical vocabulary), the simple view just sketched threatens to make the meaningfulness of claims like 'space turned out not to be Euclidean' (in contexts where no particular physical interpretation of 'point' and 'line' has been made salient) a mystery.[10] Also, it's not clear how advocates of leveling down pluralism can understand work in variant axiom systems as having genuine geometrical character.[11]

---

[9]For example, in the case of set theory (which is clearly supposed to be parallel to that of geometry) he characterizes his multiverse theory as taking variant axioms to be true of intended ('full grown') models of set theory, unlike traditional approaches which merely take these theories to have toy models.

[10]Note that the intuitive meaning of such claims about the geometrical structure of space seems to go beyond bland claims that there are *some* physical interpretations of 'point' and 'line' on which Euclid's axioms fail. Rather it seems to involve (something like) an idea that Euclid's axioms fail *on all the most natural/intended* physical interpretations of 'point' and 'line.'

[11]They could say that all consistent axioms in the language of geometry (trivially) count as having genuine geometrical character. But this doesn't fit with use of geometry to pick out a somewhat interesting and useful natural kind within mathematics.

Alternately (taking inspiration from algebra) they could say any axiom system which proves certain core sentences (any interpretation which makes these sentences come out true) in the language of geometry will have genuine geometrical character. However, they won't be able to use traditionally expected connections to physical space to characterize (or motivate) the boundaries of geometry—as Wikipedia still does by saying "Geometry is a branch of mathematics concerned with properties of

3 Geometrical Pluralism by Leveling Up

In this section, I will try to provide a concrete example of leveling up geometrical pluralism. My story takes inspiration from Hamkins' remarks (as I will indicate below), but extrapolates beyond them in ways I don't claim he would endorse.

Let's start with the question: What does it mean to see work in variant axiom systems as having genuine geometrical character/being geometrically meaningful?[12]

As we have seen, Hamkins intriguingly associates starting to accept variant geometries as "geometrically meaningful" with getting a sense of "what it is like to live [in]" them (Hamkins, 2012). Similarly, in the case of set theory (which he claims is closely analogous), he describes forcing arguments as motivating set-theoretic pluralism by giving one an experience of glancing into "new set-theoretic worlds" and studying "what it would be like to live in them" (Hamkins, 2012). Admittedly, sometimes this talk of *living in* a different geometry/set theory does just seem to be a picturesque way of speaking about *working in* (i.e., proving things in) variant systems. However, this is not always the case. For example, he claims that set theorists can "reason about a forcing extension by jumping into it and reasoning **as though** they were living in that extension" (emphasis mine). Thus, he seems to allow for a three-way contrast between

- implicitly representing a universe where $\neg$CH, by reasoning about a suitable forcing relation on the sets in your current background universe, which you take to satisfy ZFC and CH
- 'jumping in' to this forcing extension by reasoning as if you lived in a universe where CH by working with suitable axioms (e.g., proving things from ZFC $+\neg$CH)
- **actually living** in a set universe where $\neg$CH

Accordingly, a leveling up pluralist might identify coming to see variant axiom systems as fully geometrical with developing the ability to see proofs

space such as the distance, shape, size and relative position of figures." (Wikipedia Editors, 2023). For they dismiss traditional connections between geometry and physical reality (like default favored physical interpretations of 'distance' and 'shape') as irrelevant to pure mathematics. C.f. Einstein on the appeal of connecting geometry to ideas about physical measuring (rather than treating all interpretations of geometrical vocabulary on par) in Einstein (2002).

[12] One might think the obvious answer to this question is that an axiom system is geometrically meaningful iff there is some acceptable precisification of geometrical vocabulary on which all its principles come out true. But, even if correct, this leaves us with an obvious remaining question about which interpretations are acceptable and why.

from these axioms as telling us about what it would be like to 'actually live in' a world with correspondingly different geometry.

Ok so far, but what could such talk of 'actually living in' a different geometry (or imagining doing so) mean? I think considering what people mean when they say things like "Scientists originally assumed we lived in a Euclidean world, but they were wrong" (in contexts where no specific physical interpretation of 'point' and 'line' has been mentioned) suggests a possible answer to this question. Perhaps 'actually living' in a geometrically different universe means living in a universe where facts about the structure of space are different, so that different geometrical axioms could come out true *under their traditionally intended physical interpretations* (i.e., when 'point' and 'line' are given a priori/default favored physical interpretations).

In more detail, we might say that someone can positively level up their attitudes to some variant geometrical axiom systems (come to regard them as fully geometrical) via the following process.

First, do some debunking which clears the ground for geometrical pluralism as follows. Admit that there are a range of different a priori live options for what the structural constraints on physical space (studied by traditional geometry) might be. Decide, on this basis, that we should separate traditional geometry into a priori and empirical portions. Empirical study is needed to determine the kinds of facts traditional geometry hoped to discover a priori (laws constraining the physically necessary/possible configuration of physical objects, points, lines, etc.). So if we want to retain the idea that mathematics is an a priori activity, it seems natural to propose that geometry (insofar as it's a branch of mathematics) should equally include the investigation of all a priori live options for these laws of physical geometry. This opens conceptual space for work in various incompatible axiom systems to qualify as genuinely geometrical.

Second, build up a suitable positive relationship to some variant geometrical axiom systems. Begin to see the hypothesis that these variant axiom systems truly describe physical geometry (in the sense of being true under the traditionally intended interpretations of 'point,' 'line,' etc.) as (loosely) coherent and imaginable—perhaps something that creatures with our limited faculties of a priori insight can't and shouldn't rule out a priori. This change in attitude to an axiom system involves a substantive psychological shift which might (but need not) involve coming to use mental pictures in a new way.

Putting all this together yields a form of leveling up pluralism which differs from the simple (leveling down) geometrical pluralism sketched above as follows:

- Traditional expectations about the intended physical applications of geometry remain relevant to mathematics (though not in a way that picks out a unique object for geometers to study).
- Work in various axiom systems can be regarded as having "genuine geometrical character" iff it can be regarded as telling us about an (in some sense) a priori conceivable live option for the structure of physical space (specifically, telling us about what else would be true if the structure of physical space made relevant axioms true while preserving traditionally expected connections between physical reality and geometry). Thus:
  - There's (potentially) a motivated natural kind contrast between axiom systems whose theorems can vs. can't be seen as having geometrical character.[13]
  - A substantive psychological shift is involved in switching from thinking about a geometrical axiom system as merely having toy models (like the stage two pre-pluralist geometer in Hamkins' story) to regarding work in this axiom system as having genuinely geometrical character. One must acquire (something like) the ability to imagine physical space having different structures, and come to see work in variant geometrical axiom systems as informative about imaginable possibilities where physical space makes variant axioms true (under all sufficiently traditionally intended physical interpretations of terms like 'point' and 'line).

So I think this way of fleshing out Hamkins' remarks about leveling up to geometrical pluralism fits the text reasonably well. It also avoids the pair of worries for simple leveling down pluralism mentioned above (by giving a motivated account of 'genuine geometrical character' and the default meaning of claims that physical space satisfies certain geometrical axioms).

I also think this way of fleshing out geometrical pluralism via leveling up fits decently well with Hamkins' suggestion that working in different geometries can give mathematicians a sense of "what it's like to live in [different geometries]." Hamkins (2012) and thereby prompt pluralism. For it seems fairly plausible that accruing instinctive skill, speed, and familiarity with seeing consequences in a certain geometrical axiom system might significantly help enable the above gestalt switch to finding it imaginable that physical space satisfies these axioms (on natural/intended physical interpretations of terms like 'point' and 'line').

[13] However, there may be some blurriness about this boundary and a spectrum of more vs. less physically natural physical interpretations of point and line, leading to vagueness about where this boundary lies.

Now let's return to our main topic: set theory and the multiverse. In addition to being interesting in its own right, developing (and suggesting we mirror) a *leveling up* approach to geometrical pluralism promises to do argumentative work for Hamkins. In the next section, I'll review why multiverse theory initially seems better positioned than other forms of set-theoretic pluralism (like formalism) to mirror leveling up pluralism about geometry, and then question this appearance.

4 Is Hamkins' Multiverse Set-Theoretic Pluralism via Leveling Up?

Suppose you like leveling up pluralism about geometry, and want to mirror it in the case of set theory. Can we do so? Does Hamkins' multiverse theory actually provide a form of leveled up pluralism about set theory?

Hamkins' multiverse view can initially seem to provide a kind of 'leveling up' route to set-theoretic pluralism, analogous to leveling up pluralism about geometry. For acknowledging that there are universes corresponding to different set axioms sounds like granting them some kind of new positive status. And Hamkins lists taking forcing arguments to put one in contact with 'fully grown' models of set theory—rather than merely countable toy models—as an advantage of the multiverse view over traditional single universe approaches to set theory (which he compares to stage two pre-pluralist attitudes to geometry).[14]

However, Hamkins doesn't say much to explicate this claim to (distinctively) regard forcing arguments as telling us about full grown set-theoretic universes. And cashing it out turns out to be somewhat problematic. Note that Hamkins cannot simply say that 'full grown' models are supposed to be ones which satisfy the traditional iterative hierarchy conception of the width of the set-theoretic universe (by containing, at each level, sets corresponding to all possible ways of choosing from sets at lower levels). For he explicitly denies that any sets in the multiverse have this property (since he takes each universe V to exist alongside a larger universe, which adds a subset to some set already contained in V).[15]

---

[14]He writes, "The toy model perspective can ultimately be unsatisfying, however, since it is of course in each case not the toy model in which we are interested, but rather the fully grown-up universe." (Hamkins, 2013).

[15]Relatedly, we might worry about whether Hamkins can develop an attractively unified account of what's required for interpretations of theorems in various set-theoretic axiom systems to count as having

For example, yes Hamkins can say forcing arguments tell us about ensembles of abstract objects (sets) which satisfy various axioms. But so do contemporary single universe Platonists who take forcing arguments to tell us about how countable models of ZF can be extended (and whose attitude Hamkins compares to second stage pre-pluralism about geometry).[16] One might object that the models conventional universe theorists acknowledge are not fully grown, because they are seen to be countable (and thus lacking some sets) from the perspective of some larger set universe.

However, *Hamkins also maintains* that every universe is countable from the perspective of some larger universe. Indeed, work by Gitman and Hamkins (2010) shows that universe theorists should say there's a natural toy model (the collection of all countably computably saturated models of ZFC) which satisfies all the Hamkins' multiverse axioms in Hamkins (2012). So, one might say, the mainstream pre-pluralist single universe set theorist *already* accepts every positive claim about iterative hierarchies of sets Hamkins' multiverse theorist wants to make, and only differs from the latter in claiming there's a largest overall universe which contains all these hierarchies. In this way, multiverse theory seems to be pluralism via leveling down (i.e., it secures geometrical pluralism merely by debunking traditional claims to unique favored status, without also attributing some substantive positive status to alternative understandings).

To put this point more generally,[17] Hamkins maintains that every universe in the multiverse is countable (and, indeed, ill founded) from the perspective of some other universe in the multiverse. This seems to suggest that every

---

'genuine set-theoretic content.' An anonymous referee suggested Hamkins might do this by appealing to continuity with the history of prior set theory, maintaining that "what makes an axiom system of set-theoretic interest as opposed to mere formalist interest is that it is about the principles and phenomena which form the subject matter of set theory. So this would include the continuum hypothesis, forcing axioms, large cardinals, etc. This answer isn't provided by multiversism per se, but Hamkins' project is about explaining set-theoretic practice so the multiversist can freely point to ordinary work in set theory without undermining themself." However, adopting this strategy would leave questions about what is required for something to count as, e.g., giving a variant perspective on the continuum hypothesis (and therefore having genuine set-theoretic content), if not all things which syntactically look to do so count. And one might also feel that it leaves us unable to vindicate intuitions that set theory has a single unified subject matter. It also would not seem to vindicate Hamkins' claims that multiverse theory is distinctively able to see variant axioms as having full grown models.

[16]The completeness theorem ensures that all consistent first order theories have models. And according to single universe Platonism, these models are themselves sets and have abstract objects (sets) as their elements, so this means all syntactically consistent first order axioms can be interpreted as truly describing some abstract mathematical objects. Specifically single universe theorists take forcing arguments to tell us about the existence and extendability of countable models for the ZF axioms within the intended hierarchy of sets.

[17]Thanks to a referee for suggesting this way of putting things, and reference to the theorem above.

universe is a toy model, as seen by some larger universe. So how can Hamkins distinguish genuine interpretations from toy models?

Arguably, the real difference between Hamkins' pluralist multiverse theory and conventional non-pluralist understandings of set theory (which he compares to the stage two pre-pluralist attitudes to geometry) is only this. The mainstream single universe set theorist maintains that the true hierarchy of sets has certain kinds of positive status (e.g., being the fattest universe, containing sets corresponding to all possible ways of choosing subsets from sets it contains). In contrast, Hamkins' multiverse proposal denies that any model of set theory has these features.[18]

If so, then multiverse Platonism actually turns out to be a form of leveling *down* pluralism (just as much as formalism is). For, switching to multiverse theory doesn't actually involve any positive element (of coming to see variant axiom systems as telling you about fully grown set-theoretic universes rather than merely about countable models). Rather, Hamkins' multiverse proposal secures pluralism merely by *denying* that any interpretation of 'set' and 'element' had the features traditionally supposed to distinguish the intended model of set theory. So it would seem to be pluralism by debunking. Therefore we should question whether multiverse theory can get any special motivation (beyond that given to all forms of set-theoretic pluralism) from calls to mirror (leveling up) pluralism about geometry.

5 What Might Set-Theoretic Pluralism via Leveling Up Look Like?

5.1 Basic Proposal for Leveling Up Pluralism About Set Theory

If Hamkins' multiverse isn't (in any obvious way) an example of leveling up pluralism about set theory, is such pluralism possible? What might real leveling up pluralism about set theory look like?

In this (final) section I'll argue that closely mirroring the example of leveling up geometrical pluralism in Sect. 3 yields an interesting variant on Hamkins' multiverse proposal, which *would* count as leveling up pluralism about set theory and could let us attractively:

---

[18]Note that the multiverse as a whole is not supposed to be an intended interpretation for set theory, and seemingly does not let you provide one.

- explicate what's required to see a theorem as having 'genuine set-theoretic character'
- justify (a version of) Hamkins' claims to see variant axioms as telling us about "full-grown models" of set theory, in some distinctive sense

Recall that in the case of geometry, we appealed to traditionally expected physical applications of geometry to help flesh out leveling up pluralism as follows. Traditionally, geometrical facts were expected to track facts about the structure of space and the possible spatial relationships between physical objects, by way of certain a priori expected bridge laws connecting geometry to physical facts.[19] We then identified coming to see theorems as 'having genuine geometrical content' with coming to see them as telling us about a conceivable way the structure of space could be, via these traditionally expected applications.

Can we mirror this in the case of set theory? Some nice parallels between a priori expected physical applications of set theory and geometry suggest a natural way to do this. However, while in the geometrical case we appealed to an intuition that there are different (metaphysically possible) ways physical space could be, in the set-theoretic case we'll consider different (metaphysically impossible but 'weakly' a priori epistemic[20]) options for facts about 'all possible ways of choosing.'

To explain this in more detail, recall that on the traditional single universe view of set theory, there's a unique intended hierarchy of sets which satisfies the following iterative hierarchy conception. The intended hierarchy of sets has a well-ordered spine of ordinals. And at each ordinal layer there are sets available corresponding to 'all possible ways of choosing' some objects (either sets or ur-elements) available at lower layers[21]).

Accordingly, there's an a priori expected close relationship between set theory and lawlike (counterfactual-supporting) constraints on all possible ways of choosing how any properties and relations can apply to any objects. And this parallels the traditionally expected close relationship between geometry

[19] For example, someone holding this traditional view might say that geometrical claims commit us to facts about physical space because they were expected to come out true on all sufficiently intended/natural physical interpretations of 'point' and 'line. And these latter facts about spatial points and lines, in turn, imply constraints on spatial relationships between objects by way of other a priori principles.

[20] Here I specify 'weakly' a priori epistemic live options, because a perfectly rational being would presumably have direct access to all facts about logical possibility/all possible ways of choosing. In contrast, there are plausibly a range of options that cannot be ruled by the kind of logical and combinatorial principles and inference methods which we (very much non-ideal) beings are lucky enough to find a priori compelling (c.f. Berry, 2013; Williamson, 2016.

[21] C.f. Boolos, 1971; McGee, 1997.

and lawlike, counterfactual-supporting constraints on spatial possibility (how physical objects can stand in spatial relations to one another) as follows.

- Naive geometry attempts to study what we might call spatial possibility—how it's (in some sense) possible for physical objects, points, and lines to be spatially related to one another, via the idea that
  - The true principles of geometry will express truths on the intended physical interpretations of 'point' and 'line,' etc.
  - Such facts about physical points and lines imply counterfactual-supporting constraints on the structure of space and hence how physical objects can relate to each other.
- Traditional (iterative hierarchy) set theory attempts to study how it's (in some sense) possible to choose how arbitrary properties apply to some objects (be they sets or ur-elements) via the idea that
  - The true hierarchy of sets with ur-elements contains sets corresponding to 'all possible ways of choosing' (some physical objects, or some elements from a single set which this hierarchy contains).
  - And such facts about all possible ways of choosing reflect constraints on how physical properties can apply to physical objects.

Considering this parallel suggests a direct way of mirroring the leveling up pluralism about geometry advocated in Sect. 3, in the case of set theory. We can say that

- **Actually Living in** a world corresponding to some set-theoretic axioms[22] means living in a world where the intended hierarchy of sets (which contains sets corresponding to 'all possible ways of choosing' subsets from sets it contains, etc.) obeys those axioms.
- Set theory (insofar as it's a branch of pure mathematics) equally studies all (weakly) a priori epistemic possibilities for how the facts about all possible ways of choosing (and thus all intended width hierarchies of sets) could be.
- Seeing work in variant axiom systems (whether studied directly or via forcing arguments) as genuinely set-theoretic involves seeing this work as revealing facts about some weakly a priori epistemic live option for how facts about all possible ways of choosing (and hence the intended hierarchy of sets) could be.

[22] Here I mean to invoke the strong sense of 'living in' which Hamkins contrasts with merely imagining living in such a world.

An advocate of this kind of leveling up set-theoretic pluralism could endorse a version of Hamkins' claim to see forcing arguments as telling us about full grown models (in a way that the mainstream single universe set theorists cannot). For this leveling up pluralist can say that fully grown models of set theory are ones which witness facts about all possible ways of choosing in the traditionally expected way (given by the iterative hierarchy conception of sets). And we can say that forcing arguments tell us about what would be true in the metaphysically impossible scenarios where (facts about all possible ways of choosing/logical possibility were different so that) *these intended/fully grown* models of set theory had a different structure. Specifically, we can see the study of variant axiom systems via forcing arguments as telling us about metaphysically impossible scenarios where the laws of logical possibility are different (from those relevant to the actual world, or the model of set theory we work in when making the forcing argument) in the following way.

> An intrinsic duplicate of the (actual world) intended hierarchy of sets V (i.e., the thing we directly refer to and reason about when making this forcing argument) exists within a larger structure V[G]. V[G] (rather than V) is the intended model of set theory in this counterpossible scenario (so it contains sets witnessing 'all possible ways of choosing' *in the expanded sense relevant to this impossible world*).[23]

Even more specifically, we might see forcing arguments as telling us about counterpossible counterfactuals: what would be true if (per impossibile) the modal facts about logical possibility were different so that

- there are were extra logically possible ways of choosing some of the natural numbers (and sets witnessing this) than there actually are such that
- an intrinsic duplicate of the actual-world intended hierarchy of sets (which has the intended relationship to logical possibility facts in the actual world) can exist inside a larger intended hierarchy of sets (which reflects the more generous facts about 'all possible ways of choosing' in this scenario)

[23]For these purposes I'm thinking about laws in a non-Humean fashion, so that two possible (or impossible) scenarios can differ in what they make physically possible (and perhaps what they make logically possible) without differing in the actual non-model Humean mosaic of events. I'm also thinking about logical possibility/all possible ways of choosing as a fundamentally modal notion, which may or may not be elegantly and exhaustively described by any conveniently stateable list of 'laws of combinatorics.'

5.2 Prior Work on Counterpossible Counterfactuals

Appeal to conditionals with metaphysically impossible antecedents (like those above) might seem odd or problematic. But various tools for understanding such strange claims ('if the laws of logical possibility were different so as to allow extra subsets of the natural numbers as per ...forcing extension then...') have been suggested by prior work on impossible worlds and counterpossible counterfactuals (Nolan, 1997).

Traditionally, much work on counterfactuals has focused on claims with a false but metaphysically possible antecedent like 'if Nixon had pushed the button, then there would have been nuclear war.' Lewis and Stalnacker famously analyzed such claims by appealing to metaphysically possible worlds, with a counterfactual like this being true (at the actual world) if at all sufficiently close possible worlds to the actual world where the antecedent is true (Nixon pushed the button), the consequent is also true (nuclear war occurs). And different theories have been explored about how to understand the relevant closeness relation, with agreement that things like similarity to the actual world in important respects and mostly preserving general physical laws (except perhaps for minor miracles as needed to make the antecedent of the counterfactual true) tend to make for closeness.

One reason talk of metaphysically impossible worlds and counterpossibles has been unpopular is that according to Lewis' influential but infamous modal realism objects in possible worlds are just as real as objects in the actual world and have all properties attributed to them. So accepting non-trivial counterpossible counterfactuals would seem to require accepting that objects with metaphysically impossible combinations of properties (e.g., round square tables and marbles that are both red and not red) exist.

However, as (Nolan, 1997) points out, most popular ways of making sense of possible worlds and counterfactual talk differ from Lewis' infamous modal realism in this regard. They don't require us to accept objects with metaphysically possible but not actual combinations of properties like "talking donkeys, phlogiston, crystal spheres spinning around the center of their universe." Accordingly, they can often be naturally extended to allow for non-trivial counterpossible counterfactuals without commitment to any objects having metaphysically impossible combinations of properties.

For example, those who take possible worlds to be collections of propositions, or sets of sentence-like representations, can similarly identify impossible worlds with less restricted collections of propositions (which don't have to be logically coherent or metaphysically compatible). Those who take possible worlds to be sui generis abstract objects can say the same about impossible worlds. And those who do not admit the literal existence of possible worlds, but engage in talk of them all the same (fictionalists, instrumentalists, Meinongians) could do the same with impossible worlds.

Admittedly, questions arise about how to develop each approach. For example, it's not clear that taking possible worlds to be collections of sentences allows us to distinguish enough different scenarios, or make sense of talk about intrinsic duplicates of structures in one world existing within another. However, talk about intrinsic duplicates of objects at one world within another is a staple of Lewisian metaphysics. And counterfactuals like 'if Genghis Khan had had two more children than he actually did...' and 'if Napoleon had been taller than he actually was ...' are widely accepted as potentially meaningful and non-trivial. So, one might argue, this is a problem for everyone (or all friends of counterfactuals).

Similarly, taking there to be sui generis abstract objects corresponding to different metaphysically impossible scenarios involving different iterative hierarchy structures raises cardinality paradox/Burali forti worries (however many objects there are, it would seem there should be more distinct possibilities). However, one can raise the same cardinality worries about metaphysically possible worlds (if one does not think there's any metaphysically necessary limit on the number of objects that exist).

But if relevant metaphysical and technical obstacles to this project can be surmounted,[24] we might come to see forcing arguments, etc. as providing relatively concrete, precise, motivated, and rigorous way of reasoning about metaphysically impossible scenarios and counterpossible counterfactuals.

5.3 Features of This View

Thinking about forcing arguments (and other ways of switching between universes) in this way plausibly counts as set-theoretic pluralism by leveling up.

[24] Admittedly, if we see forcing arguments as telling us about counterpossible counterfactuals in the way that I've suggested, we will have to say something a bit more circuitous about how making forcing arguments can establish consistency results (like the famous result that ZFC $+\neg CH$ is consistent). For there will be (remote) impossible worlds corresponding to syntactically inconsistent theories. So merely establishing that $\neg CH$ something would be true in the closest metaphysically impossible scenario where the hierarchy of sets in the actual world is extended in certain ways does not automatically give us reason to think that a theory which combines CH and ZFC is consistent. But one might fall back on the traditional realist way of getting from forcing to syntactic consistency to ensure this (i.e., one could say that forcing arguments *also* tell one about how countable models could be extended).

One might also worry about whether the (unquestioned, unreconstructed) talk of intrinsic duplicates I've indulged in creates pressure to accept claims about isomorphic mappability which a multiverse theorist should reject. I take no position on whether this problem is solvable, here.

For it lets us attribute work in variant axiom systems as having a substantive positive status (telling us about a cogent live option for what entwined facts about logical possibility and set theory would be like) not attributed to them by common contemporary pre-pluralist set theory, which regards such theorems as merely telling us about unintended models within the intended hierarchy of sets. We can perhaps accommodate Hamkins' remarks about phenomenology, his suggestion that forcing arguments seem to consider the whole hierarchy of sets you are explicitly working in (not just some countable model of ZFC inside it) being extended.

What about the other kinds of closure principles (aside from taking forcing extensions) that Hamkins suggests? To most directly parallel Hamkins' multiverse theory[25] (while providing genuine leveling up pluralism in the way I'm suggesting), the leveling up set-theoretic pluralist could simply add that all of Hamkins' closure principles for the multiverse preserve cogency/being a weakly a priori live option in the relevant sense (e.g., for every scenario that can be so considered, there's another one from whose perspective the first scenario is countable, not well founded, etc.).[26]

Let me end by noting three things. First, although it is (ultimately) not my preferred view, I think the above proposal closely resembles Hamkins' multiverse in some interesting ways—and can be seen as the result of applying a small twist to Hamkins' original multiverse proposal which brings out the latent radicalness of the latter view.

For as we have noted, it is traditional to expect a certain close relationship between set theory and facts about all possible ways of choosing/logical possibility (and thereby constraints on how physical properties like color can apply to concrete objects like cats and spaceships). The set-theoretic leveling up pluralism I've sketched results from simply keeping this connection between math and logic intact while embracing Hamkins' pluralism about set theory and claims to see forcing arguments as telling us about 'full grown' set-theoretic universes. As a result, it sees pluralist set theory as not just exploring different set-theoretic universes, but also scenarios where facts about 'all possible ways

[25] My leveling up pluralist identifies seeing a proof as having genuine set-theoretic content with seeing it as illuminating a (weakly) a priori live hypothesis about all possible ways of choosing (and hence the intended hierarchy of sets). They could, in principle, take a range of different positions about which axioms systems theories can reasonably be so interpreted.

[26] Possibilities which differ on the *width* of the intended hierarchy of sets, like CH, will correspond to different views about modal facts about all possible ways of choosing. For facts about all possible way of choosing fix facts about the intended interpretation of second order quantification and thereby—via Zermelo's quasicategoricity theorem—facts about the structure of the intended hierarchy of sets up to height.

of choosing'/logical possibility (and thence plausibly also physical possibility) are different. The result is unsurprisingly radical and perhaps a bit psychedelic.

Just as we can make sense of (so to speak) Lovecraftian science fiction novels where protagonists' physical space has a radically different geometry, this leveling up pluralism about set theory suggests a way to make sense of Borghesian science fiction novels (Priest et al., 2016; Borges & Hurley, 1998) which depict life in (impossible) worlds with radically different set theory, logical possibility/'all possible ways of choosing,' and perhaps also physical possibilities for concrete objects. And it (i.e., the leveling up set-theoretic pluralism sketched in this section) regards the study of such strange scenarios as a legitimate part of set theory.

Second, the version of leveling up pluralism sketched in this section may fortuitously help answer a puzzle about reference and multiverse theory raised by Barton in (2016). Hamkins talks about starting by working in one set-theoretic universe and then repeatedly switching references by taking forcing extensions, considering ground models, etc. But Barton asks how we could ever succeed in referring to a unique set universe if multiverse theory is true. After all, any axioms we assert about the sets will be satisfied by many universes. I don't think it would be a disaster for Hamkins to say that talk of 'the universe you are currently working in' is not meant seriously and there isn't actually a unique universe we are currently working in. However, a fan of the leveling up pluralism sketched in this section could instead answer Barton's challenge (at least as regards the width of the hierarchy of sets) in a more positive way. They could say that we, by default, initially refer to a set universe whose width reflects facts about logical possibility/all possible ways of choosing in the actual world. Forcing arguments could then be seen as revealing counterpossible counterfactuals, about what would be true if there were more possible ways of choosing so (intrinsic duplicates of) this actual-world-intended universe could exist within a larger one.

Third, considering this kind of leveling up pluralism calls into question how much analogies with geometry (alone) can do to motivate Hamkins' multiverse proposal—specifically his claim that there's no right answer to questions like CH, aside from mathematicians' choice of which axioms to work in. For my leveling up pluralist about set theory will presumably say there *is* a mathematical axiom choice independent right answer to CH (and all other questions determined entirely by the width of the hierarchy of sets) which reflects facts about all possible ways of choosing in the actual world—just as there's a physically right answer to the parallel postulate which reflects facts about spatial possibility in the actual world (what axioms come out true under all sufficiently intended/natural physical interpretations). And the leveling up pluralist will agree with Hamkins that mathematicians can legitimately study variant axiom systems and the results they prove will have genuine set-theoretic content. But they will maintain that in at least one important sense there's a right answer to CH (reflecting actual-world facts about 'all possible ways of choosing'/logical possibility), whether or not we can ever discover it.

6 Conclusion

In this chapter, I've argued that Hamkins' (2012) special development of familiar pluralist analogy between set theory and geometry suggests a distinction between leveling up and leveling down versions of mathematical pluralism, which is interesting in its own right. On the unusual leveling up approach Hamkins seems to favor, we become pluralists partly by developing a substantive new way of interpreting work in variant axiom systems—not merely by debunking assumptions that favored certain choices of axioms.

I've suggested a way of fleshing out leveling up pluralism about geometry (inspired by Hamkins' brief remarks). I've then argued that it's not clear that Hamkins' Platonist multiverse actually delivers leveling up pluralism about set theory—calling into question his use of analogies between set theory and geometry to motivate multiverse theory (rather than mere pluralism). I've ended by giving an example of what genuine leveling up pluralism about set theory might look like (paralleling the above leveling up pluralism about geometry).

References

Barton, N. (2016). Multiversism and concepts of set: How much relativism is acceptable? In F. Boccuni & A. Sereni (Eds.), *Objectivity, realism, and proof. Filmat studies in the philosophy of mathematics* (pp. 189–209). Springer.

Berry, S. (2013). Default reasonableness and the mathoids. *Synthese, 190*(17), 3695–3713.

Berry, S. (2022). *A logical foundation for potentialist set theory.* Cambridge University Press.

Boolos, G. (1971). The iterative conception of set. *Journal of Philosophy, 68*(8), 215–231.

Borges, J. L., & Hurley, A. (1998). *Collected fictions.* Penguin Classics Deluxe Edition. Penguin Books.

Clarke-Doane, J. (2020). *Morality and mathematics.* Oxford University Press.

Einstein, A. (2002). Geometry and experience. In *The Collected Papers of Albert Einstein, Volume 7: The Berlin Years: Writings, 1918–1921* (Vol. 7, pp. 232–246). Princeton University Press.

Gitman, V., & Hamkins, J. D. (2010). A natural model of the multiverse axioms. *Notre Dame Journal of Formal Logic, 51*(4), 475–484.

Hamkins, J. (2013). *The multiverse perspective in set theory* (2013)

Hamkins, J. D. (2012). The set-theoretic multiverse. *The Review of Symbolic Logic, 5*(3), 416–449.

McGee, V. (1997). How we learn mathematical language. *Philosophical Review, 106*(1), 35–68.

Nolan, D. (1997). Impossible worlds: A modest approach. *Notre Dame Journal of Formal Logic, 38*(4), 535–572.

Priest, G., Tanaka, K., & Weber, Z. (2016). Paraconsistent logic. In E. N. Zalta (Ed.), *The stanford encyclopedia of philosophy*. Metaphysics Research Lab, Stanford University, Winter 2016 edition.

Wikipedia Editors. *Geometry—Wikipedia* (2023).

Williamson, T. (2016). Absolute provability and safe knowledge of axioms. In L. Horsten & P. Welch (Eds.), *Gödel's disjunction: The scope and limits of mathematical knowledge*. Oxford University Press.

9

Basic Mathematics Beyond Classicality

Sourav Tarafder

1 Introduction

We shall study a few models of non-classical set theories in this chapter. By a non-classical set theory we mean a set theory having a non-classical underlying logic, unlike ZFC which uses classical logic. We will look at set theories that incorporate both classical and non-classical underlying logics, in conjunction with the relevant set-theoretic axiom system. This system can either be a fragment of ZFC, an extension of ZFC, or an axiom system that is incomparable to ZFC, as long as it generates a coherent and mathematically viable framework. In this chapter, we shall try to explain what we mean by a 'viable framework' in this context.

The present work is based on a few developments made in the last few years on algebra-valued models of non-classical set theories. In contrast to other lines of research in non-classical mathematics that focus on including proper classes such as the Russell class, the main motivation of these chapters is to address the following questions:

(i) How much classicality is needed for 'basic mathematics'?

S. Tarafder (✉)
St. Xavier's College, Kolkata, West Bengal, India

C. Antos et al. (eds.), *The Palgrave Companion to the Philosophy of Set Theory*, Palgrave Companions, https://doi.org/10.1007/978-3-031-62387-5_9

(ii) The *General Comprehension Principle* says that for any property **P**, there exists a set consisting of all the objects satisfying **P** and no other objects. Even for some weak non-classical backgrounds all the ZF axioms can be derived from the General Comprehension Principle and the Axiom of Extensionality Weber [2010], Libert [2005]. Do there exist non-classical models (other than the models of Intuitionistic Zermelo-Fraenkel Set Theory, IZF) which validate all the axioms of ZFC, but not the principle of General Comprehension?
(iii) What are the independence results in non-classical set theories?
(iv) How do the different interpretations of $=$ and $\in$ produce changes in the validity of the set-theoretic sentences?

It is a debatable issue how much of abstract mathematics is needed in the development of science and technology. There are examples of different branches of mathematics which had no practical use at the time of invention, but eventually they were used to develop modern technologies. Number theory (in its full generality) can be considered as one such important example. In Hardy [1916, p. 350], the famous mathematician G. H. Hardy wrote

> The Theory of Numbers has always been regarded as one of the most obviously useless branches of Pure Mathematics. The accusation is one against which there is no valid defense; and it is never more just than when directed against the parts of the theory which are more particularly concerned with primes. A science is said to be useful if its development tends to accentuate the existing inequalities in the distribution of wealth, or more directly promotes the destruction of human life. The theory of prime numbers satisfies no such criteria. Those who pursue it will, if they are wise, make no attempt to justify their interest in a subject so trivial and so remote, and will console themselves with the thought that the greatest mathematicians of all ages have found in it a mysterious attraction impossible to resist.

But the scenario changed after a few years. Nowadays, number theory is widely used in the fields Cryptography, Computer Science, and more. The inexorable use of mathematics in everyday life revolves around what is referred to as 'basic mathematics' in (i). This chapter addresses a fundamental question: what should the foundational logic for this basic mathematics be? Conventionally, classical logic serves as the assumed foundational logic. However, during our educational journey in mathematics, we remain unaware of an underlying system that operates behind the scenes, revealed later as a formal logical axiom system—widely acknowledged as classical logic. During that stage, the reason for accepting this fact almost unquestioningly lies in the seamless alignment of all the mathematical concepts studied at the school level with classical logic. This leads us to a crucial inquiry: which aspect should we hold as more fundamental—basic mathematics or classical logic? To be more precise, should we adopt classical logic as the foundational logic and then deduce this basic mathematics from it within a suitable theory, utilizing them in our daily lives when needed? Alternatively, should we endeavor to comprehend the fundamental mathematics required for survival and flourishing first and then develop a logical system to support it? While we acknowledge that the application of a logical system in mathematics alone does not suffice to justify its validity, it undeniably plays a significant role in the validation process.

This chapter shows that there exist non-classical set theories which coincide with the classical set theory when basic mathematics is concerned, but (may) differ in more abstract matters. This leads to the question: are we justified in considering classical mathematics as the only foundation of science and technology (putting aside some branches like quantum mechanics)?

In Sect. 2, we study generalized-algebra-valued models of set theories based on the chapter Löwe and Tarafder [2015], which shows a method of getting models of certain non-classical set theories. The main goal of this chapter is discussed in Sect. 3, where we define 'basic mathematics' and prove that there exist non-classical set theories satisfying it. In Sect. 4, a few more set-theoretic properties of those set theories are studied. Finally we give some concluding remarks in Sect. 5.

2 Non-classical Set Theories and Their Models

In this chapter, we shall study algebra-valued models of set theories in a manner similar to Boolean-valued models of classical set theory.

2.1 Generalized Algebra-Valued Models

The construction of an algebra-valued model starts with a foundational model $\mathbf{V}$ of Zermelo-Fraenkel set theory within which an algebra $\mathbb{A}$ is introduced as a semantics for a specific propositional logic system. For instance, one can begin with a Boolean algebra or a Heyting algebra, which, respectively, corresponds to Classical Propositional Logic or Intuitionistic Propositional Logic. By utilizing the set-theoretic model $\mathbf{V}$ and the algebra $\mathbb{A}$, a subclass $\mathbf{V}^{(\mathbb{A})}$ of $\mathbf{V}$ is built that satisfies certain targeted set-theoretic axioms. Consequently, $\mathbf{V}^{(\mathbb{A})}$ is a model of a particular set theory. For instance, if $\mathbb{A}$ is regarded as a complete

Boolean algebra (or complete Heyting algebra), $\mathbf{V}^{(\mathbb{A})}$ becomes a model of ZF (or IZF). Similarly, if an algebra A is employed, which encompasses a non-classical logic, the aforementioned construction leads to a non-classical set theory. There is no compelling basis to assume that the propositional logics of algebra $\mathbb{A}$ and set theory $\mathbf{V}^{(\mathbb{A})}$ will necessarily coincide [Löwe et al., 2018]. However, this distinction gives rise to a separate area of investigation that falls outside the primary objective of this chapter.

Following Löwe and Tarafder [2015], we present a comprehensive technical overview of generalized algebra-valued models of set theory. Let $\mathbf{V}$ be a model of ZF and ORD be the class of all ordinal numbers in $\mathbf{V}$. Let us now assume that $\mathbb{A} = \langle \mathbf{A}, \wedge, \vee, \Rightarrow, ^*, \mathbf{1}, \mathbf{0} \rangle$ is a complete distributive lattice, augmented with one binary operator $\Rightarrow$ and one unary operator *. Then a universe of *names* is defined by transfinite recursion as follows. For any $\alpha \in$ ORD,

$$\mathbf{V}_{\alpha}^{(\mathbb{A})} = \{u : u \text{ is a function where } \operatorname{ran}(u) \subseteq \mathbf{A}$$

$$\text{and there is } \xi < \alpha \text{ with } \operatorname{dom}(u) \subseteq \mathbf{V}_{\xi}^{(\mathbb{A})}\}, \text{ and}$$

$$\mathbf{V}^{(\mathbb{A})} = \{u : \exists \alpha (u \in \mathbf{V}_{\alpha}^{(\mathbb{A})})\}.$$

After obtaining $\mathbf{V}^{(\mathbb{A})}$, extend the language $\mathcal{L}$ of ZF by adding a constant symbol for every element in $\mathbf{V}^{(\mathbb{A})}$, where $\mathcal{L}$ contains the binary predicate symbols $\in, =$, connectives $\wedge, \vee, \rightarrow, \neg, \bot$, and quantifiers $\forall, \exists$. The extended language is denoted by $\mathcal{L}_{\mathbb{A}}$.

Definition 9.1 Define an assignment function $[\![\cdot]\!]$ from the class of all closed well-formed formulas (or sentences) in $\mathcal{L}_{\mathbb{A}}$ to the set $\mathbf{A}$ of *truth values* as follows. If $u, v \in \mathbf{V}^{(\mathbb{A})}$ and φ, ψ are any two closed well-formed formulas in $\mathcal{L}_{\mathbb{A}}$, then

$$[\![\bot]\!] = \mathbf{0},$$

$$[\![u \in v]\!] = \bigvee_{x \in \operatorname{dom}(v)} (v(x) \wedge [\![x = u]\!]),$$

$$[\![u = v]\!] = \bigwedge_{x \in \operatorname{dom}(u)} (u(x) \Rightarrow [\![x \in v]\!]) \wedge \bigwedge_{y \in \operatorname{dom}(v)} (v(y) \Rightarrow [\![y \in u]\!]),$$

$$[\![\varphi \wedge \psi]\!] = [\![\varphi]\!] \wedge [\![\psi]\!],$$

$$[\![\varphi \vee \psi]\!] = [\![\varphi]\!] \vee [\![\psi]\!],$$

$$
\begin{aligned}
[\![\varphi \to \psi]\!] &= [\![\varphi]\!] \Rightarrow [\![\psi]\!],\\
[\![\neg\varphi]\!] &= [\![\varphi]\!]^*,\\
[\![\forall x \varphi(x)]\!] &= \bigwedge_{u \in \mathbf{V}^{(\mathbb{A})}} [\![\varphi(u)]\!], \text{ and}\\
[\![\exists x \varphi(x)]\!] &= \bigvee_{u \in \mathbf{V}^{(\mathbb{A})}} [\![\varphi(u)]\!].
\end{aligned}
$$

Observe that sentences containing 'meet', 'join', 'arrow', and 'negation' are assigned truth values based on conjunction, disjunction, implication, and complementation in the algebra **A**. For sentences with universal and existential quantifications, their truth values correspond to the infimum and supremum values of the set of all truth values of the sentences under the scope of the quantifiers.

Definition 9.2 A set $D \subseteq \mathbf{A}$ is called a *designated set* of $\mathbb{A}$ if it is a *filter* in $\mathbb{A}$, that is,

(i) $\mathbf{1} \in D$,
(ii) $\mathbf{0} \notin D$,
(iii) if $a \in D$ and $a \leq b$ then $b \in D$, and
(iv) for $a, b \in D$, we have $a \wedge b \in D$.

It is called an ultra-designated set, if it is an ultrafilter, that is, D is a filter and there does not exist any other filter F of $\mathbb{A}$ such that $D \subsetneq F$.

A designated set D is a subset of **A**. Informally, the elements of **A** are considered to be the truth values of the sentences of the language. In particular, if a sentence's truth value belongs to D, the sentence is deemed valid.

A sentence φ of $\mathcal{L}_{\mathbb{A}}$ is said to be D-valid or simply valid in $\mathbf{V}^{(\mathbb{A})}$, denoted by $\mathbf{V}^{(\mathbb{A})} \models_D \varphi$ or simply $\mathbf{V}^{(\mathbb{A})} \models \varphi$ (when the corresponding designated set is clear from the context), if $[\![\varphi]\!] \in D$.

It is widely recognized that when $\mathbb{A}$ constitutes a complete Boolean algebra, $\mathbf{V}^{(\mathbb{A})}$ assumes the role of an algebra-valued model of ZF, that is, $\mathbf{V}^{(\mathbb{A})} \models$ ZF. These models, known as Boolean-valued models, were specifically devised by Dana Scott, Robert M. Solovay, and Petr Vopěnka in the 1960s to facilitate the comprehension and elucidation of Paul Cohen's forcing method (cf. Bell, 2005). Similarly, for a complete Heyting algebra $\mathbb{A}$, $\mathbf{V}^{(\mathbb{A})}$ is an algebra-valued model of IZF, that is, $\mathbf{V}^{(\mathbb{A})} \models$ IZF [Grayson, 1979]. In both these cases, the designated set is $\{\mathbf{1}\}$.

Let us now consider the *bounded quantification property*:

$$[\![\forall x(x \in u \to \varphi(x))]\!] = \bigwedge_{x \in \mathrm{dom}(u)} (u(x) \Rightarrow [\![\varphi(x)]\!]). \qquad (\mathrm{BQ}_\varphi)$$

Notice the difference between the value of a universal quantified formula under $[\![.]\!]$ and BQ_φ. The infimum in BQ_φ is taken only over the domain of the element u. So the natural question is whether BQ_φ holds in all algebra-valued models. For all formulas φ in $\mathcal{L}$, BQ_φ holds in all the Boolean-valued models and Heyting-valued models. However, it was proved in Löwe and Tarafder [2015] that there are algebras $\mathbb{A}$ such that BQ_φ does not hold in general in the algebra-valued models $\mathbf{V}^{(\mathbb{A})}$.

2.2 Negation-Free Fragment

By NFF, the negation-free fragment of $\mathcal{L}$, we mean the closure of the atomic formulas of $\mathcal{L}$ under $\wedge$, $\vee$, $\to$, $\bot$, $\exists$, and $\forall$. The elements of NFF are called negation-free formulas. By NFF-ZF and NFF-ZF$^-$ we mean the negation-free fragment of ZF and the negation-free fragment of ZF excluding the Axiom of Foundation, respectively. The NFF-bounded quantification property (NFF-BQ_φ) stands for the bounded quantification property BQ_φ for all negation-free formulas φ. Classically and intuitionistically, every formula is equivalent to a negation-free formula, as $\neg\varphi$ is equivalent to $\varphi \to \bot$. Hence, in these cases, we can conclude that NFF-ZF and NFF-IZF are equivalent to ZF and IZF, respectively.

2.3 Reasonable Implication Algebra

Löwe and Tarafder [2015, p. 194] An algebra $\mathbb{A} := \langle \mathbf{A}, \wedge, \vee, \Rightarrow, \mathbf{1}, \mathbf{0} \rangle$ is called a *deductive reasonable implication algebra* if $\langle \mathbf{A}, \wedge, \vee, \mathbf{1}, \mathbf{0} \rangle$ is a complete distributive lattice and the following properties hold: for any $x, y, z \in \mathbf{A}$

P1 $(x \wedge y) \leq z$ implies $x \leq (y \Rightarrow z)$,
P2 $y \leq z$ implies $(x \Rightarrow y) \leq (x \Rightarrow z)$,
P3 $y \leq z$ implies $(z \Rightarrow x) \leq (y \Rightarrow x)$, and
P4 $((x \wedge y) \Rightarrow z) = (x \Rightarrow (y \Rightarrow z))$.

The following theorem gave rise to generalized algebra-valued models of set theories.

Theorem 9.3 (Löwe and Tarafder, 2015, Theorems 3.3 & 3.4) *If $\mathbb{A}$ is a deductive reasonable implication algebra such that NFF-BQ$_\varphi$ holds in $\mathbf{V}^{(\mathbb{A})}$, then for any choice of the designated set D, we have $\mathbf{V}^{(\mathbb{A})} \models_D$ NFF-ZF$^-$.*

The search of finding 'reasonable mathematics' in the models of non-classical set theories started from Theorem 9.3. If there are algebras $\mathbb{A}$ that do not belong to the Boolean algebra class, yet fulfill the conditions stated in Theorem 9.3, then the construct $\mathbf{V}^{(\mathbb{A})}$ constitutes a non-classical model of NFF-ZF$^-$. An example of such algebras is the complete Heyting algebra, which has been extensively studied in Heyting-valued models since 1978 by R. J. Grayson and subsequent researchers. However, we have identified a substantial number of algebras that neither belong to the Boolean algebra class nor the Heyting algebra class, yet satisfy these conditions. The detailed investigations of these cases can be found in Löwe and Tarafder [2015], Tarafder and Chakraborty [2015], Jockwich Martinez and Venturi [2021], Tarafder [2022].

3 Mathematics in the Models of Non-classical Set Theories

Certainly, mathematics is an art in its own right. In the book 'A Mathematician's Apology', Hardy pointed out that a pure mathematician generally does not create a new theory to apply it to technology. The beauty of mathematics lies in itself, independent of its application. Moreover, he added that the most beautiful mathematics are those which do not have any practical application. He rather compared a mathematician with a painter or poet [Hardy, 1940]:

> A mathematician, like a painter or poet, is a maker of patterns. If his patterns are more permanent than theirs, it is because they are made with ideas.

Mihir Kr. Chakraborty, a logician and pure mathematician, made significant strides in the exploration of (Western) logic in India. Beyond his extensive research in mainstream mathematics and logic, he authored numerous articles and books on this fascinating subject. Among his literary contributions is the enlightening work titled 'Amimangsar Alo-Andhari' (light and shade of the undecidable), thoughtfully crafted to cater to a broader audience in the Bengali language. The following quote, drawn from his remarkable book [Chakraborty, 2010], holds profound wisdom and insight.

> Usability is not an essential attribute of mathematics. Rather, it is like a side-effect
> ...
> The corner of beauty is its ultimate home. In this sense, mathematics is much closer to art or poetry. If there is any truth in beauty, then this truth is acquired by mathematics.

For the layman, abstract mathematical problems may appear inconsequential, regardless of their significance within the field of mathematics. Let us examine a couple of examples to illustrate this point. Firstly, let us consider Fermat's Last Theorem, proposed by the French mathematician Pierre de Fermat in 1637, scribbled in the margin of a copy of Arithmetica. He claimed to possess a proof, but it was too extensive to fit within the margin. For 358 years, this problem remained one of the captivating unsolved challenges in mathematics. However, in 1994, the English mathematician Andrew Wiles successfully provided a proof, an achievement that earned him the esteemed Abel Prize in 2016.

Next, we explore the Continuum Hypothesis (CH), which was the foremost among 23 unsolved open problems presented by the German mathematician David Hilbert at the International Congress of Mathematicians in Paris in 1900. In 1940, Kurt Gödel demonstrated that the negation of CH cannot be proven within ZFC, and later in 1963, the American mathematician Paul Cohen showed that CH itself cannot be proven within ZFC, thereby establishing its independence from ZFC. For this significant contribution to mathematics, Cohen was honored with a Fields Medal in 1966.

The mathematical community acknowledges the immense value of abstract mathematical results for the advancement of the discipline. These results play a crucial role in pushing the boundaries of mathematical knowledge and understanding. However, it is important to recognize that outside of academia, such abstract concepts may not hold the same practical significance for the general public: resolving Fermat's Last Theorem or determining the independence of the Continuum Hypothesis from the Zermelo-Fraenkel set theory might not have any direct impact on their daily lives or professional activities. Furthermore, some may argue whether the abstract concept of 'infinity' can truly be related to any observable phenomenon in the real world. However, it is widely acknowledged that the application of infinite sets is inevitable in very many branches of pure mathematics.

Despite the disconnect between abstract mathematical concepts and their relevance in everyday life, we cannot deny the pervasive role of basic mathematics in our daily routines. Basic mathematics, which encompasses fundamental arithmetic operations, is an integral part of our day-to-day activities.

In this chapter, by 'basic mathematics' we mean the mathematics needed in our daily life, comprising the basic arithmetic operations. But on which domain those basic arithmetic operations will be defined? This is another debatable issue. To explain it, let the domain of all those binary operators be $S \times S$. In this current context, what should the set S be? The set S should definitely be a superset of the set of all natural numbers $\mathbb{N}$. But, should it be the set of all real numbers, $\mathbb{R}$? Are really irrational numbers needed for making calculations of amounts, sizes, or other measurements in our daily life? Of course fractions are needed, but for how many decimal places? That means, we may not even need all the rational numbers, from a layman's perspective! So we are keeping the possibilities of the set S open for the future works. However, in this chapter, we are fixing S as the minimum set required for our daily life calculations as the set of all natural numbers. Hence, we define the following.

Definition 9.4 (Basic Mathematics) The mathematics consisting of basic arithmetic operations *addition, multiplication*, and *exponentiation* over the set of all natural numbers is called the basic mathematics.

So we are assuming that the way basic arithmetic operations act in classical mathematics matches with the natural phenomenon. In addition, as a fact, classical mathematics contributes in the development of science and technology through some of its branches like geometry, calculus, applications of differential equations, partial differential equations, and more.

We started the study of non-classical set theories to check whether there exist models other than classical set theory, which matches with the basic mathematics. We have succeeded to find these kind of models. In this chapter, we shall discuss these developments. Our future goal is to check whether it matches with the branches already used in the development of science and technology, the work is in progress.

Following Tarafder [2022] we shall give a class of well-ordered algebras and study their algebra-valued models, which will fulfill our requirements.

Definition 9.5 Let $\mathcal{PS}$ be the collection of all algebras $\mathbb{A} = \langle \mathbf{A}, \wedge, \vee, \Rightarrow, ^*, \mathbf{1}, \mathbf{0} \rangle$ satisfying the following conditions:

(i) $\langle \mathbf{A}, \wedge, \vee, \mathbf{1}, \mathbf{0} \rangle$ is a bounded lattice having top and bottom elements $\mathbf{1}$ and $\mathbf{0}$, respectively,
(ii) the lattice order $\leq$ is a well-ordered relation on $\mathbf{A}$, and

(iii) the operator $\Rightarrow$ is defined by

$$a \Rightarrow b = \begin{cases} \mathbf{0}, & \text{if } a \neq \mathbf{0} \text{ and } b = \mathbf{0}; \\ \mathbf{1}, & \text{otherwise,} \end{cases}$$

(iv) * is a unary operator on $\mathbf{A}$ which satisfies $\mathbf{1}^* = \mathbf{0}$ and $\mathbf{0}^* = \mathbf{1}$.

Any algebra $\mathbb{A} \in \mathcal{PS}$ will be called a $\mathcal{PS}$-algebra.

Then we have the following observations in Tarafder [2022].

Observation 9.6 There are $\mathcal{PS}$ algebras $\mathbb{A}$ and designated set D such that

(i) the logic of $(\mathbb{A}, D)$ is non-classical and
(ii) $\mathbf{V}^{(\mathbb{A})} \models_D$ NFF-ZF.

So, $\mathbf{V}^{(\mathbb{A})}$ becomes an algebra-valued model of a non-classical set theory.

There is no reason to think that only $\mathcal{PS}$ algebras are those which satisfy Observation 9.6, to explore more such algebras one may go through the papers Tarafder and Venturi [2022] and Tarafder and Venturi [2021]. In this chapter, we are studying only $\mathcal{PS}$ algebras to fulfill our goal. In $\mathcal{PS}$-algebra-valued models, we have ordinal numbers, whose first transfinite initial segment is the set of natural numbers, like classical set theory.

Definition 9.7 (Tarafder, 2022) Let $\mathbb{A}$ be a $\mathcal{PS}$-algebra. An element $x \in \mathbf{V}^{(\mathbb{A})}$ is said to be

(i) 0-like if $x(y) = \mathbf{0}$ for every $y \in \text{dom}(x)$,
(ii) α-like for some $\alpha \in$ ORD if for each $\beta \in \alpha$ there exists a β-like $y \in \text{dom}(x)$ such that $x(y) \in D$, and for any $z \in \text{dom}(x)$ if it is not β-like for any $\beta \in \alpha$ then $x(z) = \mathbf{0}$, and
(iii) ordinal-like if it is α-like for some $\alpha \in$ ORD.

Intuitively, the definition of ordinal-like elements can be explained similar to the notion of ordinal numbers in a model of classical set theory. The $\mathbf{0}$-like elements are those elements of $\mathbf{V}^{(\mathbb{A})}$ whose image sets are $\{\mathbf{0}\}$. An element $u \in \mathbf{V}^{(\mathbb{A})}$ will be called an α-like element, for some $\alpha \in$ ORD, if for any $\beta \in \alpha$ there exists a β-like element in the domain of u having image in the designated set.

From the definition, we have natural-number-like elements for any natural numbers $n \in$ ORD. Then we have the following.

Theorem 9.8 (Tarafder, 2022, Theorem 20) *For any $\mathcal{PS}$-algebra $\mathbb{A}$ and any designated set D, $\mathbf{V}^{(\mathbb{A})} \models_D$ Mathematical Induction.*

In Sections 5 and 6 of Tarafder [2022], we defined the cardinal numbers and cardinal arithmetic in $\mathcal{PS}$-algebra-valued models, which are as follows.

Theorem 9.9 (Tarafder, 2022, Theorem 48) *For any $\mathcal{PS}$-algebra $\mathbb{A}$ and any cardinal numbers $u, v, w \in \mathbf{V}^{(\mathbb{A})}$, the following are valid in $\mathbf{V}^{(\mathbb{A})}$.*

(i) Any 0-like element is an additive identity.
(ii) $(u + v) + w = u + (v + w)$.
(iii) $u + v = v + w$.
(iv) If $u \leq v$ *then* $u + w \leq v + w$.

For the next two theorems, we define for every $x \in \mathbf{V}$ an element $\check{x}$ of $\mathbf{V}^{(\mathbb{A})}$ recursively as: $\check{\varnothing} = \varnothing$ and $\check{x} = \{\langle \check{y}, \mathbf{1} \rangle : y \in x\}$.

Theorem 9.10 (Tarafder, 2022, Theorem 51) *For any cardinal numbers $u, v, w \in \mathbf{V}^{(\mathbb{A})}$, where $\mathbb{A}$ is a $\mathcal{PS}$-algebra, the following are valid in $\mathbf{V}^{(\mathbb{A})}$.*

(i) Any 1-like element is a multiplicative identity.
(ii) $u \,.\, \mathsf{zero} = \check{0}$, *where* zero *is any 0-like element.*
(iii) $(u \,.\, v) \,.\, w = u \,.\, (v \,.\, w)$.
(iv) $u \,.\, v = v \,.\, u$.
(v) If $u \leq v$ *then* $u \,.\, w \leq v \,.\, w$.
(vi) $u \,.\, (v + w) = u \,.\, v + u \,.\, w$.

Theorem 9.11 (Tarafder, 2022, Theorem 57) *If u, v, w are three cardinal numbers in $\mathbf{V}^{(\mathbb{A})}$, where $\mathbb{A}$ is a $\mathcal{PS}$-algebra, then the following are valid in $\mathbf{V}^{(\mathbb{A})}$.*

(i) $u^{\mathsf{zero}} = \check{1}$; $\mathsf{one}^u = \check{1}$; $\mathsf{zero}^u = \check{0}$ *if* $\check{0} < |u|_{\mathbf{V}^{(\mathbb{A})}}$, *where* zero *and* one *are any 0-like element and 1-like element, respectively.*
(ii) $(u \,.\, v)^w = u^w \,.\, v^w$.
(iii) $u^{v+w} = u^v \,.\, u^w$.
(iv) $(u^v)^w = u^{v \,.\, w}$.
(v) If $u \leq v$ *then* $u^w \leq v^w$.
(vi) If $\check{0} < u \leq v$ *then* $w^u \leq w^v$.

Notice that Theorems 9.9, 9.10, and 9.11 are, respectively, nothing but the generalizations of the usual addition, multiplication, and exponentiation on natural numbers. So, if we restrict the cardinal arithmetic consisting of the operators addition, multiplication, and exponentiation into the collection of natural-number-like elements then the so-called basic mathematics is restored over any $\mathcal{PS}$-algebra-valued models. Hence, we have the following theorem.

Theorem 9.12 *There exist non-classical set theories which satisfy the basic mathematics.*

Proof The proof follows directly from the Observation 9.6 and Theorems 9.9, 9.10, 9.11, restricted over the collection of all natural-number-like elements. □

The $\mathcal{PS}$-algebra-valued models not only satisfy our primary goal, they provide a few more interesting mathematical results as well.

Theorem 9.13 (Tarafder, 2022, Theorem 22 & 26) *Let* $\mathbf{V}$ *be a model of ZF. For any* $\mathcal{PS}$*-algebra* $\mathbb{A}$ *and an ultra-designated set* D *of it,*

(i) if $\mathbf{V} \models$ *Axiom of Choice, then* $\mathbf{V}^{(\mathbb{A})} \models_D$ *Axiom of Choice and*
(ii) if $\mathbf{V} \models$ Well-Ordering Theorem, *then* $\mathbf{V}^{(\mathbb{A})} \models_D$ Well-Ordering Theorem.

To compare the $\mathcal{PS}$-algebra-valued models of non-classical set theories with the other non-classical set theories, we should mention here that in the set theories studied in Weber [2010] and Libert [2005], the General Comprehension Principle was taken as an axiom of those theories. As a result, the Russell's set, the collection of all sets, the collection of all ordinals, etc. become sets in those set theories. But we proved that these collections are not sets in $\mathcal{PS}$-algebra-valued models in Tarafder [2015]. Hence, we can conclude that the General Comprehension Principle is not valid in $\mathcal{PS}$-algebra-valued models.

3.1 The Quine-Putnam Indispensability Argument

At this juncture, it becomes pertinent to explore a bit the Quine-Putnam indispensability argument for mathematical realism within the framework of this chapter's perspective. Quine and Putnam, distinguished proponents of this argument, posit that the indispensability of mathematics in the realm of empirical science furnishes compelling grounds to uphold the existence of mathematical entities. Central to their line of reasoning is the notion that invoking or quantifying over mathematical entities—ranging from sets and numbers to functions—is indispensable for the coherence of our most robust scientific theories. As a result, a compelling case emerges for embracing the existence of these mathematical entities.

In the context of this chapter's discourse, it is contended that the tenets of the Quine-Putnam indispensability argument find alignment with $\mathcal{PS}$-algebra-valued models, particularly when basic mathematics is concerned. At this foundational level, the compatibility remains intact. However, it is imperative to acknowledge that as we ascend to loftier levels of abstraction, the argument assumes a more contentious stance. Within the realm of higher abstraction, intriguingly, $\mathcal{PS}$-algebra-valued models definitely exhibit certain mathematical properties that diverge from their classical counterparts, thereby questioning the validity of the argument in general.

Therefore, while at the fundamental stratum $\mathcal{PS}$-algebra-valued models harmonize with the Quine-Putnam indispensability argument, the scenario becomes progressively intricate as we ascend the abstract hierarchy. The existence of conflicting mathematical properties within PS-algebra-valued models in comparison to the classical framework warrants a nuanced evaluation of the argument's applicability within differing mathematical contexts.

4 Independence and the Interpretations of $\in$ & $=$

4.1 Independence Results in Non-classical Set Theories

A few independence results in non-classical set theories were studied in Tarafder and Venturi [2021]. Those results are not only confined over the class of $\mathcal{PS}$-algebra-valued models, but a much larger class. The logics of the corresponding set theories of those algebra-valued models were also investigated.

To discuss a particular case, if the unary operator * (the algebraic interpretation of $\neg$ in the language of ZFC) of a $\mathcal{PS}$-algebra $\mathbb{A}$ is as follows:

$$\mathbf{1}^* = \mathbf{0},\ \ \mathbf{0}^* = \mathbf{1} \text{ and } a^* = a, \ \text{ for all a } \neq \mathbf{1}, \mathbf{0}$$

then the logic of $(\mathbb{A}, D)$ becomes a paraconsistent logic,[1] where D is the ultra-designated set. Hence, the corresponding algebra-valued model gives rise to a

[1]A logic is called *paraconsistent* if there are formulas ϕ and ψ such that $\{\phi, \neg\phi\} \not\vdash \psi$.

paraconsistent set theory. Not only that it was proved that these paraconsistent set theories are proper fragments of the classical set theory.

We know that the independence of a statement in a theory inherited by any of its subtheories. This is in general proved by the two models of the main theory which disagree the validity of the statement, as those two models are the models of the sub-theory also. But we investigated a bit more in non-classical set theories. In Tarafder and Venturi [2021, Theorem 5.35], a new technique was found which shows that, if φ is an independent statement of classical set theory and $\mathbf{T}$ is a proper sub-theory of the classical set theory, then we have two algebra-valued models $\mathbf{M_1}$ and $\mathbf{M_2}$ of $\mathbf{T}$, but not of the classical set theory, such that $\mathbf{M_1} \models \varphi$ but $\mathbf{M_2} \not\models \varphi$. Besides many other important independent results in non-classical set theories, this was one of the main results of the chapter [Tarafder & Venturi, 2021].

We have another independent result in the set theories corresponding to the $\mathcal{PS}$-algebra-valued models, which is given as follows.

Theorem 9.14 (Tarafder, 2022, Theorem 42) *For any $\mathcal{PS}$-algebra $\mathbb{A}$ and a model $\mathbf{V}$ of ZFC, $\mathbf{V} \models$ GCH if and only if $\mathbf{V}^{(\mathbb{A})} \models$ GCH, where GCH stands for the Generalized Continuum Hypothesis.*

4.2 Interpretations of ∈ and =

This subsection is dependent on the study made in Jockwich Martinez et al. [2022]. It was proved that the *Leibniz's law of indiscernibility of identicals*:

$$\forall x \forall y\big((x = y \wedge \varphi(x)) \rightarrow \varphi(y)\big) \qquad (\mathrm{LL}_\varphi)$$

is not in general valid in the $\mathcal{PS}$-algebra-valued models, that is, there are sentences φ (which are not NFF) and $\mathcal{PS}$-algebras $\mathbb{A}$ such that $\mathbf{V}^{(\mathbb{A})} \models_D \varphi$, for some designated set D. It is proved in Jockwich Martinez et al. [2022] that one of the reasons that LL_φ fails in the $\mathcal{PS}$-algebra-valued models is the algebraic interpretations of $\in$ and $=$ through the assignment function $[\![\cdot]\!]$ as defined in Definition 9.1.

Let us assign a new assignment function $[\![\cdot]\!]_{\mathrm{PA}}$ from the collection of all sentences of $\mathcal{L}_\mathbb{A}$ into $\mathbb{A}$ as same as Definition 9.1, but replace the expressions $[\![u \in v]\!]$ and $[\![u = v]\!]$ by the following:

$$[\![u \in v]\!]_{\mathrm{PA}} = \bigvee_{x \in \mathrm{dom}(v)} \big(v(x) \wedge [\![x = u]\!]_{\mathrm{PA}}\big),$$

$$[\![u = v]\!]_{\mathrm{PA}} = \bigwedge_{x \in \mathrm{dom}(u)} \big((u(x) \Rightarrow [\![x \in v]\!]_{\mathrm{PA}}) \wedge ([\![x \in v]\!]^*_{\mathrm{PA}} \Rightarrow u(x)^*)\big)$$
$$\wedge \bigwedge_{y \in \mathrm{dom}(v)} \big((v(y) \Rightarrow [\![y \in u]\!]_{\mathrm{PA}}) \wedge ([\![y \in u]\!]^*_{\mathrm{PA}} \Rightarrow v(y)^*)\big).$$

Then we have proved a generalized version of the following theorem in Jockwich Martinez et al. [2022].

Theorem 9.15 *For any $\mathcal{PS}$-algebra $\mathbb{A}$ and the ultra-designated set D, with respect to the $[\![\cdot]\!]_{PA}$ assignment function,*

(i) $\mathbf{V}^{(\mathbb{A})} \models_D LL_\varphi$, *for all sentences φ in $\mathcal{L}_\mathbb{A}$ and*
(ii) $\mathbf{V}^{(\mathbb{A})} \models_D \overline{ZF}$,

where $\overline{ZF}$ consists of all the axioms of ZF as it is, other than the Axiom of Extensionality, which is replaced by

$$\forall x \forall y \forall z\big(((z \in x \leftrightarrow z \in y) \wedge (\neg(z \in x) \leftrightarrow \neg(z \in y))) \to x = y\big). \quad (\overline{\text{Extensionality}})$$

Notice here that the replacement of the assignment function $[\![\cdot]\!]$ by $[\![\cdot]\!]_{\mathrm{PA}}$ not only validates the Leibniz's law of indescernibility of identicals, but it validates ZF (not only NFF-ZF) with a reasonable modification of Axiom of Extensionality in the $\mathcal{PS}$-algebra-valued models. Clearly, ZF and $\overline{\text{ZF}}$ are classically equivalent theories.

5 Conclusion

In the vast landscape of modern mathematics, numerous branches have been explored by dedicated mathematicians, and it is certain that many more will emerge in the future. The true elegance of pure mathematics lies in its abstraction, allowing concepts to transcend the confines of reality.

However, this chapter's focus is on the practical side of mathematics—the fundamental principles that find relevance in our daily lives. Our research endeavors to determine whether basic mathematics is solely confined to classical mathematics or if it extends beyond classicality. To tackle this inquiry, we have delved into set-theoretic approaches: seeking out non-classical set theories that still satisfy the core principles of basic mathematics. By uncovering these non-classical models, we demonstrate that classical mathematics is not the sole domain for accommodating fundamental mathematical ideas.

Furthermore, we explore additional set-theoretic properties of these intriguing models. Naturally, we expect these non-classical interpretations to diverge from classical mathematics concerning abstraction levels, yet they form their own captivating realm of abstraction.

This leads us to ponder an essential question: If a non-classical mathematical framework adheres to the fundamental principles of basic mathematics while deviating in abstraction from the classical norm, should we not afford it equal significance as classical mathematics? Rather than outright dismissing it due to its non-classical nature, we must consider embracing it as a potential avenue of mathematical exploration.

Acknowledgments Firstly, I extend my sincere gratitude to Prof. Mihir Kr. Chakraborty for his invaluable insights and discussions on the various aspects presented in this chapter. Additionally, I would like to express my appreciation to the anonymous referees whose constructive suggestions have significantly contributed to the enhancement of this manuscript. Finally, I would like to express my gratitude to Neil Barton for his assistance in articulating several paragraphs with simple words.

References

Bell, J. L. (2005). *Set theory, Boolean-valued models and independence proofs*. Oxford Logic Guides (Vol. 47, 3rd edn). The Clarendon Press Oxford University Press.

Chakraborty, M. K. (2010). *Amimangsar Alo-Andhari*. Nandimukh Samsad.

Grayson, R. J. (1979). Heyting-valued models for intuitionistic set theory. In M. P. Fourman, C. J. Mulvey, & D. S. Scott (Eds.), *Applications of Sheaves, Proceedings of the Research Symposium on Applications of Sheaf Theory to Logic, Algebra and Analysis held at the University of Durham, Durham, July 9–21, 1977*. Lecture Notes in Mathematics (Vol. 753, pp. 402–414). Springer.

Hardy, G. H. (1916). Prime numbers. In British Association for the Advancement of Science (Ed.), *Report of the Eighty-Fifth Meeting* (pp. 350–354). John Murray.

Hardy, G. H. (1940). *A mathematician's apology*. Cambridge University Press.

Jockwich Martinez, S., Tarafder, S., & Venturi, G. (2024). Zf and its interpretations. *Annals of Pure and Applied Logic, 175*(6), 103427.

Jockwich Martinez, S., & Venturi, G. (2021). Non-classical models of **ZF**. *Studia Logica, 109*, 509–537.

Libert, T. (2005). Models for paraconsistent set theory. *Journal of Applied Logic, 3*(1), 15–41.

Löwe, B., Paßmann, R., & Tarafder, S. (2018). Constructing illoyal algebra-valued models of set theory. *ILLC Prepublication Series PP-2018-15*.

Löwe, B., & Tarafder, S. (2015). Generalized algebra-valued models of set theory. *The Review of Symbolic Logic, 8*(1), 192–205.

Tarafder, S. (2015). Ordinals in an algebra-valued model of a paraconsistent set theory. In M. Banerjee & S. Krishna (Eds.), *Logic and Its Applications, Lecture Notes in Computer Science, 8923. Proceedings of the 6th International Conference, ICLA 2015, Mumbai, India, January 8–10, 2015* (Vol. 8923, pp. 195–206). Springer.

Tarafder, S. (2022). Non-classical foundations of set theory. *The Journal of Symbolic Logic, 87*(1), 347–376.

Tarafder, S., & Chakraborty, M. K. (2015). A paraconsistent logic obtained from an algebra-valued model of set theory. In J. Y. Beziau, M. K. Chakraborty, & S. Dutta (Eds.), *New Directions in Paraconsistent Logic, Springer Proceedings in Mathematics & Statistics, 152. Proceedings of the 5th WCP, Kolkata, India, February 2014* (pp. 165–183). Springer.

Tarafder, S., & Venturi, G (2021). Independence proofs in non-classical set theories. *The Review of Symbolic Logic, 16*, 979–1010. https://doi.org/10.1017/S1755020321000095

Tarafder, S., & Venturi, G. (2022). Zf between classicality and non-classicality. *Studia Logica, 110*, 189–218.

Weber, Z. (2010). Transfinite numbers in paraconsistent set theory. *The Review of Symbolic Logic, 3*(1), 71–92

Part III

Ontology

10

Iteration and Dependence Again

Luca Incurvati

As late as 1971, George Boolos could say of the iterative conception that although 'well known among logicians', '[a]uthors of set-theory texts either omit it or relegate it to back pages; philosophers, in the main, seem to be unaware of it, or of the preeminence of **ZF**, which may be said to embody it' (Boolos, 1971, 218, fn. 3).

Much has changed since then. Mainstream set theory textbooks such as Devlin (1993) and Jech (2003) characterize the cumulative hierarchy in their initial sections. And—no doubt partly thanks to Boolos's own influence—the iterative conception has come to be regarded by philosophers of mathematics as the conception of set lying behind the axioms of at least several of the axioms of first-order Zermelo-Fraenkel set theory (**ZF**). Moreover, **ZF**—perhaps with

Earlier versions of this material were presented at the *Metaphysical and Mathematical Explanation: New Perspectives* workshop in Paris, the *Foundations of Mathematics: Modern Views* conference in Munich, the *Campinas-Konstanz Seminar in the Philosophy of Mathematics* and the *Barcelona Set Theory Seminar*. Many thanks to members of these audiences and two anonymous referees for comments and feedback. Special thanks to Carolin Antos, Neil Barton and Giorgio Venturi for organizing the Campinas-Konstanz Seminar and putting this volume together. This work has received funding from the European Research Council under the European Union's Horizon 2020 research and innovation programme (grant agreement No 101086295) within the project *From the Expression of Disagreement to New Foundations for Expressivist Semantics*.

L. Incurvati (✉)
Department of Philosophy and Institute for Logic, Language and Computation, University of Amsterdam, Amsterdam, Netherlands
e-mail: L.Incurvati@uva.nl

C. Antos et al. (eds.), *The Palgrave Companion to the Philosophy of Set Theory*, Palgrave Companions, https://doi.org/10.1007/978-3-031-62387-5_10

the addition of the Axiom of Choice, thus giving rise to **ZFC**—is now widely regarded by philosophers as the standard axiomatic system for set theory.

Indeed, after the publication of Boolos's article, things moved quickly. Two central works on the iterative conception, namely Charles Parsons's (1977) 'What is the iterative conception of set?' and the chapter on the concept of set of Hao Wang's (1974) *From Mathematics to Philosophy*, appeared in the subsequent years and were reprinted, together with Boolos's article, in a special section on the iterative conception in the famous *Philosophy of Mathematics* collection (1983), edited by Paul Benaccerraf and Hilary Putnam.[1] A new paradigm had emerged.

With the paradigm in place, philosophers began to inquire into its metaphysical underpinnings. Early accounts of the iterative conception by Boolos (1971), Joseph Shoenfield (1965, 1977) and Wang (1974) presented the conception appealing to the idea of sets being *constructed* in stages. Dissatisfaction with this constructivist understanding of the iterative conception, however, led to the emergence of a different approach. While the roots of this new approach can already be found in Parsons's 1977 paper, its explicit articulation appears for the first time in the 1980s, in the work of James van Aken (1986), who formulates a set theory based on the idea of a set *presupposing* its elements in a metaphysical sense. Then in the 1990s and the 2000s, partly thanks to Kit Fine's (1995) work on essence, this idea of metaphysical presupposition starts being spelled out in terms of the notion of metaphysical dependence, and it becomes customary in the philosophical literature to understand the iterative conception in terms of this notion.[2]

In previous work (Incurvati, 2012, 2020), I have defended a minimalist understanding of the iterative conception, which dispenses with the idea that a set presupposes or is metaphysically dependent upon its members. I will not rehearse my arguments here. My aim is instead to clarify what is at stake in the debate between the dependency account and the minimalist account of the iterative conception. In particular, I will distinguish this debate from the debate between actualist and potentialist accounts of the iterative conception. These two debates, I will argue, concern different aspects of

[1] By contrast, the previous edition of the collection only contained one paper on the iterative conception, namely Kurt Godel's 'What is Cantor's continuum problem' (1947).

[2] Philosophers who subscribe to an understanding of the iterative conception in terms of dependence include Potter (2004), Linnebo (2008), Schaffer (2009), Trogdon (2013), Gasser (2015), Wigglesworth (2015) and Mount (2019). Notably, Potter identifies a platonist understanding of set theory with the dependency account of the iterative conception, and in more recent work goes as far as saying that '[m]uch of the interest in the philosophy of set theory consists in trying to make [the notion of dependency of a set upon its members] clearer' (Potter, 2020, 391).

the conception. I will then use this distinction to assess some arguments that Mark Gasser (2015) has advanced against *ante rem* structuralism. In particular, I will show that this distinction means that there are two different objections in Gasser's article. The first objection is that the iterative conception conflicts with dependence claims made by structuralists. As I have suggested in previous work, *ante rem* structuralists can address this objection by endorsing a minimalist account of the iterative conception. The second objection is that the indefinite extensibility of the set concept conflicts with the idea that the cumulative hierarchy is exhausted by the **ZFC** axioms. I chart various possible ways of addressing the second objection and show that disentangling *ante rem* structuralism from the idea that mathematical structures ought to be given by implicit definitions opens up a novel way of dealing with the objection. Having defended *ante rem* structuralism against Gasser's objections, I will, however, close the chapter by contending that although my arguments show that *ante rem* structuralism is compatible with the iterative conception, there are still reasons to favour a more deflationary understanding of structuralism, advocated by John Burgess (2009, 2015), Joel Hamkins (2021, 32ff.) and Gasser himself.

1 The Cumulative Hierarchy Structure

According to the iterative conception of set, sets are obtained by iterated application of the *set of* operation starting with some basic objects. The *set of* operation takes any given objects and delivers a set. The objects that are given are the basic objects and any objects obtained before or at a given state of the iteration process. The basic objects are usually called *individuals* or, in honour of the German founders of set theory, *Urelemente*. Anything which is not a set can be taken to be one of the *Urelemente*. In the abstract study of sets, however, it is customary to dispense with *Urelemente* altogether and assume that there are no basic objects. As a result, all sets will be pure—their transitive closure contains nothing but sets—and the domain of quantification only includes sets.

This informal description of the iterative conception can be made more formal. Note, first, that the process of iterating the *set of* operation is assumed to take place in an orderly manner and can therefore be indexed by ordinals. And, second, observe that, at any stage of the iteration process, the *set of* operation collects together any totality of things previously obtained. It is easy to see that this is tantamount to forming all members of the powerset of the preceding stage at a successor stage and taking unions of things formed at earlier stages at a limit stage. The upshot is that, on the iterative conception, sets appear at some level of the *cumulative hierarchy of sets*, **V**:

Definition 10.1 (The Cumulative Hierarchy) Let On be the class of all ordinals. Then the *cumulative hierarchy* is the transfinite sequence $\langle V_\alpha : \alpha \in \mathrm{On}\rangle$ such that:

$$V_0 = \emptyset;$$
$$V_{\alpha+1} = \mathcal{P}(V_\alpha);$$
$$V_\lambda = \bigcup_{\alpha<\lambda} V_\alpha \text{ if } \lambda \text{ is a limit ordinal.}$$

Once thus sharpened, the central tenet of the iterative conception is in fact a theorem of **ZFC**:

Proposition 2 *Every set belongs to some* V_α.

Indeed, Proposition 2 is equivalent to the Axiom of Foundation over the other **ZFC** axioms.

The cumulative hierarchy picture of the set-theoretic universe provides important structural insights. At the same time, it raises two fundamental questions about the structural richness of the universe. The first question concerns the height of the hierarchy, the second its width. Such questions, however, are in principle separate from the key structural element of the iterative conception that every set appears at some V_α (Proposition 2). We can illustrate this point with the help of some formal results from set theory which deserve to be better known.

The distinction between the key structural element of the iterative conception that every set appears at some V_α and the structural question concerning the height and width of the hierarchy took some time to emerge. This is probably due to certain features of the Zermelo-style axiomatization which are worth pausing on. Let us start with a first-order setting, i.e. by considering **ZFC**. We have noted that this theory proves that every set belongs to some V_α and that this statement is equivalent to the Axiom of Foundation over the other **ZFC** axioms. It can be shown that Replacement is needed to derive from Foundation the assertion that every set belongs to some V_α. This is why $\mathbf{ZF_2}$ and $\mathbf{ZFC_2}$ (the second-order versions of **ZF** and **ZFC**, obtained by replacing the Replacement Schema with the corresponding second-order axiom) are quasi-categorical—given any two models, one is isomorphic to an initial segment of the other—but the result no longer holds if we drop Replacement (Uzquiano, 1999).

Replacement, therefore, leads a double life within Zermelo-style axiomatizations: it contributes both to enforcing the cumulative hierarchy structure upon the set-theoretic universe and to increasing the lower bound of the cumulative hierarchy. In particular, whilst any V_λ with λ a limit ordinal greater than ω is a model of Zermelo set theory (**ZF** without Replacement) and its second-order cousin, some such V_λs, for instance, $V_{\omega+\omega}$, are not models of **ZFC** or ZFC_2. Indeed, it can be shown that $V_\kappa \models \mathsf{ZFC}$ if and only if κ is a worldly cardinal, where a cardinal is worldly just in case it is uncountable and strong limit. Moving to the second-order context, we have that $V_\kappa \models \mathsf{ZFC}_2$ if and only if κ is an inaccessible cardinal, where a cardinal is inaccessible just in case, as well as being worldly, it is regular.

Can we do better? Yes, we can. Our starting point will be an axiomatization of set theory presented by Dana Scott (1974). Scott's axiomatization is intended to be directly based on the cumulative hierarchy picture of the universe of sets. It makes use of a two-sorted first-order language: we have variables $x, y, \ldots$ ranging over sets and variables $V, V', \ldots$ ranging over levels ordered by membership. (Scott's term is 'partial universes', but we shall refer to them as 'levels' to emphasize the connection with the cumulative hierarchy.) In addition to Extensionality and the Separation Schema, we have three axioms: an Axiom of Accumulation, which states the members of a level are the members or subsets of earlier levels; an Axiom of Restriction, which asserts that every set belongs to some level (and embodies, therefore, the key tenet of the iterative conception); and an Axiom of Reflection, which says that every property expressible in the language of first-order set theory is completely reflected in some level. The precise form of the latter does not really matter for our purposes, but, as Scott himself shows, it is equivalent to Infinity and Replacement over the other axioms of his system.

Scott-Potter set theory (**SP**) is a descendant of Scott's system which has a one-sorted language and in which the notion of a level is explicitly defined in terms of membership (Potter, 2004, see also van Aken, 1986). Moreover, **SP** does not explicitly include the Accumulation Axiom, since this now follows directly from the definition of a level in terms of membership. Finally, in **SP** the Reflection Axiom is replaced by an Axiom of Creation, stating that there is no highest level, and an Axiom of Infinity, asserting the existence of a limit level.

Now, let SP^- be **SP** without its existence axioms, that is, the Axioms of Creation and Infinity; and let SP_2^- be SP^- with the Separation Schema replaced by its second-order counterpart. Then we have the following result, first explicitly noted in Incurvati (2010, 130) and proved in detail by Tim Button and Sean Walsh (2018, 184):

Theorem 10.3 *The models of* SP_2^- *are all and only the* V_α*s.*

This result shows that by switching to a Scott-style axiomatization—a theory more directly based on the iterative conception of set—we have clearly separated issues concerning the key structural component of the hierarchy from questions concerning its height, that is, questions about how far the iteration of the *set of* operation goes. In particular, the result shows that, by making it part of our initial axiomatization that every set belongs to some level, we can completely fix the cumulative structure of the hierarchy whilst leaving it completely open how far the hierarchy goes. Indeed, the cumulative hierarchy structure enforced by SP_2^-, a theory which has the central tenet of the iterative conception as an axiom, has a model in V_0.

2 Distinguishing Between Two Debates

Having now separated issues concerning the key structural component of the hierarchy and questions concerning its height, we can put this distinction to philosophical use. In particular, it allows us to clarify what is at stake in two central debates in the philosophy of set theory and the relationship between them. These debates are the debate between dependency and minimalist accounts of the iterative conception and the debate between actualism and potentialism.

Let us start from the first debate. According to the dependency account, the iterative conception is to be understood as the conception of set which takes sets to be ontologically or metaphysically dependent upon their members. This account of the iterative conception is explicitly endorsed by Michael Potter (2004) and Øystein Linnebo (2008), but, as mentioned, earlier developments of the account can also be found in work by van Aken (1986) and Parsons (1977). On the dependency account, the key structural element of the iterative conception that every set belongs to some level of the cumulative hierarchy is to be established by appealing to structural features of the metaphysical dependence relation. For this to be possible, the dependence relation must behave structurally like the ancestral of the membership relation in iterative set theory. Thus, the dependence relation must be antisymmetric, irreflexive, transitive and well founded.[3]

According to the minimalist account (Incurvati, 2012, 2020), by contrast, the content of the iterative conception is exhausted by saying that sets are the objects obtained by iterating the *set of* operation. Note that while the minimalist account is not committed to the thesis that sets are dependent on their members, it is not committed to the negation of this thesis either. Thus, minimalism is strictly speaking compatible with the possibility that sets are dependent upon their members, although this, on the minimalist account, is not part of the *content* of the iterative conception of set.[4] Furthermore, on the minimalist approach, the correctness of the iterative conception—the fact that every set appears at some level of the cumulative hierarchy—is defended by appealing to an *inference to the best conception*: the iterative conception should be preferred to its rivals because it fares better with respect to certain *desiderata* on conceptions of set.

Thus, the debate between the dependency account and the minimalist account concerns how to understand and motivate the key structural component of the iterative conception. But what about the 'iterative' in the 'iterative conception', captured by the idea that sets are the things obtained by *iterating* the *set of* operation?

This component is not what is at stake in the debate between the dependency and the minimalist account of the iterative conception and does not even appear in the dependency account. Yet, this component of the iterative conception is key to its claim to explain the set-theoretic paradoxes. For according to a plausible diagnosis, these paradoxes arise because on the naïve

---

[3]The dependency theorist's reliance on the dependency relation having certain structural features is critically discussed in Incurvati (2012, 75–79) and Incurvati (2020, 53–58). Which structural features the dependence relation must be assumed to have at the outset and which ones can be derived is controversial. Following the treatment of set theory in Scott (1974), to which we shall return below, van Aken (1986) motivates axioms for the set-theoretic hierarchy by appealing to structural features of the notion of metaphysical presupposition. Like Scott, van Aken is able to derive a principle of induction on membership, which immediately entails that the sets are well founded, without explicitly imposing a well-foundedness requirement. However, the derivation uses the Separation Schema, which would be contentious in certain non-well-founded settings, since it allows one to derive the existence of the Russell set from the existence of a universal set. For more on the significance of Scott's result, see Boolos (1984) and Incurvati (2020, 80).

[4]An analogy with minimalism about truth might help. Minimalism about truth is sometimes understood as being committed to the thesis that truth is not a substantive property (see, e.g., Horwich, 1998; Horsten, 2009). However, it is possible to understand minimalism about truth as simply holding that the *content* of the truth predicate is exhausted by the T-Schema or the T-rules. On this understanding, minimalism about truth is committed neither to truth being substantive nor to truth not being substantive. For a defence of this understanding of minimalism about truth (and of minimalist accounts of other notions such as proposition and attitude), see Incurvati and Schlöder (2023).

conception of set we have both indefinite extensibility and universality of the set concept (Russell, 1906; Incurvati, 2020). According to the indefinite extensibility of the set concept, whenever we have succeeded in defining a set *u* of objects falling under the concept of set, there is an operation which, given *u*, produces an object which intuitively falls under the concept of set but is not already contained in *u*. According to the universality of the set concept, there is a set of all objects falling under the concept of set. These two principles are classically inconsistent. The iterative conception deals with the conflict between indefinite extensibility and universality by retaining the former at the expense of the latter.

Minimally, this means that the hierarchy is *unbounded* in the sense that there is no highest level.[5] This is exactly what **SP**'s Axiom of Creation says and delivers Powerset and Pairing in the context of the SP^- axioms.

But one might want to go further. We are told that the process of iterating the *set of* operation is indexed by ordinals, but how far does this process go? An initial option is to say that it goes as far as the ordinals. This sanctions an axiom stating that for every α there is a V_α, which is, in effect, an instance of the Replacement Schema (Incurvati, 2020, 92–93). But how far do the *ordinals* go? A frequent answer is that the process of iterating the *set of* operation should be iterated *as far as possible*. In terms of the hierarchy, this amounts to endorsing the following principle:

Height Maximality There are as many levels of the hierarchy as possible.

What Height Maximality delivers is a matter of dispute: according to some, it licenses suitable lower bounds in the form of axioms of (higher) infinity or, more directly, reflection principles, with further disputes about exactly what forms of reflection are sanctioned by Height Maximality. We need not enter these disputes here. See Incurvati (2017) for an overview.

The discussion about questions of height brings us to the second debate in the philosophy of set theory to be considered, namely that between actualists (see, e.g., Uzquiano, 2003; Burgess, 2004) and potentialists (Hellman, 1989; Linnebo, 2013; Studd, 2019). *Actualism* holds that there is a definite totality or plurality of all sets: the hierarchy is actual. *Potentialism*, on the other hand, denies that there is such a definite totality or plurality: the hierarchy is potential in nature. In the context of iterative set theory, potentialism and actualism are

[5] When using the terminology of ranks, this is tantamount to saying that every set has lower rank than some set.

often understood as having specifically to do with the *height* of the hierarchy. Thus, *height* actualism holds that there is a definite totality or plurality of all *ordinals*, which means that the hierarchy is actual in height, whilst height potentialism holds that there is no such definite totality or plurality, which means that the hierarchy is potential in height.

Note here that the debate between actualists and potentialists in height ultimately concerns how to understand the process of iterating the *set of* operation. Actualism holds that the ordinals indexing this iteration process form a definite totality or plurality, whereas potentialism holds that they do not.

The upshot is that the debates between the dependency account and the minimalist account on one side and the debate between actualist and potentialists on the other concern different aspects of the iterative conception of set and the associated cumulative hierarchy picture. Indeed, one can find all combinations of views in the literature. Potter (2004) holds that there is a primitive relation of dependence holding between a set and its members, but is an actualist about both height and width.[6] Linnebo (2008) holds there to be a relation of metaphysical dependence between a set and its members, to be analysed in terms of the notion of essence articulated by Fine (1994, 1995) and E.J. Lowe (2003, 2005) among others,[7] but is one of the main proponents of a potentialist understanding of the cumulative hierarchy of sets (Linnebo, 2013). My minimalist account of the iterative conception was partly inspired by remarks by Boolos (1989), who was a thoroughgoing actualist and has been described as having 'no use' for the notion of indefinite extensibility (Burgess, 2004, 205). Finally, in the paper in which the cumulative hierarchy structure is presented for the first time (Zermelo, 1930), Ernst Zermelo, in describing the universe of sets as an open-ended sequences of normal domains, has often been interpreted as endorsing or at least suggesting some form of potentialism about the universe of sets, which he characterizes as never reaching true completion (Zermelo, 1930, 431). At the same time, Zermelo's position in the same paper seems to fall within the minimalist camp, as has been argued, in effect, by Daniel Isaacson (2010). The combination of minimalism and potentialism is even clearer in the case of Charles Parsons. Throughout his career, Parsons has

[6] For the view that the dependency relation should be considered a primitive and not further analysable in the metaphysics literature, see, e.g., Thomasson (1998, Ch. 2) and Barnes (2012, 879).

[7] In particular, Linnebo takes the dependence relation holding between a set and its members to be a relation of *identity* dependence, in the sense that a set depends on its members for its identity. In fact, Fine has used the case of the relation between sets and their members to criticize a modal account of dependence and motivate his own essentialist account. For a response to Fine's arguments, see Wildman (2013).

been a key proponent of a potentialist understanding of set theory. And while, as we noted earlier, a dependency account was suggested by remarks in his 1977 paper, Parsons has moved away from such an understanding of the iterative conception, endorsing instead the idea that the iterative conception does not involve any substantial notion of presupposition or dependence between a set and its members (see, especially, Parsons, 2008, Ch. 4).

How about the debate between actualists and potentialists *in width*? As we have seen, when properly formulated, second-order axiomatizations of set theory based on the iterative conception settle, at least at the semantic level, questions concerning the width of the hierarchy. For this reason, the appropriate setting for width potentialism is a first-order one. In this setting, the Separation Axiom is formulated as a schema, with one instance for each property expressible in the language of first-order set theory $\mathcal{L}_{\in}$. But at each level, one should like to collect together *all* subsets of sets already formed, not only those defined by properties expressible in $\mathcal{L}_{\in}$. Even here, however, the debate between actualists and potentialists in width concerns whether there is a definite totality of all subsets of a given level, not how to understand or motivate the key structural component of the iterative conception.

3 Structuralism and Dependence

In this section, I consider an objection recently advanced by Gasser (2015) against a certain brand of mathematical structuralism. The distinctions drawn in the previous section will put us in a position to see that, in fact, Gasser is putting forward two distinct objections.

Before turning to Gasser's objection, it is important to delimit its intended target. As is well known, many varieties of mathematical structuralism have been formulated and discussed. Gasser's target is the version of structuralism known as *ante rem* structuralism and famously defended by Stewart Shapiro (1997). At a first pass, we can characterize *ante rem* structuralism as the thesis that mathematics is about a realm of structures existing independently of any system instantiating them and that mathematical objects are positions in such structures. As we shall see, one issue that we will have to confront is precisely what *ante rem* structuralism amounts or should amount to. For the time being, however, this characterization will do.

Gasser argues that the iterative conception is incompatible with *ante rem* structuralism. Having introduced the iterative conception, he writes:

> Now note that there is more to the sets suggested by the iterative conception than to the sets construed by a structuralist. For instance, the notion that sets are "built up" from certain elements suggests a form of dependence absent from a purely structuralist point of view—on the iterative conception, a set's elements must somehow "already be there" for the set to be formed, while the structuralist view says nothing at all about the dependence of sets on their elements. (Gasser, 2015, 15)

As formulated, the objection is related to one advanced by Linnebo (2008), according to whom the *ante rem* structuralist is best understood as subscribing to two dependence claims. The first states that every object in a structure depends on every other object in that structure. The second says that every object in a structure depends on the structure itself. Because *ante rem* structuralism is committed to these two dependence claims, Linnebo argues, it conflicts with the iterative conception, since on that conception a set depends on its members but not *vice versa*, and a set, far from depending on **V**, depends *only* on its members.

Following Parsons, Gasser claims that a committed *ante rem* structuralist should regard the iterative conception only as 'part of an explanation by analogies, not necessarily part of the literal truth about sets' (Parsons, 1995, 79).[8] But, he says, there is evidence that the iterative conception is not used only as an analogy. Hence, he concludes, we have a conflict between *ante rem* structuralism and mathematical practice. It is of course open to *ante rem* structuralists to bite the bullet and insist that this conflict has to be resolved in favour of *ante rem* structuralism. Using a distinction by Burgess (1983) and familiar from the debate on mathematical fictionalism, this way of resolving the conflict would make the *ante rem* structuralist approach a revolutionary one (as opposed to a hermeneutic one, aiming to provide a faithful interpretation of mathematical discourse and practice). However, such a revolutionary approach would seem to undercut one of the main motivations for *ante rem* structuralism, namely the indifference in mathematical discourse and practice to non-structural features of mathematical objects.

[8]Note that, as observed above, Parsons now endorses something akin to a minimalist account of the iterative conception, and it will be argued below that on a minimalist account the iterative conception is not reduced to an explanation by analogies. Thus, there appears to have been a shift in Parsons's views on the matter over the years.

In support of the claim that the iterative conception is not only used as an explanation by analogies, Gasser gives an oft-quoted passage by Gödel:

> [T]he axioms of set theory by no means for a system closed in itself, but, quite on the contrary, the very concept of set [i.e. the iterative conception] on which they are based suggests their extension by new axioms which assert the existence of still further iterations of the operation "set of". (Gödel, 1964, 260)

Gasser's claim is that since the iterative conception is used to describe the universe of sets and provides a richer conception than what is codified in ZFC, this is incompatible with *ante rem* structuralism.

The conflict between the iterative conception and *ante rem* structuralism does not arise on the minimalist account (see Incurvati, 2012; Incurvati, 2020, Ch. 2). In particular, since the minimalist account is not committed to there being a metaphysical dependence relation holding between a set and its members but not *vice versa*, it allows the *ante rem* structuralist to hold on to the view that every set in the hierarchy is dependent on every other set in the hierarchy and on the hierarchy itself. Gasser (2015, 16) is correct that Gödel does not use the iterative conception simply as an analogy: according to Gödel, the iterative conception describes a universe of sets in which the standard axioms of set theory hold. But the minimalist account does not reduce to saying that the iterative conception is just part of an explanation by analogies; on the minimalist account, the iterative conception commits someone who asserts it unconditionally to believing that sets are part of a cumulative hierarchy arranged into levels and that the universe of sets satisfies, for instance, the Powerset Axiom. What the minimalist account rejects is the idea that the iterative conception ought to be understood in terms of metaphysical dependence.

Gasser thinks that structuralism is better understood as a hypothesis about mathematical practice. Thus, he insists that whether or not the iterative conception can be understood according to the minimalist account, 'some notion of dependence mattered to Gödel's take on the iterative conception' (Gasser, 2015, 15, fn. 31), and that this suffices to show that structuralists are wrong to think that mathematical practice does not concern itself with non-structural aspects of mathematical objects. Such non-structural aspects are important in the foundations of mathematics, Gasser argues, as in Gödel's case. Now, I am sympathetic with Gasser's way of understanding structuralism—or, at the very least, its core motivation—as a hypothesis about mathematical practice, and I will return to this below. Moreover, we can set aside, for current purposes, the possible reply that Gödel's remarks on the iterative conception could be taken to belong not to mathematical practice, but to philosophical aspects of the foundations of mathematics, which should be considered distinct from mathematical practice. For if that reply were viable, one might then argue that as long as there is no conflict between the practice of set theorists studying the mathematical properties of the cumulative hierarchy of sets and structuralism, then structuralists are off the hook: all that we have is a conflict between two different interpretations of mathematical practice, namely the structuralist one and Gödel's. Instead, I would like to contend that the passage quoted above provides no evidence that Gödel understood the iterative conception as involving metaphysical dependence.

Gasser insists that Gödel understood the iterative conception in a way that a structuralist cannot allow. He again quotes Gödel, when he remarks that the iterated powerset operation

> is to be understood so as to include also transfinite iteration, the totality of sets obtained by finite iteration forming again a set and a basis for a further application of the operation. (Gödel, 1964, 259, fn. 13)

Thus, Gasser concludes, 'Gödel may not have had any robust notion of dependence in mind here, but this is nonetheless more than a structuralist can allow' (2015, 15). Again, however, there appears to be no evidence in the quoted passage that Gödel took sets to be metaphysically dependent on their members. Thus, Gasser has failed to provide the required argument that there is a conflict between *ante rem* structuralism and the iterative conception based on the notion of metaphysical dependence.

Nonetheless, the fact that Gasser focuses on passages in which Gödel is concerned with iteration of the *set of* operation going beyond what is sanctioned by the axioms of set theory suggests that Gasser might have a different objection in mind. We should distinguish between two possible objections to *ante rem* structuralism. The first is that the iterative conception, at least as understood by Gödel, encodes a notion of dependence, which *ante rem* structuralism cannot allow. The second is that the iterative conception, at least as understood by Gödel, describes a structure which is not exhaustively captured by the ZFC axioms. The discussion in Sect. 1 shows that these are potential conflicts between the iterative conception and *ante rem* structuralism which concern *different components* of the iterative conception (or understanding thereof). The first is based on how the basic cumulative hierarchy structure is to be understood. This is what the debate between dependency and minimalist accounts of the iterative conception is about. The second is based on the indefinite extensibility of the set concept on the iterative conception. How this indefinite extensibility is to be understood, we saw, is what the debate between actualists and potentialists in height is about. The minimalist account provides a way for the *ante rem* structuralist to deal with the first objection to *ante rem* structuralism and, as we have seen, there is no evidence that Gödel endorsed more than a minimalist understanding of the key structural component of the iterative conception. The question is whether there is nonetheless a conflict between *ante rem* structuralism and the indefinite extensibility of the set concept, which the iterative conception endorses.

In the next section, we will explore the options for *ante rem* structuralists for dealing with the second objection based on an alleged conflict between *ante rem* structuralism and the iterative conception of set. It should be emphasized, however, that the conflict, if it exists, would appear to be much more general than the objection would seem to suggest when formulated on the basis of the iterative conception. For, as we noted earlier, the iterative conception retains the indefinite extensibility at the expense of universality in its solution of the set-theoretic paradoxes, but it is by no means the only conception which does so. For instance, the graph conception of set (Incurvati, 2014) also endorses the claim that the concept of set is indefinitely extensible. Indeed, the conception sanctions a theory whose universe can be characterized by means of an iterative process, albeit one allowing for non-well-founded sets (see Incurvati, 2020, §7.7).

4 Structuralism and Indefinite Extensibility

I have argued that, in his discussion of Gödel, Gasser appears to have in mind a different objection to *ante rem* structuralism than the one based on the alleged dependence of a set upon its members on the iterative conception. For, when discussing Gödel, Gasser seems to focus on the supposed conflict between *ante rem* structuralism and a different aspect of the iterative conception, namely the indefinite extensibility of set concept. Gasser presents the objection in the following passage:

> On the structuralist view of sets, it's not clear what this richer conception [i.e. a conception that is not exhausted by the **ZFC** axioms] could possibly tell us. For a structuralist, sets are essentially just positions in the structure of the set-theoretic universe—that is, as Parsons puts it, positions in the structure common to any "domain of 'objects' related in a relation called 'membership' satisfying conditions that can then be stated in the language of set theory," for instance the conditions stated by the **ZFC** axioms. (Gasser, 2015, 15)

We can offer the following preliminary reconstruction of the argument in the passage, intended to establish that there is indeed a conflict between *ante rem* structuralism and the indefinite extensibility of the set concept, as sanctioned by the iterative conception.

1. According to *ante rem* structuralism, sets are just positions in a structure characterized by some set of axioms (e.g. the **ZFC** axioms).
2. On the iterative conception, the concept of set is indefinitely extensible.
3. Hence, the cumulative hierarchy is not exhausted by any set of axioms. (From 2)
4. This contradicts (1).

How could *ante rem* structuralists respond? At first sight, premise (1) is simply a statement of the *ante rem* structuralist view, so we can set it aside for the time being. We shall have occasion to return to it below. Similarly, we can set aside premise (2), since we are assuming with Gasser that *ante rem* structuralists want to hold on to the iterative conception as part of mathematical practice and that, therefore, rejecting it would clash with the original motivation of *ante rem* structuralism, namely that of giving its due to mathematical discourse and practice. Given this, the only possibility left for *ante rem* structuralists would seem to be to challenge the entailment from (2) to (3).

One way to do this would be to focus on second-order axiomatizations of set theory, as indeed many *ante rem* structuralists have done (see, e.g., Shapiro, 1997; Leitgeb, 2020a,b; for a defence of *ante rem* structuralism in a first-order setting, see Resnik, 1997). For in a second-order setting, we must distinguish between two readings of 'exhausted' in 'the cumulative hierarchy is not exhausted by any set of axioms', *model-theoretic* and *proof-theoretic*. Some structuralists have argued that, once we focus on second-order axiomatizations, the cumulative hierarchy *is* model-theoretically exhausted by some set of axioms. In the literature, we find two strategies which aim to accomplish this.

The first strategy holds that the hierarchy is model-theoretically exhausted by some set of axioms because it is genuinely *indeterminate* how far the hierarchy goes beyond what is ensured by those axioms. For instance, Isaacson considers the ZFC_2 axioms, whose models are all V_κ with κ an inaccessible cardinal. He then writes:

> [W]hat is undecided in virtue of this degree of non-categoricity is genuinely undecided, in the same way that the fifth postulate of Euclid's geometry is genuinely undecided by the other axioms. (Isaacson, 2010, 53)

The idea is that, say, the assertion that there is an inaccessible cardinal does not have a determinate truth value because it is true in some models of ZFC_2 and false in others. The result that the models of ZFC_2 are all V_κ with κ an inaccessible cardinal was famously proved by Zermelo (1930),[9] and a position similar to Isaacson's may be reasonably ascribed to Zermelo himself. For as, we noted above, Zermelo, describes the universe of sets as an open-ended sequences of normal domains, exhausted by the ZFC_2 axioms.

The second strategy holds that the hierarchy is model-theoretically exhausted by some set of axioms because there *is* a set of axioms that categorically characterizes it. For instance, Gabriel Uzquiano (2002) appeals to Vann McGee's (1997) result that ZFCU_2 plus the axiom that the *Urelemente* form a set and the assumption that the quantifiers range over everything (i.e. the theory can only have models which contain everything) is categorical with respect to the pure sets.

Does this mean that *ante rem* structuralists must reject Gödel's claim that the iterative conception suggests further iterations of the *set of* operation and so sanctions the indefinite extensibility of the set concept? Not necessarily. In particular, remember that we should distinguish between two readings of 'exhausted' in 'the cumulative hierarchy is not exhausted by any set of axioms'. Helping themselves to this distinction, *ante rem* structuralists could insist that although *model-theoretically* the hierarchy is exhausted by the axioms of ZFC_2 or ZFCU_2 plus the Urelemente Axiom and the assumption that the quantifiers range over everything, it is not so exhausted *proof-theoretically*, that is, axioms of higher infinity allow us to prove more about what holds in the cumulative structure as characterized by the axioms. Thus, *ante rem* structuralists can try to resolve the conflict between their position and the indefinite extensibility of the set concept by resisting the entailment from (2) to (3) in Gasser's argument.

This is not to say, however, that the two strategies we have outlined on the *ante rem* structuralist's behalf are not without problems. In the case of the first strategy, the role of axioms asserting the existence of levels V_κ for κ an inaccessible cardinal appears problematic. For, one the one hand, the truth values of these axioms is said to be indeterminate. And, on the other, these axioms must be suggested by the iterative conception itself, to give its due to the indefinite extensibility of the set concept. The problem does not arise for the second strategy, which holds that the cumulative hierarchy has been fully characterized and so the role of axioms of higher infinity is clear: they make

[9]The theory Zermelo actually considered in his paper was a version of ZF_2 without Infinity. For this reason, the theory also has models in V_ω.

proof-theoretically explicit what was already model-theoretically secured by the axioms of set theory, motivated by the iterative conception of set.

The challenge for the second strategy is rather whether McGee's categoricity result does deliver what it purports to. As I argued in previous work (Incurvati, 2016), the result's assumption that the quantifiers range over everything jeopardizes its significance: there is no clear sense in which the result shows that the axiomatic theory fixes a single structure in a way that can be explained by appealing solely to our use and practice using that axiomatic theory. In other words, it is easy enough to specify an axiomatic theory that singles out the intended model using a meta-axiom that rules out unintended models by *fiat*; what is needed is an explanation as to why those models are unintended and, in ruling out models that do not encompass everything by assuming that the quantifiers do range over everything, McGee's result fails to provide the required explanation.

Thus, both strategies to deal with the apparent conflict between *ante rem* structuralism and the indefinite extensibility of the set concept on the iterative conception appear to have problems. However, I am now going to argue that challenging the entailment from (2) to (3) in our earlier reconstruction of Gasser's argument is not the only option for *ante rem* structuralists. For consider our reconstruction again. In that reconstruction, we followed Gasser in taking structuralism to be committed to the thesis that sets are just positions in a structure characterized by some set of axioms (e.g. the **ZFC** axioms). However, this thesis is best seen as a combination of two theses, leading to the following reconstruction of the argument:

1. According to *ante rem* structuralism, sets are just positions in a structure.
2. According to *ante rem* structuralism, structures are always characterized by some sets of axioms.
3. Hence, according to *ante rem* structuralism, sets are just positions in a structure characterized by some set of axioms (e.g. the **ZFC** axioms). (From 1 and 2)
4. On the iterative conception, the concept of set is open-ended.
5. Hence, the cumulative hierarchy is not exhausted by any set of axioms. (From 4)

That sets are just positions in the structure of the set-theoretic universe is part and parcel of the *ante rem* structuralist doctrine. And a recurrent theme in the writing of structuralists *is* that structures are fully characterized by the axioms, which can then function as implicit definitions of those structures. For instance, Shapiro writes:

> Ultimately, mathematical structures are characterized axiomatically (as implicit definitions), or they are defined in set theory. (Shapiro, 1997, 258)

Given that the cumulative hierarchy cannot be found as a set within itself, it must be given by a coherent axiomatic characterization, which then works as an implicit definition of the cumulative hierarchy.

> Structuralism entails a relativity of objects. Rather than objects simpliciter, we speak of objects *of a structure*. Each mathematical object is implicitly or explicitly tied to the structure of which it is a place. Structures, in turn, are characterized by coherent theories, via devices like implicit definition. (Shapiro, 1997, 258)

> In other words, with ordinary (pure) mathematics, a given theory is an implicit definition of a freestanding structure or a class of freestanding structures. (Shapiro, 1997, 260)

Although Shapiro himself (1997, 165) interprets Hilbert as an eliminative structuralist, the Hilbertian origins of the view that mathematical structures are given by coherent sets of axioms which work as implicit definitions are clear. And the view is, of course, an important and influential one in mathematics and its philosophy. However, it should be stressed that, unlike the thesis that mathematical objects are positions in a structure, this view is *not* part and parcel of the *ante rem* structuralist doctrine. It is an additional commitment that *ante rem* structuralists may but need not undertake. And rejecting this commitment allows *ante rem* structuralists to turn the apparent conflict between their view and the indefinite extensibility of the set concept on its head. It is precisely our grasp of the fact that the *set of* operation is to be iterated indefinitely and that, as a result, the concept of set is indefinitely extensible that makes it the case that iterative set theory is about the cumulative hierarchy of sets. Even if the indefinite extensibility of the set concept as embodied by the iterative conception cannot be fully captured by the axioms of iterative set theory, this does not mean that it cannot contribute to our understanding of the structure iterative set theory is meant to be about. As Benacerraf once put it:

> Even if we agree [...] that the determinants of mathematical meaning must lie somewhere in our *use* of mathematical language, why think use to be captured

(or capturable) by axioms (we *do* argue about new axioms)? (Benacerraf, 1985, 111)

Thus, distinguishing between the thesis that sets are positions in a structure and the thesis that structures are characterized axiomatically allows us to see another possible avenue of response to Gasser's objection. In particular, the *ante rem* structuralist can hold on to the thesis that sets are positions in a structure but accept that the indefinite extensibility of the set concept on the iterative conception shows that *some* structures cannot be fully characterized by any set of axioms. This, however, is compatible with thinking that remarks such as Gödel's on the indefinite extensibility of the set concept on the iterative conception have a role to play in our grasp of the mathematical structure iterative set theory is about, namely the cumulative hierarchy. Thus, although standard presentations of *ante rem* structuralism do often say that structures are *always* given by set of axioms via implicit definitions, this aspect should be distinguished from the thesis that mathematics is the science of structures and that mathematical objects are positions in structures.[10]

5 Conclusion

The debate between minimalist and dependency accounts and the debate between potentialist and actualist accounts concern different aspects of the iterative conception, and indeed all four positions in the resulting logical space have been occupied in the literature. Distinguishing between these two aspects of the iterative conception allows us to distinguish clearly two objections against *ante rem* structuralism based on the iterative conception. The first objection is that the *ante rem* structuralist cannot allow for the possibility of a set asymmetrically depending on its members. I have argued that the *ante rem* structuralist can address this objection by endorsing the minimalist account of the iterative conception. The second objection is that the *ante rem* structuralist cannot allow for the indefinite extensibility of the set concept, as embodied by the iterative conception. I have charted various avenues of response to the second objection, and argued that, by distinguishing between the thesis that

[10] Shapiro distinguishes between algebraic (e.g. group theory, field theory and ring theory) and non-algebraic theories (e.g. arithmetic and set theory). The latter are about a class of structures whereas the former are intended to be about a particular structure. Algebraic theories are typically understood to be about all and only their models—a group is nothing but a model satisfying the group axioms. One natural possibility for the *ante rem* structuralist would therefore be to hold on to the idea that structures are characterized axiomatically, but to reject it for non-algebraic theories.

mathematics is the science of structures and mathematical objects are positions in those structures on the one hand and the idea that mathematical structures are given by axioms as implicit definitions on the other, we can identify a so far neglected avenue of response to the second objection.

Having criticized *ante rem* structuralism, Gasser ends his paper by defending an understanding of structuralism which does not involve substantial claims about the ontology of mathematics. Instead, on this understanding, structuralism is simply a thesis about the practice of mathematics: ordinary mathematicians pursue mathematics by ignoring non-structural features of mathematical objects. Burgess and Hamkins too are critical of metaphysically loaded understandings of structuralism, and defend versions of structuralism which similarly concern the way mathematics is to be practised. Burgess (2015) endorses what he calls *Hardy's Principle* (1940) that in mathematics it is enough for there to be *some* interpretation of symbols on which the axioms come out as true, but it does not matter whether there are *many* interpretations, since we can adopt any of them. The only questions about the choice of interpretation are pragmatic, concerning which interpretation is most convenient for certain purposes. Similarly, Hamkins (2021) defends what he calls *structuralism in practice*, which is again not an ontological thesis but rather a thesis about how mathematics is to be pursued: by ignoring non-structural features of mathematical objects and by stating theorems and formulating mathematical concepts so that they are invariant under isomorphism.

Having spent the bulk of this chapter defending *ante rem* structuralism against Gasser's objections, I want to now end it on a reconciliatory note by expressing my sympathy for a non-ontological understanding of structuralism and outlining my reasons for this sympathy. As we saw, Gasser holds that there are aspects of set-theoretic practice that allegedly appeal to non-structural aspects of mathematical objects. I have argued that set-theoretic practice does not appeal to dependence relations between a set and its members, and that one can avoid having to regard the indefinite extensibility of the set concept as resting on non-structural features of mathematical objects by rejecting the idea that structures must be fully characterizable by axioms. Burgess and Hamkins, for their part, argue against structuralism on the basis of the existence—indeed, abundance—of non-rigid structures in mathematics. This is the well-known Burgess–Keränen objection (Burgess, 1999; Keränen, 2001), which has been extensively debated (see, e.g., Button, 2006; Shapiro, 2006; Leitgeb & Ladyman, 2008; Shapiro, 2008), and I will not enter this debate here.

For my reasons for being sympathetic with a non-ontological understanding of structuralism are different. I agree with Gasser, Burgess and Hamkins that the structuralist insight that mathematical practice ignores non-structural features of mathematical objects is an important one, which should have a central place in one's philosophy of mathematics. Moreover, and unlike Gasser, Burgess and Hamkins, I agree with *ante rem* structuralists that this insight does not only have consequences for the understanding of mathematical practice. In particular, *ante rem* structuralists are right when they typically contend that the semantics of mathematical talk should be sensitive to mathematicians' indifference to non-structural features of mathematical objects. This is because, from a broadly pragmatist perspective, one should avoid assigning to expressions semantic values that are not needed to explain their functional role—the function they play in our practices (see Incurvati & Schlöder, 2023, esp. Ch. 1).

Thus, in general, it would seem correct to assign to mathematical expressions meanings that do not appeal to non-structural features of mathematical objects. This makes the *ante rem* structuralist position *prima facie* very appealing. For consider the two major alternatives from the structuralist camp, which dispense with structures as *sui generis* entities (and hence are eliminativist in nature). The first alternative consists in taking the meaning of mathematical expressions to be given by a background ontology, typically a set-theoretic one. On this view, the meaning of, say, arithmetical expressions can be given by the von Neumann ordinals. However, this meaning assignment is arbitrary—the meaning of arithmetical expressions could have as well been given by identifying their referents with the Zermelo ordinals. The reference of numerical expressions is therefore relative to an initial choice of reference assignment, which is why this kind of structuralism is sometimes called *relativist* (Reck & Price, 2000). The problem for relativist structuralism is that whatever choice one makes, the semantics ends up ascribing to mathematical expressions semantic values that are not needed to explain their functional role, since the assignment settles questions such as whether the number 1 is a member of the number 3, which appear to be extraneous to mathematical practice.

The second alternative aims to improve on relativist structuralism by not focusing on an arbitrarily chosen assignment of possible referents but rather universally quantifying over all of them. Arithmetical statements, for instance, are to be reconstructed as generalizations over all systems of objects instantiating 'the natural number structure', though there is no need to appeal to a *sui generis* natural number structure over and above the particular systems of objects instantiating it. The problem for this *universalist* version of structuralism (Reck & Price, 2000) is that it cannot take mathematical statements at face value, which would seem to count against it. More importantly, the particular way in which arithmetical statements are semantically analysed on the universalist structuralist view appears to be non-compositional (Leitgeb, 2020a, 325).

However, from the natural language semantics fact that one should assign to mathematical expressions semantic values which do not appeal to non-structural features of mathematical objects and that are compositional, one need not draw the metaphysical conclusion that mathematical objects do not have non-structural features. This conclusion is only forced upon us if we regard the assignment of meaning to mathematical terms to proceed in a broadly referentialist manner, so that when we assign meaning to mathematical expressions we fix all of their properties by assigning to them a referent. From an inferentialist perspective, in particular, objects can have properties going beyond what they must have on the basis of their semantics. There may be *extra*-semantics reasons for regarding mathematical objects as having non-structural features. The fact that the functional role of mathematical talk does not appeal to non-structural features of mathematical objects does not imply that mathematical objects do not have such features (or, of course, that they do have them). These remarks are no doubt sketchy, but I hope that they indicate a promising direction for the development of a version of structuralism which is not metaphysically loaded and gives its due to mathematicians' indifference to non-structural features of mathematical objects.[11] For now, I conclude that there is no conflict between structuralism and the iterative conception of set.

References

Barnes, E. (2012). Emergence and fundamentality. *Mind, 121,* 873–901.

Benacerraf, P. (1985). Skolem and the skeptic. *Proceedings of the Aristotelian Society, 59,* 85–115.

Benacerraf, P., & Putnam, H. (1983). *Philosophy of mathematics. Selected readings* (2nd ed.). Cambridge University Press.

Boolos, G. (1971). The iterative conception of set. *Journal of Philosophy, 68*, 215–231. Reprinted in Boolos, G. (1998). *Logic, logic, and logic* (pp. 13–29). Harvard University Press.

Boolos, G. (1984). The justification of mathematical induction, *PSA: Proceedings of the Biennial Meeting of the Philosophy of Science Association, 1984*, 469–475.

[11] Incurvati and Schlöder (2023) develop an approach to semantic theorizing which combines expressivism and inferentialism, and defends the general account of the relationship between semantics and metaphysics outlined here. In future work, I plan to apply this approach to mathematical discourse and structuralism.

Boolos, G. (1989). Iteration again. *Philosophical Topics, 17,* 5–21. Reprinted in Boolos, G. (1998). *Logic, logic, and logic* (pp. 88–104). Harvard University Press.
Burgess, J. (1983). Why I am not a nominalist. *Notre Dame Journal of Formal Logic, 24,* 93–105.
Burgess, J. (1999). Book review: Stewart Shapiro. Philosophy of mathematics: Structure and ontology. *Notre Dame Journal of Formal Logic, 40,* 283–291.
Burgess, J. (2004). *E Pluribus Unum.* Plural logic and set theory. *Philosophia Mathematica, 3,* 193–221.
Burgess, J. P. (2009). Putting structuralism in its place. Unpublished Manuscript.
Burgess, J. P. (2015). *Rigor and structure.* Oxford University Press.
Button, T. (2006). Realistic structuralism's identity crisis: A hybrid solution. *Analysis, 66,* 216–222.
Button, T., & Walsh, S. (2018). *Philosophy and model theory.* Oxford University Press.
Devlin, K. (1993). *The joy of sets. Fundamentals of contemporary set theory* (2nd ed.). Springer.
Fine, K. (1994). Essence and modality. In J. Tomberlin (Ed.), *Philosophical perspectives 8: Logic and language* (pp. 1–16). Ridgeview.
Fine, K. (1995). Ontological dependence. *Proceedings of the Aristotelian Society, 95,* 269–290.
Gasser, M. (2015). Structuralism and its ontology. *Ergo, 2,* 1–26.
Gödel, K. (1947). What is Cantor's continuum problem? *American Mathematical Monthly, 54,* 515–525. Reprinted in Gödel, K. (1990). *Collected works II* (pp. 176–187). Oxford University Press.
Gödel, K. (1964) What is Cantor's continuum problem? In Benacerraf, P., & Putnam, H. (Eds.), *Philosophy of mathematics. Selected readings* (2nd ed., 470–485). Cambridge University Press (1983). Revised and expanded version of Gödel, K. (1947).
Hamkins, J. D. (2021). *Lectures on the philosophy of mathematics.* MIT Press.
Hardy, G. H. (1940). *A mathematician's apology.* Cambridge University Press.
Hellman, G. (1989). *Mathematics without numbers: Towards a modal-structural interpretation.* Clarendon Press.
Horsten, L. (2009). Levity. *Mind, 118,* 555–581.
Horwich, P. (1998). *Truth* (2nd ed.). Clarendon Press.
Incurvati, L. (2010). *Set theory: Its justification, logic and extent,* PhD Thesis, University of Cambridge.
Incurvati, L. (2012). How to be a minimalist about sets. *Philosophical Studies, 159,* 69–87.
Incurvati, L. (2014). The graph conception of set. *Journal of Philosophical Logic, 43,* 181–208.
Incurvati, L. (2016). Can the cumulative hierarchy be categorically characterized? *Logique et Analyse, 59,* 367–387.
Incurvati, L. (2017). Maximality principles in set theory. *Philosophia Mathematica, 25,* 159–193.

Incurvati, L. (2020). *Conceptions of Set and the Foundations of Mathematics.* Cambridge University Press.

Incurvati, L., & Schlöder, J. J. (2023). *Reasoning with attitude. Foundations and applications of inferential expressivism.* Oxford University Press.

Isaacson, D. (2010). The reality of mathematics and the case of set theory. In Z. Novak & A. Simonyi (Eds.), *Truth, reference and realism* (pp. 1–76). Central European University Press.

Jech, T. (2003). *Set theory. The third millenium edition.* Springer.

Keränen, J. (2001). The identity problem for realist structuralism. *Philosophia Mathematica, 9,* 308–330.

Leitgeb, H. (2020a). On non-eliminative structuralism. unlabeled graphs as a case study, part a. *Philosophia Mathematica, 28,* 317–346.

Leitgeb, H. (2020b). On non-eliminative structuralism. unlabeled graphs as a case study, part b. *Philosophia Mathematica, 29,* 64–87.

Leitgeb, H., & Ladyman, J. (2008). Criteria of identity and structuralist ontology. *Philosophia Mathematica, 16,* 388–396.

Linnebo, Ø. (2008). Structuralism and the notion of dependence. *Philosophical Quarterly, 58,* 59–79.

Linnebo, Ø. (2013). The potential hierarchy of sets. *Review of Symbolic Logic, 6,* 205–228.

Lowe, E. J. (2003). Individuation. In M. Loux & D. Zimmerman (Eds.), *Oxford handbook of metaphysics* (pp. 75–95). Oxford University Press.

Lowe, E. J. (2005). Ontological dependence. In E. N. Zalta (ed.), *Stanford encyclopedia of philosophy (fall 2008 edition).* Available at http://plato.stanford.edu/archives/fall2008/entries/dependence-ontological/

McGee, V. (1997). How we learn mathematical language. *Philosophical Review, 106,* 35–68.

Mount, B. M. (2019). Antireductionism and ordinals. *Philosophia Mathematica, 27,* 105–124.

Parsons, C. (1977). What is the iterative conception of set? In R. E. Butts & J. Hintikka (Eds.), *Logic, Foundations of Mathematics, and Computability Theory. Proceedings of the Fifth International Congress of Logic, Methodology and the Philosophy of Science (London, Ontario, 1975),* Reidel, Dordrecht. Reprinted in Benacerraf, P., & Putnam, H. (1983). *Philosophy of mathematics. Selected readings* (2nd ed., pp. 503–529). Cambridge University Press.

Parsons, C. (1995). Structuralism and the concept of set. In W. Sinnott-Armstrong, D. Raffman, & N. Asher (Eds.), *Modality, morality, and belief: Essays in honor of Ruth Barcan Marcus* (pp. 74–92). Cambridge University Press.

Parsons, C. (2008). *Mathematical thought and its objects.* Cambridge University Press.

Potter, M. (2004). *Set theory and its philosophy.* Oxford University Press.

Potter, M. (2020). Intuitive and regressive justifications. *Philosophia Mathematica, 28,* 385–394.

Reck, E. H., & Price, M. P. (2000). Structures and structuralism in contemporary philosophy of mathematics. *Synthese, 125*, 341–383.

Resnik, M. (1997). *Mathematics as a science of patterns*. Clarendon Press.

Russell, B. (1906). On some difficulties in the theory of transfinite numbers and order types. *Proceedings of the London Mathematical Society, 4*, 29–53.

Schaffer, J. (2009). On what grounds what. In D. Chalmers, D. Manley, & R. Wasserman (Eds.), *Metametaphysics* (pp. 347–383). Oxford University Press.

Scott, D. (1974). Axiomatizing set theory. In T. J. Jech (ed.), *Axiomatic set theory II. Proceedings of symposia in pure mathematics* (Vol. 13, pp. 207–214). American Mathematical Society.

Shapiro, S. (1997). *Philosophy of mathematics: Structure and ontology*. Oxford University Press.

Shapiro, S. (2006). Structure and identity. In F. MacBride (Ed.), *Identity and modality* (pp. 109–145). Oxford University Press.

Shapiro, S. (2008). Identity, indiscernibility, and ante rem structuralism: The tale of i and $-i$. *Philosophia Mathematica, 16*, 1–26.

Shoenfield, J. R. (1965). *Mathematical logic*. Addison-Wesley.

Shoenfield, J. R. (1977). Axioms of set theory. In J. Barwise (Ed.), *Handbook of mathematical logic* (pp. 321–344). North-Holland.

Studd, J. P. (2019). *Everything, more or less: A defence of generality relativism*. Oxford University Press.

Thomasson, A. L. (1998). *Fiction and metaphysics*. Cambridge University Press.

Trogdon, K. (2013). An introduction to grounding. In M. Hoeltje, B. Schnieder, & A. Steinberg (Eds.), *Varieties of dependence* (pp. 97–122). Philosophia Verlag.

Uzquiano, G. (1999). Models of second-order Zermelo set theory. *Bulletin of Symbolic Logic, 5*, 289–302.

Uzquiano, G. (2002). Categoricity theorems and conceptions of set. *Journal of Philosophical Logic, 31*, 181–196.

Uzquiano, G. (2003). Plural quantification and classes. *Philosophia Mathematica, 11*, 67–81.

van Aken, J. (1986). Axioms for the set-theoretic hierarchy. *Journal of Symbolic Logic, 51*, 992–1004.

Wang, H. (1974). *From Mathematics to philosophy*. Routledge and Kegan Paul.

Wigglesworth, J. (2015) Set-theoretic dependence, *The Australasian Journal of Logic* **12**, 159–176.

Wildman, N. (2013). Modality, sparsity, and essence. *The Philosophical Quarterly, 63*, 760–782.

Zermelo, E. (1930), Über Grenzzahlen und Mengenbereiche: Neue Untersuchungen über die Grundlagen der Mengenlehre, *Fundamenta Mathematicae* **16**, 29–47. Reprinted and translated in Zermelo, E. (2010). In H.-D. Ebbinghaus, C. G. Fraser, & A. Kanamori *Collected works* (Vol. I, pp. 400–431). Springer.

11

What Is Forcing Potentialism?

Ethan Brauer

Potentialism about some domain of objects, as I will understand it, is the view that that domain should be regarded as *potentially infinite* rather than as an actually completed infinite totality. For instance, arithmetic potentialism is the view that the natural numbers are merely potentially infinite and do not form an actual infinity, and geometric potentialism is the view that geometric objects form a potential infinity. Some forms of potentialism might be coupled with a denial that there are any actual infinities, while other forms of potentialism might hold that some infinities are actual, while others are merely potential. For instance, the Fefermanian predicativist might hold that the natural numbers form an actual infinity, while the sets of natural numbers are a potential infinity. Set-theoretic potentialism is the view that the universe of sets V is (in some respect) potentially infinite.

The essential characteristic of a potential infinity is that it can always be *extended*. To say the natural numbers are potentially infinite is to say that no matter which numbers you give me, I can always add more numbers to that collection. To say that the geometric objects are potentially infinite is to say that no matter which geometric objects you give me, I can always construct more geometric objects. As a corollary, if a domain of objects is potentially infinite, there can be no definite collection of *all* the entities in that domain;

E. Brauer (✉)
Department of Philosophy, Classics, History of Art and Ideas, University of Oslo, Oslo, Norway

C. Antos et al. (eds.), *The Palgrave Companion to the Philosophy of Set Theory*, Palgrave Companions, https://doi.org/10.1007/978-3-031-62387-5_11

for otherwise, if I were given that collection of *all* the objects, I could not find any new objects to add to it.

One version of set-theoretic potentialism is *forcing potentialism*. This is the view that the universe of sets is potentially infinite because, given any universe V we can add new sets to it by forcing, thus finding ourselves in the generic extension $V[G]$. While the technical tools of potentialism have been applied to study forcing extensions of universes, leading to something that has been called 'forcing potentialism', the philosophical content of such a view has not been clearly laid out. In this chapter, I pose a challenge for forcing potentialism, arguing that a philosophical account of forcing potentialism must revise the familiar iterative conception of set. After developing this challenge, I will propose one way of meeting it by distinguishing between the iterative combinatorial conception of set and the iterative logical conception of set. Importing the intuitionistic idea of a *lawless sequence* into the iterative logical conception of set yields a view akin to forcing potentialism.

I begin in Sect. 1 by explaining potentialism in more detail. In the set-theoretic case there is an important distinction between height potentialism and width potentialism. Forcing potentialism is a form of width potentialism. Section 2 then rehearses the story behind the iterative conception of set and explains how it can motivate height potentialism. In Sect. 3 I adapt some ideas from Sam Roberts to develop the challenge for forcing potentialism, arguing that it is in tension with the iterative conception of set. Hence an account of forcing potentialism requires some new way of thinking about the cumulative hierarchy of sets. In Sect. 4 I develop such an account of forcing potentialism by adding a lawless sequence to the iterative process of set formation.

1 Potentialism

If potentialism is the view that a given domain is potentially infinite, I should say something about potential infinity and its contrast with actual infinity. A collection is actually infinite if it forms a definite, completed totality of objects larger than any finite cardinality. By contrast, a collection C is potentially infinite when any definite, completed subcollection of C can be expanded to a more inclusive subcollection of C. Potential infinity is an ancient concept, tracing its origins back to Aristotle, who wrote:

> [T]he infinite has this mode of existence: one thing is always being taken after another, and each thing that is taken is always finite, but always different. (*Physics* 206a27-29)[1]

For instance, consider the natural numbers and suppose we start by being given the number 1. Then we may also 'take' the number 2, then 3, and so on. At each stage we have a finite collection of natural numbers, but it is always a different, larger collection of natural numbers.

Recent work on potential infinity has focused on the modal nature of potential infinity: *necessarily*, given any subcollection of C, it is *possible* to find an object x that is in C but is not in the original subcollection.[2] Or, in Aristotle's temporal terms, one thing is *always* being taken *after* another. That is, *whenever* we have taken some C's, we can *later go on* to take some more C's. Thus, for instance, while an actualist might express the infinitude of the natural numbers as:

$$\forall x \exists y (x < y),$$

the potentialist might express the *potential* infinitude of the natural numbers as:

$$\Box \forall x \Diamond y (x < y).$$

Similarly, where the actualist about geometry might say that for every triangle ABC there is a line segment AD bisecting the leg BC, the potentialist about geometry would only grant that necessarily, for every triangle ABC, it is *possible* for there to be a line segment AD that bisects BC. But the arithmetical potentialist will *deny* that it is possible for every number to *actually* have a successor, since that would require the collection of natural numbers to be a completed, infinite totality. Likewise, the geometric potentialist will deny that it is possible to have actually performed *all* geometric constructions, since that would require the collection of geometric objects to be a completed, infinite totality.

Potentialist approaches to set theory can be broadly divided into height potentialism and width potentialism. Height potentialism is the view that for any universe V of set theory, it is possible to extend V to a taller universe

---

[1] See also *Physics* 207a6, Aristotle (1941).

[2] Linnebo (2013) developed this modal approach to potential infinity in the context of set theory, building on Parsons (1983); Studd (2013) had also used modal logic to explicate the iterative conception of set, though without explicitly making the connection to potential infinity. Linnebo and Shapiro (2017) subsequently generalized these ideas to give a modal account of potential infinity as such.

V'. One simple version of this view is that the cumulative hierarchy of V_θ's is potentially infinite. If we have plural resources available, we can express this view as follows, where $x \prec xx$ means that x is among the plurality xx and $Ord(x)$ abbreviates the formula that says x is an ordinal:

$$\Box\forall xx(\forall x \prec xx\, Ord(x) \rightarrow \Diamond\exists\kappa\forall x \prec xx\, (x \in V_\kappa))$$

Width potentialism is the view that for any universe V of set theory there is a 'wider' universe V'. Forcing potentialism is a version of width potentialism, holding that the generic extensions of a model of set theory M are potentially infinite. Suppressing some details, it may be expressed modally like so:

$$\Box\forall M\Diamond\exists G(M \subset M[G])$$

The technical side of forcing potentialism involves using modal logic to study the properties of forcing extensions. This side of the idea has been fairly extensively pursued.[3] But potentialism has a philosophical side, as well. It is a claim about the nature of a certain domain of objects, and while using modal logic provides a formal framework for explicating the concept of potential infinity, the formalism has to be motivated and interpreted in a given setting. Why should we think of a given domain as being potentially infinite? And what is the nature of the modality in that domain? In this chapter, my attention will primarily be on the first question. How can we understand the nature of set theory so as to motivate forcing potentialism? Or in other words, what philosophical picture of the set-theoretic universe can the forcing potentialist be working with?

To appreciate the significance of such questions, contrast forcing potentialism with multiversism. Multiversism is the view that there are many different universes of set theory which each has some claim to legitimacy.[4] Take, for example, the version of multiversism introduced in Woodin (2011), which requires that the multiverse be closed under generic extensions and generic refinements.[5] On this approach, "The truths of Set Theory are the sentences

[3] Hamkins (2003), Hamkins and Löwe (2008), Hamkins et al. (2015), Hamkins and Linnebo (2022), Scambler (2021).

[4] There are a variety of versions of multiverse views in the literature. See, for example, Hamkins (2012), Steel (2014), Maddy and Meadows (2020), Antos et al. (2015). Different versions of multiversism will take different views on what makes each of the universes 'legitimate'. The claim that they are *in some sense* legitimate is what distinguishes multiversism as a philosophical position from the simple mathematical fact that **ZFC** has many different models (assuming it is consistent).

[5] N is a generic refinement of M iff M is a generic extension of N. In the so-called set-theoretic geology, generic refinements are also known as *grounds*.

which hold in each universe of the multiverse" (Woodin, 2011, 14). Modal logic has a role to play in studying the generic multiverse, because it provides tools for analyzing the relationships between the theory of a universe and the theories of its forcing extensions. But the fact that multiversism and forcing potentialism both use modal logic to understand the relations between models and their forcing extensions does not make them the same view. To the contrary, they are prima facie different views.[6] On its face, multiversism is a pluralistic thesis about the objects of set theory, whereas forcing potentialism is seemingly a thesis about the character of *the* universe of sets. So put, the difference between the two views is intuitive, but vague. To examine the difference a little further, consider two possible motivations for multiversism.

One motivation for multiversism is that any universe in the generic multiverse is a perfectly good candidate for capturing our intuitive notion of set or, alternatively, provides a perfectly good interpretation of our set-theoretic discourse.[7] This stands in contrast to the potentialist view which is based on the idea that *no* single element of the potentialist system, no 'universe fragment', is an adequate representation of our set-theoretic discourse, because our ideas about sets compel us to recognize possible extensions of each such universe fragment. Where the multiversist says "for all that we have said about sets, we could be living in any one of these universes", the potentialist says "given all that we have said about sets, we cannot be restricted to any one of these so-called universes". This motivation for the multiverse view, then, seems to be at odds with the basic potentialist view.

Another approach to multiversism is that we are compelled to recognize a plurality of universes of set theory because given any universe we are able to move to a richer universe, for instance, by forcing. This idea admittedly has a potentialistic flavor to it, since we can always 'take' one new universe after another, in the Aristotelian idiom.[8] An important question for this point of view, though, is why or how it is always possible to move to a richer universe. Where did those richer universes come from? If they were already there in some sense, then that smacks of actualism. But if the richer universe had to be created or generated in some sense, then we are entitled to ask for some clarification of *how* they were so generated. And this is exactly the question

[6] Barton (2021) also emphasizes the need for a clearer philosophical distinction between multiverse views and potentialist views.

[7] This is, I take it, the basic idea behind (Hamkins, 2012) analogy with geometry: just as our ordinary notions of point, line, etc. can be perfectly well captured in different geometries, so our ordinary notions of set, membership, etc. can be perfectly well captured in different set-theoretic universes.

[8] Thanks here to Robin Solberg.

I am after in this chapter. How are forcing extensions generated? Once we have an answer to this question, we might find that it does provide us with a way to understand certain multiverse views in potentialist terms. But also, we might not find this. Ultimately, we desire more clarity on the nature of different multiverse and potentialist views. What does each of them require? What does each presuppose? And what are the relations between the different views?

I have two goals in this chapter. First, I want to show that there is a serious challenge in giving a philosophical account of forcing potentialism. Setting forth this challenge will be the task of Sects. 2 and 3. Second, I will try to meet this challenge in Sect. 4 by sketching a philosophical story about set theory and then building a model of forcing potentialism based on that story. I want to emphasize from the outset that my goal is not to try to capture everything that has been said in the name of forcing potentialism. Rather, I want to see how much of what has gone under the name 'forcing potentialism' has a coherent philosophical rationale. More specifically, I want to see how much of forcing potentialism can find a coherent philosophical rationale *in the notion of potential infinity*. Anything that does not have such a rationale should not bear the name potentialism. To be painfully clear, the fact that a view does not have a rationale based on the notion of potential infinity does not mean that that view must be rejected. It does mean that other grounds must be provided for that view. Perhaps a theory has other philosophical motivations, or perhaps it is motivated simply by its mathematical interest. Nothing in this chapter is an argument against multiverse views or pluralist views or any other way of understanding the modal logic of forcing. What I am concerned with is whether the notion of potential infinity provides a good way of understanding the modal logic of forcing.

2 The Iterative Conception of Set

Above, I phrased the question at hand in terms of the forcing potentialist's conception of set. The dominant conception of set is the *iterative conception*. In this section I will briefly review the iterative conception of set, before arguing in the next section that this conception is inconsistent with forcing potentialism.

The iterative conception of set is best known through the work of George Boolos.[9] He aptly summed up the idea:

[9] Shoenfield (1977) is also an influential exposition.

> A set is any collection that is formed at some stage of the following process: Begin with individuals (if there are any). An individual is an object that is not a set; individuals do not contain members. At stage zero (we count from zero instead of one) form all possible collections of individuals. If there are no individuals, only one collection, the null set, which contains no members, is formed at this 0th stage. [...] At stage one, form all possible collections of individuals and sets formed at stage zero. If there are any individuals, at stage one some sets are formed that contain both individuals and sets formed at stage zero. Of course some sets are formed that contain only sets formed at stage zero. At stage two, form all possible collections of individuals, sets formed at stage zero, and sets formed at stage one. At stage three, form all possible collections of individuals and sets formed at stages zero, one, and two. At stage four, form all possible collections of individuals and sets formed at stages zero, one, two, and three. Keep on going in this way, at each stage forming all possible collections of individuals and sets formed at earlier stages. (Boolos, 1971, 220-1)

There are several pieces to the intuitive story behind the iterative conception of set, but we can isolate four ideas at the core of the conception. First is a claim about priority: the members of a set are prior to the set itself. Second, sets are nothing but collections of these prior existing objects. The third idea is only implicit in Boolos' description quoted above—and indeed it is not properly the domain of the iterative conception but is essential to any conception of set. This idea is that sets are extensional objects: a set's identity is entirely determined by its members. There is no intensional criterion of identity. The fourth idea is a maximality claim: at each stage of set formation, *all* possible collections of objects are brought into a set. (Unlike the first three claims, this maximality claim is not constitutive of the idea of iteratively forming sets of previously existing objects. It is, however, an essential part of characterizing the iterative process as a whole in such a way that justifies standard set-theoretic axioms.)

When we tell this story so that we begin with no individuals, the result is the cumulative hierarchy of pure sets. I will assume that there are no such individuals. Of course, something needs to be said about the existence of a stage ω after all the finite stages, and about limit stages more generally. And there is some controversy about whether this conception of set can underwrite all the axioms of **ZFC** (in particular, Replacement is tricky).[10] Nevertheless,

[10] See Boolos (1971, 228), also Boolos (1989), Potter (1993), Potter (2004), Paseau (2007), Rin (2015), Incurvati (2020). It is illuminating that Linnebo (2013) and Studd (2013) both need to appeal to axioms other than those governing the iterative process in order to recover the axioms of Infinity and Replacement.

it provides a nice clear and intuitive way to apprehend what set theory is all about.

What's more, this story about forming sets at stages by collecting together existing objects can easily lead to height potentialism, by arguing that the stages are potentially infinite. Given any definite totality of stages at which some sets have been formed, I could extend that collection to include a new stage, collecting together all the sets that were formed at those earlier stages. In other words, there is no limit to how far the iterative process of set formation can go. This line of thinking is of a piece with Cantor's diagnosis of the Buralli-Forti paradox that the ordinals form an 'inconsistent multiplicity', meaning that "the assumption that *all* of its elements 'are together' leads to a contradiction, so that it is impossible to conceive of this multiplicity as a unity, as 'one finished thing'" (Cantor, 1899, 114; cf. Linnebo, 2013, 206-8). Since each stage can be naturally associated with an ordinal—indeed, each stage can simply be thought of as a V_α—it is a short step to height potentialism from Cantor's description of the ordinals as not being able to be brought together into a completed definite totality. The aim here, however, is to find a motivation for forcing potentialism, which is a version of *width* potentialism, not height potentialism. So is there another way to use the iterative conception of set to motivate forcing potentialism?

3 The Challenge for Forcing Potentialism

I will argue in this section that the iterative conception of sets cannot motivate forcing potentialism. Indeed, the iterative conception of sets creates a prima facie difficulty for any version of width potentialism.

Width potentialism is the claim that the universe of sets is potentially infinite in width: for any collection of sets you give me, I can find a more inclusive collection of sets that lies no higher in the set-theoretic universe than the original collection of sets. And of course, this entails that there can be no collection of all sets of a given height. If there were, then if you gave me that collection of sets, I would not be able to find a more inclusive collection of sets of that height, contradicting the nature of width potentialism.

There is an immediate tension between width potentialism and the iterative conception of set, specifically with the maximality claim that at each stage

Studd uses a reflection principle, and Linnebo uses a reflection principle and a version of replacement applied to pluralities.

in the iterative process *all* possible collections of previously existing objects are formed into sets. Imagine we find ourselves in some set-theoretic universe that an oracular being has created by an iterative process of forming sets out of collections, and the width potentialist claims "I can add some new sets to this universe without adding any higher levels!" I would like to ask the width potentialist why those sets were not already formed by the oracle.

There are two general ways the width potentialist could reply. First, they could explain that the oracle is not forming sets out of all possible collections but is rather operating by some more restricted procedure, and for any collection of sets formed by this procedure it is possible to form a more inclusive collection of sets of a given height. Second, they could explain that the oracle is forming all possible sets at each stage, but that as we go on new things keep becoming possible at each level. Different ways of fleshing out these answers might involve more or less radical ways of elaborating or modifying the standard story behind the iterative conception of sets, as recounted by Boolos above.[11] (To foreshadow, the theory I propose in Sect. 4 is a version of the first approach. Sets are not formed out of all possible collections, but merely out of the collections which satisfy some formula.)

What I will argue is that there is no way for the width potentialist, working broadly within the iterative conception of set, to resolve this prima facie tension while maintaining that the possible extensions of a universe are all and only the forcing extensions.

I will begin by drawing a distinction between convergent and branching extensions of a collection. In brief, the argument will be that the iterative conception of set only motivates convergent extensions, but the forcing extensions of a ground model exhibit branching behavior, so the iterative conception is inconsistent with the claim that the possible extensions of a universe are all and only the forcing extensions. When I refer simply to forcing potentialism, I will mean the view that the possible extensions of a universe are all and only its forcing extensions. (This is of a piece with the terminology in Hamkins & Linnebo, 2022.)

[11] For instance, an anonymous referee suggested that instead of forming sets out of *all* collections of objects, we could consider only sets formed from *definite* collections. This would amount to a version of the first strategy for the width potentialist. To flesh out this view, one would have to (a) provide a philosophically satisfying formal explication of definiteness, (b) explain why not all possible collections are definite in the relevant sense, and then (c) show how the resulting story of iterative set formation entailed width potentialism.

3.1 Convergent vs. Branching Extensions

A domain of objects is potentially infinite when any definite collection of objects of that domain can be extended to a more inclusive collection of objects of that domain. We can ask whether there are ever mutually incompatible ways of extending such a collection or if the possible extensions are always mutually compatible.[12] When the possible extensions are always mutually compatible, we can say we have *convergent extensions*, and when there are possible extensions that are mutually incompatible with each other, we can say we have *branching extensions*. As an example of convergent extensions, consider geometric potentialism with a Euclidesque constructive gloss. Take geometry to be the study of certain types of constructions using compass and straightedge in an idealized Euclidean space. For any constructions that have been performed, it is possible to perform further constructions. For instance, if you construct a line segment you can draw a second line segment bisecting the first, or you can draw a circle tangent to the line segment, and so on. But whichever construction you perform next, it is always open to you to go on to perform one of the other constructions after that. None of these ways of extending the collection of existent geometric objects precludes any of the others, they are all mutually compatible.

On the other hand, as an example of branching extensions, consider a potentially infinite sequence of natural numbers. This sequence is always finite, but growing. Similar to imagining a being who is constructing geometric objects, we can imagine a being who chooses the values of this sequence successively, one argument at a time.[13] If values of this sequence have been chosen up to the argument n, then it is possible to extend the sequence by choosing the value for $n + 1$ to be 0 or 1 or 2 and so on. But unlike the geometrical case, these possibilities are all mutually exclusive. If the value for $n + 1$ is 0, then it cannot be 1 or 2, if it is 1 it cannot be 0 or 2, and so on.

Forcing potentialism provides another example of branching extensions. If we begin with a countable transitive model M of **ZFC**, there are forcing extensions $M[G]$ and $M[H]$ which do not have any mutual forcing extension N. To see why, suppose that r is a real that codes the countability of M. Then we can construct Cohen reals G and H that each contains partial information

[12] This distinction is discussed in Hamkins (2018) and Brauer (2020).

[13] On this picture, a potentially infinite sequence of natural numbers is basically a Brouwerian free choice sequence, cf. Brouwer (1952), Troelstra (1977), Troelstra (1982). I emphasize, however, that we can take the idea of a potentially infinite sequence of natural numbers to be perfectly comprehensible without accepting the baggage of Brouwerian intuitionism more generally.

about r, so that r cannot be obtained from either of G or H in isolation, but can be obtained from the combination of G and H. Then either G or H can be added consistently to M by forcing, but if both were added then the resulting model would have to include r and hence would realize that M is countable. But since forcing preserves ordinals, no forcing extension of M will think M is countable.[14] Hence no forcing extension includes both G and H, and $M[G]$ and $M[H]$ are mutually incompatible as forcing extensions. Thus, if the possible extensions of a universe are all and only the forcing extensions, then we have a form of branching extensions.

3.2 The Iterative Conception and Convergence

Recall the four claims I isolated as lying at the heart of the iterative conception of set. First was the priority claim that members of a set are prior to the set itself. Second, sets are nothing but collections of these 'prior' existing objects. Third was the claim that sets are extensional objects, with no intensional criterion of identity. Fourth was the maximality claim that at each stage of set formation, *all* possible collections of objects are brought into a set.

Because it was this maximality claim that led to the basic tension with width potentialism, and I do not want to beg the question against width potentialism, I will assume here only a weaker version of this maximality claim. For present purposes I will not assume that *all* collections of prior objects *must* be brought together to form a set. I will only make the weaker assumption that *any* collection of prior objects *can* be brought together into a set.

I now want to argue on the basis of these four ideas that the iterative conception is only consistent with convergent extensions of the universe of sets. The basic idea of the argument is straightforward: if you have a universe V_0, you can get different extensions V_1 and V_2 when V_1 is formed by bringing together some collections X into sets and V_2 is formed by bringing some other collections Y into sets; but nothing in the process of forming sets out of the X collections would then prevent you from later forming sets out of the Y collections. In order to show that this argument relies only on the iterative conception and does not smuggle in any further ideas, I will now flesh it out in pedantic detail.

[14] This argument can be found in Fuchs et al. (2015), where Hamkins says he heard it from Woodin in the 90s. The result actually goes back to Mostowski (1976). (Thanks to Neil Barton and Julia Kameryn Williams for providing the historical reference).

Suppose we begin with a universe V_0, and it can be extended to either V_1 or V_2. The question is whether there is a further extension V_3 that includes both V_1 and V_2. Let V_i^s be the sth stage in the construction of V_i over V_0. Since, in the iterative conception, stages are well-founded, we can argue by induction on s that V_1^s and V_2^s have a mutual extension $V_1^s \cup V_2^s$.[15] Consider an arbitrary object x in V_1^s. Since we are concerned with pure sets, x is a set. By the assumption that sets are just collections of objects, x in particular is nothing more than a collection of objects, and by the priority assumption those objects must have come from V_0 and $V_1^{<s}$. By the i.h., all those objects exist in an extension $V_1^{<s} \cup V_2^{<s}$ of V_0. So by the assumption that any collection can be brought together into a set, those objects can also be collected together in a set y in an extension $V_1^s \cup V_2^s$ of V_0. The set y will consist of exactly the same objects as x, so by extensionality y is just x itself, which thus exists in $V_1^s \cup V_2^s$. Since x was arbitrary, every element of V_1^s is in $V_1^s \cup V_2^s$. Similarly, every element of V_2^s must be in $V_1^s \cup V_2^s$. Thus, I conclude, the iterative conception is only consistent with convergent extensions, not branching extensions.

This argument is essentially due to Roberts (2020). The key premise in Roberts' version of the argument is that, if you can form a set x, then necessarily, for any set y, you can form the subset $x \cap y$ of the objects in y that are or would be in x if x existed.[16] This is a generalization of the claim that I relied on above, namely that, if you can form sets out of some collections X, then forming some other sets out of collections Y could not prevent you from *also* forming sets out of the X collections. By formalizing these ideas in a bi-modal logic, Roberts is able to isolate exactly the formal features needed for the argument to go through. This leads to a very general inconsistency result showing that certain postulates about potentialism and set formation are formally inconsistent. My claim that the iterative conception of set is inconsistent with branching extensions is one instance of this formal inconsistency. Roberts further argues that his result shows that height potentialism and width potentialism per se are incompatible. While I demur from that more general interpretation of Roberts' result—and in fact, one of the main contributions of this chapter is to develop a version of height and width potentialism—the particular instance of Roberts' argument presented above seems to me to be on solid ground.

In sum, forcing potentialism is a form of potentialism with branching extensions, as we saw in the last subsection. But the iterative conception of set

[15] The use of set-theoretic union is strictly speaking an abuse of notation, since V_i^s may not necessarily be a set, for all that we have said so far. Certainly V_i^s is not a set in the universe V_i^s.

[16] This is a gloss on what Roberts calls $\mathsf{Separation}^{\Diamond-}$.

only allows for convergent extensions, not branching extensions. So forcing potentialism is inconsistent with the iterative conception of set.

3.3 Objections

I will consider four objections to this argument.

3.3.1 What Matters Is the Logic

The first objection is that what really matters for the potentialist is not how the universe grows, but what the deductive modal theory looks like. The potentialist uses modal logic to formally explicate the intuitive idea of a domain being extendable, but what really matters is the mathematical content of the modalized theory, not the pre-formal gloss. And in the case of forcing potentialism, the right modal logic to use is **S4.2**, as proved in Hamkins and Löwe (2008). **S4.2** results from adding to **S4** the G axiom:

$$\Diamond\Box\phi \rightarrow \Box\Diamond\phi$$

In general, **S4.2** is characterized by Kripke frames where the accessibility relation is transitive, reflexive, and convergent in the following sense:

$$wRv_1 \wedge wRv_2 \Rightarrow \exists u(v_1 Ru \wedge v_2 Ru)$$

This convergence property makes it look from the perspective of Kripke frames that possible extensions v_1 and v_2 are always mutually compatible because they can have a further joint possible extension u.

The situation here is somewhat subtle. To say that **S4.2** is characterized by the (transitive, reflexive and) convergent Kripke frames is to say that the formulas valid on every such frame are exactly the validities of **S4.2**. But for ϕ to be valid on every such frame means that it is valid in every model based on that frame—in other words, for every such frame F and valuation V, ϕ is true at every world in the model $\langle F, V\rangle$. But when we treat a set of forcing extensions of M as the worlds in a Kripke model, the formulas in question are not uninterpreted atoms to which any consistent valuation can be applied. Instead, they are *interpreted* formulas whose values are determined by the true membership relation. In $M[G]$, $\tau_G \in \sigma_G$ is true iff τ_G really is a member of σ_G. And of course, what is true in a forcing extension $M[G]$ depends in part on what is true in M. Treating the system of forcing extensions as a Kripke model, there is a conspiracy among the worlds about which formulas they make true. As a result, even though the system of worlds is not convergent in the relevant sense, substantive mathematical facts about forcing entail that $\Diamond\Box\phi \rightarrow \Box\Diamond\phi$ is a valid principle of the modal logic of forcing.

In light of this, the forcing potentialist might claim that as far as the actual deductive content of their modalized theory is concerned, their universes might just as well have convergent extensions. The branching behavior is undetectable and may be written off as just a quirk of the informal motivation behind the formal theory. Once we have arrived at the formal theory, the informal story can be kicked away like Wittgenstein's ladder.

In reply to this, I want to ask why we should study the modal logic of forcing in the first place. If it is because of the theory's inherent mathematical interest, fine. But that is consistent with the theory having nothing whatsoever to do with potentialism. Forcing potentialism—if it is a coherent view at all—is presumably the view that forcing somehow reveals the potentially infinite character of the universe of sets. As such, the theory is *precisely about* the universes, their forcing extensions, and the relations between these. Talk of the extensions of different models of set theory is not an informal story we use as a heuristic to find the right modal logic, it is the very content of the philosophical view in question. Dismissing questions about the relations between different extensions of the set-theoretic universe in favor of a deductive modal theory is not so much kicking away a ladder that has served its purpose as it is sawing through the branch one sits on.

3.3.2 The Iterative Conception Is Consistent with Branching Extensions

The second objection is that the iterative conception of set is in fact consistent with branching extensions as well as convergent extensions. This can be proved by example, using the universal finite sequence. The universal finite sequence is a sequence σ with a Σ_1 definition that is provably finite, but for any countable model M, if σ is s in M and t is any finite extension of s, there is an end-extension of M in which σ is t.[17] A model M is an *end-extension* of N when

[17] See Hamkins and Williams (2021). An arithmetical universal finite sequence is studied in Hamkins (2018). A related example that could be used for this objection is the universal finite set, see Hamkins and Woodin (2018). The reply to that example would be exactly the same as the reply to the example of the universal finite sequence.

$N \subseteq M$, $\in^N = (\in\upharpoonright N)^M$, and for all $b \in N$, if $a \in^M b$ then $a \in N$. In other words, M and N agree on $\in$ for objects in N, and M does not add any new members to sets from N.

Obviously this example will lead to branching extensions, just as we noted in Sect. 3.1 that a potentially infinite sequence of natural numbers will give rise to branching extensions. Furthermore, the objection goes, end-extensions are consistent with the iterative conception of set. An end-extension of N will add new sets, but will not mess with any of the pre-existing sets. An end-extension only forms new sets by collecting together already-existing objects, which is exactly how sets are formed according to the iterative conception of set. So we have a proof by example that the iterative conception of set is consistent with branching extensions. Although this example does not reveal how the argument of Sect. 3.2 went wrong, it shows that it must have gone wrong somewhere.

In reply, I acknowledge that there are end-extensions that are allowed by the iterative conception of set, and I acknowledge that there are end-extensions that are branching extensions. But, the end-extensions that are allowed by the iterative conception of set are *not* the end-extensions that give rise to branching extensions. On the iterative conception, recall, the elements of a set are prior to the set itself. The set is merely the collection of previously existing objects. But the elements of a set are themselves also sets, and so the same is true of them: their elements must exist prior to them. So those elements must also be in the domain. And of course, *their* elements must in turn also exist prior to *them*, so *those* elements must be in the domain. The iterative conception of set requires that the universe be transitive. But, as proved in Hamkins and Williams (2021), the universal finite sequence is the empty sequence in any transitive model. So as long as we remain in transitive models—as the iterative conception requires—the universal finite sequence will be empty and thus will not witness any branching behavior.

3.3.3 Countable Models Are Non-standard

This leads to the third objection. The forcing potentialist might object that my argument relies on non-standard models and hence does not fairly represent their position. This point is similar to the objection that Brauer (2020) raises against the argument in Hamkins (2018) that arithmetic potentialism exhibits branching extensions. Hamkins' argument relied on an arithmetic version of the universal finite sequence which results in branching behavior for models of arithmetic, similar to the set-theoretic version discussed in the previous paragraphs. Brauer (2020) noted, however, that the universal finite sequence is the empty sequence in the standard model. Since arithmetic potentialism—at least on the most natural reading—is the claim that the true natural numbers are a potential infinity, it should not be saddled with baggage about non-standard numbers. (This is all the more true given that non-standard models of arithmetic contain the natural numbers as a completed sub-part, in contrast to the idea that the collection of natural numbers cannot be completed.)

Similarly, the forcing potentialist might object, the example of branching forcing extensions relies on the ground model being countable. Such models are non-standard, the forcing potentialist may protest, and their view should not be saddled with baggage about these non-standard models which, by hypothesis, they did not mean to be talking about to begin with.[18]

In reply, I want to make two points. The first is that forcing over countable transitive models is the typical way to approach forcing. Thus, drawing on facts about forcing extensions of countable models is not to saddle the forcing potentialist with non-standard baggage, but is just to appeal to the basic, standard tools of forcing. (As a side note, the forcing potentialist cannot reply by invoking the Boolean-valued model approach to forcing, since Boolean-valued models are also non-standard. This would amount to rejecting one approach to forcing because it requires non-standard models and therefore adopting a different approach that also requires non-standard models.)

The second part of my reply is to ask in what sense countable models are non-standard. Critically, we can focus on countable *transitive* models. In these models, '∈' is interpreted as the real membership relation.[19] So the non-standardness of these models is not a matter of re-interpreting the vocabulary of the theory. If a countable transitive model is non-standard in a problematic way it must be because of its size: the model thinks that some sets are uncountable even though they are 'really' countable, and the model just omits the function witnessing their countability. By this criterion, any model that by design omits sets which really exist would be considered non-standard. But that would completely vitiate forcing potentialism, because no standard model would have a forcing extension. To add a generic to a model requires

[18]This objection will not be attractive to all forcing potentialists. The theory in Scambler (2021) is specifically designed so that for every set there is a forcing extension where that set is countable. And Hamkins (2012) proposes that every universe is countable from the perspective of another universe (although this is in the context of the multiverse, not forcing potentialism).

[19]Or at least, a countable transitive model is isomorphic to one in which '∈' is the real membership relation, via the Mostowski collapse.

that there be some sets not in that model; the model must, by design, have omitted some objects that we can see from the perspective of the metatheory.

3.3.4 Some Forcing Extensions Are Convergent

The final objection is that the forcing potentialist is not forced to recognize all forcing extensions as being on par. In some restricted settings, forcing extensions are convergent rather than branching.[20] If the forcing potentialist can motivate some restriction so that all forcing extensions meeting that restriction are mutually compatible, then they will avoid the argument of §3.2.

In reply, I grant that this strategy *would* avoid the argument of §3.2. But the strategy has to actually be executed, and it is far from obvious that the forcing potentialist can actually motivate any philosophical restriction that has the desired technical effect.[21] I leave it as a task for the objector to show that this can in fact be done. The example of non-convergence from above was a very simple, even paradigmatic, case of forcing: $M[G]$ and $M[H]$ each arises from M by adding a single Cohen real. For the objector's strategy to succeed, then, they must rule out the adding of a single Cohen real as not being the sort of forcing that the forcing potentialist is concerned with. There is a danger that the forcing potentialist will be left with a very restricted conception of forcing potentialism.[22]

3.4 Conclusion

The expression 'forcing potentialism' has found its way into work both on the modal logic of forcing and modal accounts of potential infinity. But what view is supposed to answer to this name is not clear. In some uses, forcing potentialism seems to be little more than multiversism by another name. (This

[20] For example, Hamkins (2016) shows that if there is a transitive model of **ZFC** then there is a transitive model with a forcing notion $\mathbb{Q}$ such that any two generic filters G and H are mutually generic, and hence $M[G]$ and $M[H]$ have a mutual extension $M[G][H]$.

[21] In Steel's multiverse, it is assumed as an axiom that any two universes have a mutual generic extension. This is justified largely on extrinsic reasons, however: Steel wants to have an axiomatic multiverse theory that can serve a foundational role. The existence of mutual extensions is assumed in order for multiverse truth to be axiomatizable. It does not seem to me that a similar rationale is available to the forcing potentialist. Even in the original multiverse setting, we can ask why we should insist on axiomatizability. As (Maddy and Meadows, 2020, 48) conclude: "Without a satisfactory answer to this question, we have no reason to adopt the axiomatizability requirement, and we're left without a principled argument for Amalgamation [i.e., the convergence assumption]."

[22] This point is due to Neil Barton.

situation is complicated by the fact that the most prolific writer on the modal logic of forcing is also an avowed multiversist, see Hamkins, 2012). What we would really like is an explanation of forcing potentialism that brings into focus why it is a form of *potentialism*. I hope to have shown the challenges involved in this task.

I first noted that the width potentialist needs to explain why the sets added in the width extension were not already present in the universe. I then argued that any way of meeting this challenge that relies on a broadly iterative conception of set will only admit convergent extensions. On the other hand, forcing extensions exhibit branching behavior. Thus, if forcing potentialism is the view that the universe can always be expanded by all and only the forcing extensions, this view is inconsistent with the most prominent account of the subject matter of set theory. The task facing the forcing potentialist seems to be nothing less than a reconceptualization of set theory itself.

On the other hand, if forcing potentialism is not the view that the universe can be expanded by all and only the forcing extensions, then what is forcing potentialism? If not all forcing extensions are allowed, then which forcing extensions are genuine ways to extend the universe, and why those? Or, if some expansions of the universe are allowed besides forcing extensions, which ones, and why do possible expansions by forcing have to be coupled with possible expansions by other means?

4 What Forcing Potentialism Could Be?

In this section I aim to provide an account of forcing potentialism that partially meets the challenge laid down in the last section. My account will be a height- and width potentialist theory that includes forcing extensions among the possible width extensions. It will not, however, include *only* forcing extensions—this is the sense in which I only partially meet the challenge of the last section. If one were looking for a version of forcing potentialism that said the possible width extensions of a universe were all and only the forcing extensions, this result would be disappointing. On the other hand, I will be telling a story that provides a philosophical account of width extensions, including forcing extensions, in terms of potential infinity. So what we end up with is certainly a version of potentialism which includes forcing extensions.

There will be two pieces to my account. First, I will suggest a slight modification to the iterative conception of set, which may be called the iterative logical conception of set. Second, alongside the cumulative hierarchy of sets there will be a lawless sequence of sets.

4.1 The Iterative Logical Conception of Set

There is a familiar distinction between the *logical* and *combinatorial* views of sets. On the logical view of sets, sets are extensions of properties. From the property of being a dog we get the set of dogs by considering its extension. Sets are not 'mere' collections, but have a certain unity. That an object is a member of some set is explained by the object's having the relevant property. On the combinatorial way of thinking, by contrast, sets are mere combinations of possibly unrelated objects. There is nothing that makes the set hold together as one. It is simply the result of saying 'yes' or 'no' to each object in the universe, independently of one another.[23]

The logical conception of sets is generally regarded as untenable, or at least questionable, since it is a short step from there to naive comprehension and paradox. As a result, sets today are typically thought of combinatorially. This is evident in Boolos' parable of the iterative conception: sets are not formed by taking extensions of properties, but by gathering collections of objects together, and any possible collection will do, irrespective of whether the objects in the collection have anything in common.

This has led many authors to be insufficiently careful in distinguishing between the combinatorial conception of *set* and the iterative conceptions of *sets*, when in fact the two views crosscut each other.[24] The combinatorial conception is a view about the nature of individual sets: they are nothing other than collections of possibly unrelated objects. The iterative conception, on the other hand, is not really a view about the nature of sets taken individually. (What would it even mean to say that a single individual set is iterative?) Rather, the iterative conception is a view about the nature of the set-theoretic universe and the process of set formation that creates that universe: the universe of sets is generated by forming sets of pre-existing objects and then iterating this process. Where the combinatorial conception says something about the nature of sets taken individually, the iterative conception says something about the nature of the sets taken collectively.

---

[23]The logical conception of set is typically traced back to Frege's theory of extensions in Frege (2013); the combinatorial conception is, to the best of my knowledge, first explicitly presented in Bernays (1983).

[24]To offer just one example, in a review of Incurvati (2020), Linnebo writes: "The iterative conception, especially as understood by Gödel, is a combinatorial one" (Linnebo, 2020). Incurvati himself, however, is a little more exact: "even if this combinatorial conception cannot quite be *subsumed* under the iterative conception, it does seem to harmonize quite well with it: the iteration process is naturally understood in a combinatorial manner" (Incurvati, 2020, 44, emphasis original).

With this distinction in mind, what I want to do is tell a story of iterative set formation, but thinking of sets in the logical way rather than in the combinatorial way.

The conception of sets that I will work with is still iterative in that each set has as elements other sets that were formed at earlier stages. The universe can be thought of as the result of iteratively gathering together previously existing objects to form sets, just as the iterative conception would have it. However, on the account I am proposing, sets are not thought of combinatorially, obtained by independently picking some objects from the universe, but rather are picked out as the extension of some property had by objects in the universe. If we imagine the creator being faced with some universe of objects, they form sets of those objects by gathering the extensions of properties (within that universe) together into a single collection. The result will in general be a larger, more inclusive universe. Imagine this process being iteratively applied, beginning with an empty universe. The result is an iterative hierarchy of logical sets, rather than an iterative hierarchy of combinatorial sets.[25]

If we are thinking of sets as extensions of properties, an important question to address is which properties there are. Since we are concerned with pure sets, it is natural to focus on the intrinsic set-theoretic properties of objects, such as *being an ordinal* or *having 7 members*, rather than extrinsic properties such as *being Fred's favorite cardinal*. We can stipulate, then, that by property we mean property definable in the language of set theory. It is most natural also to take 'definable' to mean 'definable with parameters'. Under these stipulations, the iterative hierarchy of logical sets will be a rank-initial segment of Gödel's L. Indeed, it will be the entirety of L, if the iterative logical sets include all the 'real' ordinals.

4.2 Lawless Sequences

The iterative logical conception of set was the first part of my account of forcing potentialism. The second part involves adding a lawless sequence to the story. Lawless sequences are a form of *free choice sequences*—also known as *infinitely proceeding sequences*—which are studied in intuitionistic analysis. Free choice sequences are potentially infinite sequences of numbers, whose

[25]Perhaps the clearest precedent of this iterative-logical view of sets is the view of predicative analysis stemming from Feferman (1964). There is also an obvious affinity to the cumulative hierarchy of constructible sets; this is not surprising given that Gödel (1990, 26) and Cohen (1966, 85) both remark on the connections between Gödel's L and predicativity. See also Feferman (2005) in this connection.

values are individually chosen by the subject who creates this sequence. Brouwer (1952) described the matter thus:

> [Intuitionism] recognizes the possibility of generating new mathematical entities: firstly in the form of infinitely proceeding sequences $p_1, p_2, \ldots$, whose terms are chosen more or less freely from mathematical entities previously acquired …(p. 140)

In saying that the terms are chosen more or less freely, Brouwer means that the choice of each new value for the sequence may or may not be subject to any constraints. If a sequence is not subject to any constraints, it is *lawless*; its behavior is completely undetermined. On the other extreme, the creating subject may determine the values of a free choice by following a recursive rule. In this case the values of the sequence would be completely predetermined and the sequence is *lawlike*. In between these two extremes are intermediate degrees of freedom and indeterminacy. For instance, the creating subject may restrict themselves to only choosing values 0 or 1, so that the future values of the sequence are not completely predetermined, but are still subject to some constraints.

Brouwer's mathematics was based on an idealistic metaphysics in which mathematical objects are all mental objects.[26] As such, there was nothing unique or problematic about introducing as a new mathematical object a sequence that is actively created by some agent. As a result, free choice sequences have not gained much purchase outside of intuitionistic circles. The essence of a free choice sequence, however, is just that of a potentially infinite sequence whose values may not be predetermined. Availing ourselves of the modal approach to potential infinity, it is easy to make sense of free choice sequences without the baggage of Brouwer's metaphysics.[27]

For our purposes here, a few simple postulates about the nature of a single lawless sequence ℓ will suffice. The goal is not to develop a comprehensive axiomatic theory of lawless sequences in set theory, but just to identify the essential properties of a lawless sequence.[28] Formally, ℓ is a two-place relation and can be thought of informally as a growing list (hence the choice of 'ℓ').

---

[26]See Brouwer (1913), Brouwer (1952), and Iemhoff (2020) for overviews of Brouwer's thought. See also Heyting (1971) for an introduction to intuitionistic mathematics more generally.

[27]See Brauer et al. (2022) and Brauer (2023).

[28]For a comprehensive axiomatic theory of lawless sequences—albeit in the context of analysis rather than set theory—see Brauer (2023).

Also, let Ord abbreviate a definition of the ordinals as transitive sets linearly ordered by $\in$.

L1 $\Box\forall x\forall y(\ell xy \rightarrow x \in Ord)$

This says that ℓ associates ordinals with sets.

L2 $\Box\forall x\forall y\forall z(\ell xy \wedge \ell xz \rightarrow y = z)$

For each ordinal, ℓ has a *unique* value (if any).

L3 $\Box\Diamond\exists x \in Ord\neg\exists y\ell xy$

This says ℓ is always incomplete.

L4 $\Box\forall x \in Ord\Diamond\exists y\ell xy$

ℓ can always be extended so that it can be defined on any ordinal. Combined with the last axiom, this captures the potentially infinite character of ℓ.

L5 $\Box\forall x\forall y(\ell xy \rightarrow \Box\ell xy)$

Once values of ℓ are chosen, they are fixed and cannot be changed.

L6 $\Box\forall x(\exists y\ell xy \rightarrow \forall z \in x\exists w\ell zw)$

ℓ has no gaps: if it is defined up to x then it is defined on all ordinals less than x.

L7 $\Box\forall x(\neg\exists y\ell xy \rightarrow \forall y\Diamond\ell xy)$

If ℓ is not yet defined on x, then any value is possible for x. This is the basic postulate about the unconstrained character of the lawless sequence. We can safely take the modality here to be governed by the logic **S4**.[29]

4.3 A Model of Height and Width Potentialism

Putting together the iterative logical conception of sets and a potentially infinite lawless sequence of sets, we have the following picture of the set-theoretic universe. Sets are created through an iterative process of taking the extensions of definable properties, while simultaneously a lawless sequence of sets is being created. The critical assumption for making this both a height and width potentialist view is that the lawless sequence ℓ can figure in the formulas

[29]Brauer (2020) argues that a potentialist theory of infinite sequences requires a new modal operator, which he calls 'inevitability'. Obviously, I endorse those arguments. But while a general theory of potentially infinite sequences would need such an operator, it does not follow that every application of potentially infinite sequences requires that extra operator.

that are used to define the sets. This can be simplified to the assumption that one of the definable sets is the graph of ℓ, so that there is a set x such that $z \in x \leftrightarrow \exists z_1 \exists z_2 (z = \langle z_1, z_2 \rangle \wedge \ell z_1 z_2)$.

This picture of the process of set creation can be suggestively summarized as follows, where the steps are indexed by ordinals from our background metatheory.

- **Step 0**: Create the empty set.
- **Step** $\alpha + 1$:

  (a) If you have already created some ordinal γ on which ℓ is currently undefined, then, if you want to, you can extend ℓ by choosing new elements of the sequence from already existing sets.
  (b) If you have extended ℓ, create the set $\{z : \exists z_1 \exists z_2 (z = \langle z_1, z_2 \rangle \wedge \ell z_1 z_2)\}$.
  (c) Going back through the sets created at steps $\leq \alpha$, add any sets that are newly definable without extending the universe vertically.[30] Go back to step (a).
  (d) If you have chosen not to extend ℓ, or if ℓ is already defined on all the currently existing ordinals, then create new sets as the extensions of all currently definable properties (i.e., take the definable powerset of the present universe.)

- **Limit Step** $\lambda = \bigcup_i \alpha_i$: If you have gone through each of steps $\alpha_1, \alpha_2, \alpha_3, \ldots$, take their union. (That is, we postulate the existence of a limit stage beyond each of $\alpha_1, \alpha_2, \alpha_3, \ldots$.)

This view is height potentialist because the creator can always continue the iterative process of forming definable powersets in step (d). And it is width potentialist because the lawless sequence can be extended in a way that makes new sets definable at the already-existing levels. For instance, suppose that the creator has so far decided not to add any sets to the lawless sequence, but has iterated the definable powerset some large countable number of times γ so that the universe thus far created is a countable model L_γ of **ZFC**. And then suppose the creator decides to choose the first ω elements of ℓ, and they do so in such a way that the graph x of ℓ up to ω is a generic set over some partial order in L_γ. Then the result of adding this set and closing under definable operations will be $L_\gamma(x)$, which is a width extension of L_γ.

[30] We can say that you have extended the universe vertically if you have added a new set y such that previously there existed a set x which was not a member of any other set but which is now a member of y.

There is a simple model that elegantly captures the potentialist system associated with this view. The idea of a 'potentialist system' is to define a class $\mathfrak{M}$ of structures in a given language—in our case, the language $\{\in, \ell\}$—and a relation R between those structures. Then using R as a Kripkean accessibility relation, the whole system $\mathfrak{M}$ is then a structure for a modal extension of the original language using the standard clause:

$$\forall \mathcal{M} \in \mathfrak{M} : \ \mathcal{M} \models \Box\phi \text{ iff } \forall \mathcal{N} \in \mathfrak{M}, \text{ if } \mathcal{M}R\mathcal{N} \text{ then } \mathcal{N} \models \phi$$

For our potentialist system, the class of structures will be the ordered pairs $\langle \alpha, \xi \rangle$ satisfying the following conditions:

- $\alpha \geq \omega$
- ξ is a sequence of sets of length $\beta < \alpha$
- For every $\gamma \leq \beta$, there is some $\gamma' < \gamma$ such that $\xi(\gamma) \in L_\alpha(\xi \restriction \gamma')$

To define truth in these structures, interpret ξ as the graph of ℓ, so that any sets a and b satisfy the formula ℓxy at the world $\langle \alpha, \xi \rangle$ iff $\xi(a) = b$. Then for set-theoretic formulas ϕ, put:

$$\langle \alpha, \xi \rangle \models \phi \ :\leftrightarrow \ L_\alpha(\xi) \models \phi$$

The reason for taking the worlds to be pairs $\langle \alpha, \xi \rangle$ rather than simply the sets $L_\alpha(\xi)$ is that we want to recognize the difference between worlds where $\ell = \xi$ and worlds where $\ell = \zeta \neq \xi$, even if the sets $L_\alpha(\xi)$ and $L_\alpha(\zeta)$ are identical. For instance, we want to recognize the difference between a world where the lawless sequence ℓ has been assigned the single value $\emptyset$ and a world where ℓ has been assigned the single value $\{\emptyset\}$. Although the sets $L_\omega(\langle \emptyset \rangle)$ and $L_\omega(\langle \{\emptyset\} \rangle)$ are identical, we are able to respect the intuitive difference between these two stages of our potentialist system by representing these stages as $\langle \omega, \langle \emptyset \rangle \rangle$ and $\langle \omega, \langle \{\emptyset\} \rangle \rangle$.

The assumption that $\alpha \geq \omega$ is adopted for simplicity. Since the distinctive parts of set theory concern the transfinite, we may as well take the hereditarily finite sets as given. The relation R is defined in the obvious way: $\langle \alpha, \xi \rangle R \langle \beta, \zeta \rangle$ iff $\alpha \leq \beta$ and ξ is an initial segment of ζ. Denote this potentialist system $\mathfrak{L}$.

4.3.1 Formal Properties of $\mathfrak{L}$

It is clear that this potentialist system is not convergent, but exhibits branching behavior. For instance, from $\langle \omega, \emptyset \rangle$, both $\langle \omega, \langle \emptyset \rangle \rangle$ and $\langle \omega, \langle \{\emptyset\} \rangle \rangle$ are accessible, but there is no mutual extension of these worlds in our potentialist system because $\langle\emptyset\rangle$ and $\langle\{\emptyset\}\rangle$ are mutually incompatible initial segments of ℓ.

On the other hand, because $L_\omega(\langle\emptyset\rangle)$ and $L_\omega(\langle\{\emptyset\}\rangle)$ are identical qua sets, exactly the same set-theoretic formulas will be true at the two worlds. Generalizing this observation, for any two worlds $L_\alpha(\xi)$ and $L_\alpha(\zeta)$, there will be respective extensions $L_\beta(\xi^\frown\zeta)$ and $L_\beta(\zeta^\frown\xi)$ that satisfy the exactly same set-theoretic formulas as each other. Thus, as we saw was the case with the system of forcing extensions of a model of set theory, even though the potentialist system is not convergent, the G axiom $\Diamond\Box\phi \rightarrow \Box\Diamond\phi$ is still valid.

Proposition 11.1 *The modal validities of $\mathfrak{L}$ for formulas in the language of set theory are exactly those of* ***S4.2****.*

(A proof is sketched in the appendix.)

Within the potentialist system $\mathfrak{L}$ we can also find various models interpreting **ZFC** by defining 'vertical modalities'. Let $\overline{\ell}$ denote the graph of ℓ; then given ξ, define $\Box_\xi\phi :\leftrightarrow \Box(\overline{\ell} = \xi \rightarrow \phi)$ and $\Diamond_\xi\phi :\leftrightarrow \Diamond(\overline{\ell} = \xi \wedge \phi)$. Now it is easy to verify that, conditional on $\overline{\ell} = \xi$, these modalities $\Box_\xi$ and $\Diamond_\xi$ will satisfy the logic **S4.2**; moreover, atomic formulas in the language of set theory will be *stable*. That is, the following hold:

$$\Box_\xi\forall x\forall y(x \in y \rightarrow \Box_\xi x \in y) \text{ and } \Box_\xi\forall x\forall y(x \notin y \rightarrow \Box_\xi x \notin y)$$

The vertical modalities $\Box_\xi$ and $\Diamond_\xi$ have the effect of restricting our attention to the linearly ordered $L_\alpha(\xi)$'s. Thus, contained within $\mathfrak{L}$ there is a height potentialist system that approximates $L(\xi)$ in the limit. To reason about this height potentialist system as a whole, then, we want to talk not necessarily about what is true at any given stage, but what can happen in general; in particular, we want to generalize not only over the sets that exist at a given stage, but over the *possible sets* that may be created at some stage in the potential construction of $L(\xi)$. This motivates the potentialist translation familiar from Linnebo (2013).

Definition 11.2 Given a formula ϕ in the language of set theory, let $\phi^{\Diamond_\xi}$ be the result of replacing each instance of $\forall$ with $\Box_\xi\forall$ and each instance of $\exists$ with $\Diamond_\xi\exists$.

Then we have:

Proposition 11.3 $\mathfrak{L} \models \Box_\xi\, (\mathbf{ZFC} + V = L(\xi))^{\Diamond_\xi}$

4.3.2 Philosophical Remarks

The potentialist system $\mathfrak{L}$ is both height and width potentialist, as desired. Moreover, it will include all the forcing extensions of any $L_\alpha(\xi)$ in the system that is a model of **ZFC**. If $\langle g_i \rangle$ enumerates a generic G for some partial order in $L_\alpha(\xi)$, then the extension $L_\alpha(\xi^\frown\langle g_i\rangle)$ will also be in $\mathfrak{L}$. It is easy to see that $L_\alpha(\xi^\frown\langle g_i\rangle)$ is the same as the forcing extension $L_\alpha(\xi)[G]$.[31]

On the other hand, $\mathfrak{L}$ also includes width extensions that are not forcing extensions. For instance, $L_{\omega+1}(\emptyset)$ has the extension $L_{\omega+1}(0^\#)$, which is not a forcing extension. $\mathfrak{L}$ includes all, but not only, the forcing extensions. However, we do have an explanation at hand for *why* there must exist non-forcing extensions alongside the forcing extensions. The very lawlessness of ℓ that allows it to enumerate generics is also what allows it to enumerate non-generics. We can no more prevent ℓ from carrying a lot of information, say by enumerating $0^\#$, than we can prevent it from being very random, say by enumerating a generic.

The potentialist system $\mathfrak{L}$ is a nice toy model of the height and width potentialist picture of sets I have proposed, based on the iterative logical conception of sets and a lawless sequence. The definition of $\mathfrak{L}$, however, was given in a background theory that is prima facie highly actualist. So while this model gives a proof of concept for the potentialist, it is not actually meaningful by the potentialist's own lights. If they were to point to $\mathfrak{L}$ as an articulation of their view, they would be open to the revenge objection that they have implicitly assumed actualism in their articulation of potentialism. A solution for the potentialist would be to offer an axiomatization of the height and width potentialist picture that is able to give a proof-theoretic interpretation of $\mathbf{ZFC} + V = L(\xi)$ in place of the model-theoretic Proposition 11.3.[32] While it would not be difficult to find some axioms that did the job technically, the

[31] This is basically just the fact that when G is a generic for a poset in L_α, the forcing extension $L_\alpha[G]$ is the same as $L_\alpha[G]$ in the sense of constructible with a predicate for G. That the forcing extension $L_\alpha[G]$ is included in the relativized constructible set $L_\alpha[G]$ is immediate from the minimality of forcing extensions. Conversely, the Forcing Theorem entails that every set that is definable using the parameter G will also be definable in the forcing extension $L_\alpha[G]$.

[32] Thus, the present view is in roughly the situation of Meadows (2015) before Scambler (2021).

challenging part would be to find an axiomatization that can be intrinsically motivated from the potentialist perspective.

Finally, let me return to the question of how potentialism relates to multiversism. The model I have sketched of height and width potentialism reveals that there is a coherent rationale for a potentialist system that includes forcing extensions, and this rationale did not rely on any distinctively pluralistic or multiversesque ideas. Width potentialism does not necessarily rely on multiverse views. However, there is a connection between my height and width potentialist system and a form of multiverse view, albeit not the generic multiverse. Proposition 11.3 shows that the universes $L(\xi)$ can be found within $\mathfrak{L}$ by restricting one's point of view to a vertical perspective. If you only look up, you will see $L(\xi)$. But take a step to the left or right and you might find yourself looking up at $L(\xi^\frown\zeta)$ instead. Rather than following from multiverse views, my picture of height and width potentialism leads to a certain multiverse view.

5 Conclusion

In this chapter, I have asked what philosophical view, if any, might motivate forcing potentialism. Studying the modal logic of forcing, and related topics, is an interesting mathematical project and is perfectly legitimate in its own right. Nevertheless, we can also ask whether there is a particular philosophical motivation for this study based on the concept of potential infinity. And if so, what light does this shed on the relation between potentialism, pluralism, and multiversism?

But there is an obstacle to answering these questions. I have argued that the standard iterative, combinatorial conception of sets only sanctions versions of convergent potentialism. Forcing extensions, by contrast, exhibit branching possibilities. If there is a view of potentialism where the possible extensions of a universe are all and only the forcing extensions, then it must be based on a different conception of set than the iterative combinatorial conception.

To meet this challenge, I introduced the *iterative logical* conception of sets. Then, alongside the iterative process of set formation, we imagine a lawless sequence of freely chosen sets. Together, this leads to a picture of set theory that is both height and width potentialist. The width extensions will include all but not only the forcing extensions. In this respect, my account may fall somewhat short in recapturing *forcing potentialism*. However, the account does allow an explanation of *why* it includes width extensions that are not forcing extensions. And although my account does not rely on any multiversesque ideas, a certain multiverse-style view naturally falls out of the account.

I have thus offered one way to reconcile height and width potentialism and to find a genuine rationale for forcing potentialism rooted directly in the concept of potential infinity. I am not, however, claiming that this is the only way to do so. Exploring alternative approaches to these questions remains an interesting and largely unexplored topic. What other ways are there of melding height and width potentialism? Are there other conceptions of set that cohere with a potentialist theory of forcing extensions and other width extensions? Working within the iterative logical conception of sets, I used a lawless sequence to facilitate an increase in expressive resources and thereby width extensions. What other ways are there to increase the expressive resources, and what width extensions do they lead to?

Acknowledgments Thanks are due to an audience at the UNICAMP-Konstanz philosophy of set theory seminar where I presented this material in April 2021. Thanks also to Chris Scambler, two anonymous referees, and especially Neil Barton for reading and discussing drafts of this chapter. This work was supported by the Research Grants Council of Hong Kong SAR, China, grant PDFS2021-3H05.

A Proofs

A.1 Proposition 11.1

Define the *modal validities* of a potentialist system in a language $\mathcal{L}$ to be the set of formulas $\psi(p_1, \ldots, p_n)$ in the language of modal logic such that for every sentence $\psi_1, \ldots, \psi_n$ in $\mathcal{L}$, $\phi(\psi_1, \ldots, \psi_n)$ is true at every world W in the potentialist system. (For concreteness, I will assume that the ψ_i's do not have parameters from W.)

To prove Proposition 11.1 we need to show, first, that **S4.2** is a lower bound to the modal validities of $\mathfrak{L}$ in the language of set theory and, second, that **S4.2** is also an upper bound.

To show that **S4.2** is a lower bound is straightforward. That the modal validities include the K, T, and 4 axioms all follow from the fact that the accessibility relation in $\mathfrak{L}$ is transitive and reflexive. That the modal validities are closed under necessitation and modus ponens is obvious.

It remains to show that the G axiom is valid. If ϕ is any formula in the language of set theory plus modal operators, but with no occurrences of ℓ, then by an easy induction on the modal degree of ϕ one can prove:

(*) If $L_\alpha(\xi)$ and $L_\gamma(\zeta)$ are identical as sets, then $\langle \alpha, \xi \rangle \models_{\mathfrak{L}} \phi$ iff $\langle \gamma, \zeta \rangle \models_{\mathfrak{L}} \phi$.

Now suppose that $\Diamond\Box\phi$ holds in $\langle \alpha, \xi \rangle$, so that there is some $\langle \beta, \xi^\frown\zeta \rangle$ where $\Box\phi$ holds. Then consider an arbitrary extension $\langle \gamma, \xi^\frown\eta \rangle$ of $\langle \alpha, \xi \rangle$. Then because $\Box\phi$ holds in $\langle \beta, \xi^\frown\zeta \rangle$, ϕ will also hold in $\langle \sup(\beta + \gamma, \gamma + \beta), \xi^\frown\zeta^\frown\eta \rangle$. Hence by (*) ϕ also holds in $\langle \sup(\beta + \gamma, \gamma + \beta), \xi^\frown\eta^\frown\zeta \rangle$. So $\Diamond\phi$ holds in $\langle \gamma, \xi^\frown\eta \rangle$, and $\Box\Diamond\phi$ must hold in $\langle \alpha, \xi \rangle$.

Second, we want to observe that **S4.2** is an upper bound on the modal validities of $\mathfrak{L}$. This follows from several ideas and results in Hamkins and Löwe (2008), Hamkins et al. (2015), and Hamkins and Linnebo (2022). The argument turns on the concepts of a *button* and a *dial*. A button is a sentence ϕ such that $\Diamond\Box\phi$ holds at every world; if $\Box\phi$ holds at a world, then the button is 'pushed'. A dial is a set of sentences $\phi_1, \ldots, \phi_n$ such that at every world exactly one of the ϕ_i's holds, while all the ϕ_j's are possible. If ϕ_k is the one that holds, then the dial is 'set to k'. Then (adapting an earlier theorem from Hamkins et al. (2015)) Hamkins and Linnebo (2022, Theorem 5) show that, given a potentialist system with arbitrarily large collections of independent buttons and an arbitrarily large dial that is independent of those buttons, at a world where none of those buttons are pushed the modal validities are bounded above by **S4.2**.

Finding an arbitrarily large dial in $\mathfrak{L}$ is easy. Intuitively, let ϕ_i say that $L_\alpha(\xi)$ has been extended to the height of an ith successor ordinal, with ϕ_0 saying that α is a limit ordinal. For a dial of size n, let ϕ_n say that α is a $\geq n$ successor ordinal.

Finally, we need to find an arbitrarily large set of buttons which are independent of each other and of the dial. For this, we can draw on (Hamkins and Löwe, 2008, Lemma 6.1), who observe that any model of $V = L$ contains the following infinite family of independent buttons:

- Let button b_n be "ω_n^L is not a cardinal".[1]

They are working in the potentialist system of consisting of forcing extensions, but it is easy to see that b_n's are also buttons in the potentialist system $\mathfrak{L}$. A cardinal collapsed by moving to a width extension (such as by adding a generic set witnessing the collapse) will remain collapsed in any height extension therefrom. Now, since we can find a world in $\mathfrak{L}$ that is a model of $V = L$—for instance, $\mathfrak{L}$ will include the Sherpherdson-Cohen minimal model—it follows

[1]Note that ω_n^L is a definite description in the language of set theory, not a parameter from the model.

that the modal validities at that world are exactly **S4.2**, and hence **S4.2** is an upper bound for the modal validities of the potentialist system $\mathfrak{L}$.

A.2 Proposition 11.3

By a straightforward induction on complexity, one can prove that for any α, ξ and for any $a_1, \ldots, a_n \in L_\alpha(\xi)$, we have:

$$L(\xi) \models \phi(a_1, \ldots, a_n) \text{ iff } \langle \alpha, \xi \rangle \models_{\mathfrak{L}} \phi^{\Diamond_\xi}(a_1, \ldots, a_n)$$

Since the theorems of $\mathbf{ZFC} + V = L(\xi)$ are all closed sentences with no parameters, it follows that for every α, $\langle \alpha, \xi \rangle \models_{\mathfrak{L}} (\mathbf{ZFC} + V = L(\xi))^{\Diamond_\xi}$, assuming $\langle \alpha, \xi \rangle$ is a member of $\mathfrak{L}$ in the first place. Thus, for every $\langle \gamma, \zeta \rangle \in \mathfrak{L}$, we have $\langle \gamma, \zeta \rangle \models \Box_\xi (\mathbf{ZFC} + V = L(\xi))^{\Diamond_\xi}$.

References

Antos, C., et al. (2015). Multiverse conceptions in set theory. *Synthese, 192*, 2463–2488.

Aristotle (1941). In R. McKeon (Ed.), *The basic works of aristotle*. Random House.

Barton, N. (2021). Indeterminacy and 'The' universe of sets: Mulitversism, potentialism, and pluralism. In M. Fitting (Ed.), *Selected topics from contemporary logics* (pp. 105–182). College Publications.

Bernays, P. (1983). Platonism in mathematics. In P. Benacceraf & H. Putnam (Eds.), *Philosophy of mathematics: Selected readings*. Trans. by Charles Parsons. 2nd. Originally published 1935 in *L'enseignement mathématique*. Cambridge University Press.

Boolos, G. (1971). The iterative conception of set. *Journal of Philosophy, 68*(8), 215–231.

Boolos, G. (1989). Iteration again. *Philosophical Topics, 17*(2), 5–21.

Brauer, E. (2020). The modal logic of potential infinity: Branching vs. convergent possibilities. *Erkenntnis, 87*, 2161–2179.

Brauer, E. (2023). A classical modal theory of lawless sequences. *Bulletin of Symbolic Logic, 29*(3), 406–452.

Brauer, E., Linnebo, Ø., & Shapiro, S. (2022). Divergent potentialism: A modal analysis with an application to choice sequences. *Philosophia Mathematica, 30*(2), 143–172.

Brouwer, L. E. J. (1913). Intuitionism and formalism. *Bulletin of the American Mathematical Society, 20*(2), 81–96.

Brouwer, L. E. J. (1952). Historical background, principles and methods of intuitionism. *South African Journal of Science, 49*, 139–146.

Cantor, G. (1899). Letter to dedekind. In J. van Heijenoort (Ed.), *From frege to Gödel: A sourcebook in mathematical logic* (pp. 113–117). Trans. by S. Bauer-Mengelberg & J. van Heijenoort. Harvard University Press.

Cohen, P.J. (1966). *Set theory and the continuum hypothesis*. W.A. Benjamin.

Feferman, S. (1964). Systems of predicative analysis. *Journal of Symbolic Logic, 29*(1), 1–30.

Feferman, S. (2005). Predicativity. In S. Shapiro (Ed.), *Oxford handbook of philosophy of mathematics and logic* (pp. 590–624). Oxford University Press.

Frege, G. (2013). *Basic laws of arithmetic*. Trans. by P. A. Ebert & M. Rossberg. Oxford University Press.

Fuchs, G., Hamkins, J. D., & Reitz, J. (2015). Set theoretic geology. *Annals of Pure and Applied Logic, 166*(4), 464–501.

Gödel, K. (1990). The consistency of the axiom of choice and the generalized continuum hypothesis (1938). In S. Feferman et al. (Eds.), *Collected works* (Vol. 2). Oxford University Press.

Hamkins, J. D. (2003). A simple maximality principle. *Journal of Symbolic Logic, 68*(2), 527–550.

Hamkins, J. D. (2012). The set-theoretic multiverse. *Review of Symbolic Logic, 5*, 416–449.

Hamkins, J. D. (2016). Upward closure and amalgamation in the generic multiverse of a countable model of set theory. In *RIMS Kyôkyu ̂roku* (pp. 17–31). http://wp.me/p5M0LV-1cv

Hamkins, J. D. (2018). The Modal Logic of Arithmetic Potentialism and the Universal Algorithm. *arXiv:1801.04599v3*.

Hamkins, J. D., Leibman, G., & Löwe, B. (2015). Structural connections between a forcing class and its modal logic. *Israel Journal of Mathematics, 207*(2), 617–651.

Hamkins, J. D., & Linnebo, Ø. (2022). The modal logic of set-theoretic potentialism and the potentialist maximality principles. *Review of Symbolic Logic, 15*(1), 1–35.

Hamkins, J. D., & Löwe, B. (2008). The modal logic of forcing. *Transactions of the American Mathematical Society, 360*(4), 1793–1817.

Hamkins, J. D., & Williams, K. J. (2021). The Σ_1 definable universal finite sequence. *Journal of Symbolic Logic, 87*(2), 783–801.

Hamkins, J. D., & Woodin, W. H. (2018). The Universal Finite Set. *arXiv:1711.07952v2*.

Heyting, A. (1971). *Intuitionism* (3rd ed.). North Holland.

Iemhoff, R. (2020). Intuitionism in the philosophy of mathematics. In E. N. Zalta (Ed.), *The stanford encyclopedia of philosophy*. Fall 2020.

Incurvati, L. (2020). *Conceptions of set and the foundations of mathematics*. Cambridge University Press.

Linnebo, Ø. (2013). The potential heirarchy of sets. *Review of Symbolic Logic, 6*(2), 205–228.

Linnebo, Ø. (2020). Review of Luca incurvati. In *Conceptions of set and the foundations of mathematics*. Notre Dame Philosophical Reviews. https://ndpr.nd.edu/reviews/conceptions-of-set-and-the-foundations-of-mathematics/

Linnebo, Ø., & Shapiro, S. (2017). Actual and potential infinity. In *Nous*.

Maddy, P., & Meadows, T. (2020). A reconstruction of steel's multiverse project. *Bulletin of Symbolic Logic, 26*(2), 118–169.

Meadows, T. (2015). Naive infinitism. *Notre Dame Journal of Formal Logic, 56*(1), 191–212.

Mostowski, A. (1976). A remark on models of the Gödel-Bernays axioms for set theory. In G. H. Müller (Ed.), *Sets and classes: On the work by Paul Bernays*. Studies in Logic and the Foundations of Mathematics (Vol. 84, pp. 325–340). North Holland.

Parsons, C. (1983). Sets and modality. In *Mathematics in philosophy* (pp. 298–341). Cornell University Press.

Paseau, A. (2007). Boolos on the justification of set theory. *Philosophia Mathematica, 15*(1), 30–53.

Potter, M. (1993). Iterative set theory. *Philosophical Quarterly, 43*(171), 178–193.

Potter, M. (2004). *Set theory and its philosophy: A critical introduction*. Oxford University Press.

Rin, B. (2015). Transfinite recursion and computation in the iterative conception of set. *Synthese, 192*, 2437–2462.

Roberts, S. (2020). Ultimate V. *Unpublished manusript*.

Scambler, C. (2021). Can all things be counted? *Journal of Philosophical Logic, 50*, 1079–1106.

Shoenfield, J. (1977). The axioms of set theory. In J. Barwise & H. J. Keisler (Eds.), *Handbook of mathematical logic*. Studies in Logic and the Foundations of Mathematics (Vol. 90, pp. 321–344). North Holland.

Steel, J. R. (2014). Gödel's program. In J. Kennedy (Ed.), *Interpreting Gödel: Critical essays*. Cambridge University Press.

Studd, J. (2013). The iterative conception of set: A (Bi-)modal axiomatization. *Journal of Philosophical Logic, 42*, 697–725.

Troelstra, A. (1977). *Choice sequences: A chapter of intuitionistic mathematics*. Clarendon Press.

Troelstra, A. (1982). On the origen and development of Brouwer's concept of choice sequence. In A. Troelstra & D. van Dalen (Eds.), *L.E.J. Brouwer Centenary Symposium* (pp. 465–486). North Holland.

Woodin, W. H. (2011). The continuum hypothesis, the generic multiverse of sets, and the Ω conjecture. In *Set theory, arithmetic, and foundations of mathematics*. Lecture Notes in Logic (Vol. 36, pp. 13–42). Cambridge University Press.

12

On the Consistency of Height and Width Potentialism

Chris Scambler

1 Overview

Recent work in philosophy of set theory has furnished some arguments that height and width potentialism are inconsistent with one another. One such argument can be found in this volume (Brauer, 2024); another is forthcoming Roberts (2024).

At the same time, others have suggested there may be some merit in the combination of height and width potentialism. Such authors have presented views that appear to manifest the combination in a non-trivial way, and have defended philosophical claims on their basis (e.g. that all sets are ultimately countable—cf. Builes & Wilson, 2022; Meadows, 2015; Scambler, 2021; Pruss, 2020).

Clearly there is a tension here. The business of this chapter is to explain (what I take to be) its solution. I will argue that height and width potentialism are compatible, and hence that the programmes cited above are in good standing. I will do this by developing some axiomatic theories that explicate height and width potentialism, and explaining how a proponent of such theories might reply to the arguments alleging inconsistency. Along the way, we will see how the two arguments at issue do nevertheless serve a useful purpose, in helping to bring out distinctive philosophical commitments of height and

C. Scambler (✉)
All Souls College, University of Oxford, Oxford, UK
e-mail: chris.scambler@all-souls.ox.ac.uk

C. Antos et al. (eds.), *The Palgrave Companion to the Philosophy of Set Theory*, Palgrave Companions, https://doi.org/10.1007/978-3-031-62387-5_12

width potentialism, commitments in set theory and plural logic. I will also present some new results relating height and width potentialism to extensions of second order arithmetic by regularity principles.

The plan is as follows. Section 2 gives some definitional background and context. Section 3 then presents some axiomatic systems that purport to combine height and width potentialism and discusses issues relevant to consistency, citing new results on strength and equivalence to standard 'actualist' set theories. Section 4 presents the inconsistency argument of Roberts and explains how a proponent of height and width potentialism should reply. Section 5 does the same for Brauer.

2 Background

Let me start by explaining the terms.

For present purposes, *potentialism* in set theory will be the idea that there could always be more sets than there in fact are. In slogan form, for the potentialist, the universe of sets is 'indefinitely extendible'.

Height potentialism is the idea that the universe is always extendible 'upwards' to include new sets of higher rank than any given ones. *Width potentialism*, on the other hand, is the idea that the universe is always extendible 'outwards', to contain new sets of no greater rank than the max of any given ones (assuming of course the 'given ones' contain some of rank $> \omega$).

One natural source of motivation for height potentialism comes from considerations involving the set-theoretic paradoxes. Russell's paradox shows there is no set of all non-self-membered sets, and hence that the cumulative hierarchy of all sets is itself not a set. But the height potentialist complains that any 'stopping point' for the cumulative hierarchy would be arbitrary. There is no conceptual obstacle to any ranks of the cumulative hierarchy admitting extensions: indeed this sort of thing is witnessed by the ranks of the cumulative hierarchy within a model of set theory. The potentialist's idea is that, like the set-sized initial segments of the cumulative hierarchy *within* ZFC, any initial segment of the cumulative hierarchy (set-sized or not) can be found as a set in still larger universes.

Of course, standard set theories themselves cannot manifest this idea, since according to them the whole hierarchy is not a set-sized structure and does not admit extensions. But it is natural to explicate this idea of an indefinitely extendible universe of sets in a modal logic. Indeed, modal axiom systems based around core height potentialist ideas are known to exhibit tight forms of equivalence with standard iterative set theories like ZFC, even though they do accommodate the idea that *any* ranks of the cumulative hierarchy *can be* extended.

Width potentialism has been historically less popular, although it has been attracting some attention over the last few decades as a way to understand the independence phenomenon in set theory. This is not surprising, since there is a close structural similarity in the motivation for height and width potentialism, one that ought to recommend width potentialism as a natural view to consider for those interested in height potentialism. Just as the height potentialist begins with the intuition that it is arbitrary that there should be some ranks of the cumulative hierarchy that somehow inherently can't be extended to include further sets, the width potentialist may begin with the idea that it is arbitrary that there should be some universe of sets that cannot be extended by forcing over its partial orders. Just as the height potentialist could support the coherence of their view by appeal to the set-sized initial segments of the cumulative hierarchy according to ZFC, so too can the width potentialist support the coherence of their view by appeal to the countable transitive models of ZFC: If any given universe is a countable transitive model from the point of view of another, larger one, it is easy to see (and formally spelled out below) that forcing is always possible.

Again, as with height potentialism, ZFC tells us it cannot be right: some forcing notions do not admit generics. But as before, the width potentialist can formalise their view using modal logic: one can formulate an axiom to the effect that, for any partial order $\mathbb{P}$, it is possible to find a generic for $\mathbb{P}$. Such theories are known, although not as well as their height potentialist cousins. Part of this chapter's contribution is in examining their mathematical (consistency) strength, and relation to familiar non-modal mathematical theories.

Of course, the width potentialist's view is more radical than the standard height potentialist's view, in that it goes further against standard set-theoretic orthodoxy in various ways. Indeed, it is so far from orthodoxy that, as I said above, it has recently been suggested that it strays into logical or at least conceptual inconsistency. The purpose of this chapter is to lay out some height and width potentialist theories, and defend them against such charges.

3 Height and Width Potentialism Combined

In this section, I will present some axiom systems that seem to explicate the intuitive idea of height and width potentialism, as described above.

3.1 Core Logical Principles

All the versions of height and width potentialism we will consider will be built around two principles: the first is the height potentialist principle that any things can be the elements of a set; the second is the width potentialist 'forcing axiom', that a generic filter can be found for any partial order.

Both potentialist principles involve modality; the first also involves plural quantification. Accordingly, the language $\mathcal{L}_0$ we use to formulate these theories will have at least a modal operator $\Box$, singular variables x_n, plural variables X_n, the propositional connectives and quantifiers, the identity symbol $=$, and the symbol $\in$ for set membership. The circumstance that some thing x is one of some things X will be represented by the concatenation Xx. Identity is only well formed between terms of the same type (singular/plural).

The following core axioms will be included in all the potentialist theories we will go on to discuss.

FQ Standard rules for free quantifier logic with identity for each type of variable.

Mod The modal logic **S4.2** with converse Barcan formula, necessitation, and standard rules for universal quantification within the scope of $\Box$.

P-ext $\Box\forall x[\Diamond Xx \equiv \Diamond Yx] \supset X = Y$

PR The plural rigidity axioms

1. $\Diamond Xx \rightarrow \Box Xx$
2. $\Diamond(\exists x[Xx \wedge x = y]) \rightarrow \exists x[Xx \wedge x = y]$.

Comp For any formula $\varphi(x, y, Y)$, $\Box\forall y\Box\forall Y\Box\exists X\forall x[Xx \leftrightarrow \varphi(x, y, Y)]$ is an axiom.

P-Choice A plural version of the axiom of choice.

M-Rep A modal version of the axiom scheme of replacement.

Inf An axiom saying that the set of natural numbers exists.

Set-Ext The axiom of extensionality in the form $\Box\forall x[\Diamond x \in y \equiv \Diamond x \in z] \supset y = z$

Sets The axiom of foundation, and an axiom $\Box\forall x\exists X\Box\forall y[Xy \leftrightarrow y \in x]$ asserting the rigidity of set membership.

These axioms provide the basic principles of plural logic and set theory that will support the exploration of various potentialist axioms of set existence. The axiom of infinity is assumed in its standard (rather than potentialist) form just as a simplifying measure; all the phenomena we are interested in occur at the level of infinite sets.

We now turn to the development of potentialist axiom systems over the core logic.

3.2 Simple HWP

In this section I will present what seems to me to be the simplest formal combination of height and width potentialism. The theory is not mathematically strong, being exactly equivalent in strength to second order arithmetic. But it does offer a clean and simple proof of concept for the combination of height and width potentialism, as well as a useful starting point for further extensions.

The first axiom we add is:

HP $\Box\forall X\Diamond\exists x\forall y[y \in x \leftrightarrow Xy]$

- *Any things can be the elements of a set.*

HP enshrines the core idea of height potentialism, since it implies (given the usual notion of rank) that given any sets one can find others of still higher rank.

The second requires a little more preparation. In our background logic we can define the notion of a partial order, the notion of being a filter on a partial order, and the notion of being a dense set in a partial order in the standard way. Let blackboard variables $\mathbb{P}, \mathbb{Q}$ range over partial orders. (These are singular variables.) Let $D(X, \mathbb{P})$ mean that the Xs are all the dense subsets of $\mathbb{P}$. Finally, let $F_\cap(x, \mathbb{P}, X)$ mean that x is a filter on $\mathbb{P}$ that has non-empty intersection with every one of the X. Then our width potentialist axiom can be stated as:

WP $\Box\forall\mathbb{P}\forall X[D(X, \mathbb{P}) \rightarrow \Diamond\exists g(F_\cap(g, \mathbb{P}, X))]$

- *Any partial order can be used in forcing.*

WP is one way to flesh out width potentialism, since it implies that any sufficiently rich plurality of sets may be extended to include ones of rank no greater than the max of their ranks by forcing.

The following proposition will be useful to us going forward. In it, $f : \mathbb{N} \twoheadrightarrow X$ is an abbreviation for the assertion that f is a function on $\mathbb{N}$ with every one of the Xs in its range.

Proposition 12.1 *Let Count be the principle:*

$$\Box\forall X\Diamond\exists f[f:\mathbb{N}\twoheadrightarrow X]$$

then Count is equivalent to WP over the core logic + HP.

Proof See Scambler (2021). □

Besides its technical use to us further down the line, this equivalence result also points in the direction of a possible motivation for WP, grounded in the intuitive idea that there should be only one size of infinity. For more on this connection, see Meadows (2015) and Builes and Wilson (2022).

In any event, let us call the result of adding HP and either WP or Count to the core logic Simple Height and Width Potentialism, or SHWP.

Turning now to questions of consistency, one can prove that SHWP is consistent relative to Second Order Arithmetic (SOA), and that in fact SHWP itself interprets SOA back under a natural translation. The theories are (close to) 'mutually interpretable', as the jargon goes.

(Here and below, we understand SOA under the guise of ZFC without power + all sets are countable. This is definitionally equivalent to the more standard arithmetical formulations; see, e.g., Krapf (2017, p97 ff).)

Let $\mathcal{L}_\in$ be the first order language of set theory. Then:

Theorem 12.2 *There is a map* $\cdot^\Diamond : \mathcal{L}_\in \to \mathcal{L}_0$ *that preserves theoremhood from SOA to SHWP.*

Theorem 12.3 *There is a map* $\cdot^\exists : \mathcal{L}_0 \to \mathcal{L}_\in$ *that preserves theoremhood from SHWP to SOA.*

Theorem 12.2 says that SHWP interprets SOA. The translation $\varphi \mapsto \varphi^\Diamond$ proceeds by prepending every universal quantifier in φ by a $\Box$ and every existential by a $\Diamond$. A result due to Linnebo[1] says that for $\varphi \in \mathcal{L}_\in$,

$$\Gamma \vdash \varphi \Leftrightarrow \Gamma^\Diamond \vdash_{Core} \varphi^\Diamond$$

where $\Gamma^\Diamond$ has the obvious meaning of $\{\gamma^\Diamond : \gamma \in \Gamma\}$. It thus suffices to prove $\varphi^\Diamond$ for each axiom of SOA in SHWP, which (using Proposition 12.1) is not difficult.

[1] See (Linnebo, 2018, p74).

It is worth emphasising that SOA is formulated here as a *first order set theory*, somewhat confusingly; and if we extend that first order set theory to a second order language in the usual way, the result then fails to hold. The reason for this is that second order set theories will typically have comprehension axioms for the second order variables

$$\exists X \forall y [Xy \equiv \varphi(y)]$$

while the instance

$$\Diamond \exists X \Box \forall y [Xy \equiv y = y]$$

of the translation of this schema is demonstrably false in SHWP and its extensions.

This incompatibility suggests that the equivalence between these theories is not deep. Ultimately, they disagree on questions about second order existence, such as whether there can be a plurality of all (possible) sets. Nevertheless, it is useful in that it means we can help ourselves to the modal translations of arithmetic truths (and the corresponding truths of the fragment of set theory) when reasoning in height and width potentialist theories.

Theorem 12.3 is of more direct significance to us. It says that SOA interprets SHWP and hence secures the consistency of the latter relative to the former. Full details of the ideal proof are a little fiddly and would go beyond our needs here. But some idea of how things go down will be useful.

The key idea behind the translation $\varphi \mapsto \varphi^{\exists}$ is to factor out use of modal operators in favour of quantification over possible worlds. Ultimately what the actualist aims to do is to define a notion of a world, and of a formula's holding at a world, in such a way that when the potentialist says something is possible the actualist has a corresponding truth at a possible world.

The introduction of talk of worlds raises some questions about the potentialist view that are not settled by the axioms up to now, but that are settled under translation, in various ways, depending on the choice of definition for 'possible world'. For example, supposing the Xs are everything and the Ys are some arbitrary things, is it possible that everything is either one of the Xs or the set whose elements are the Ys?

Various answers to this question and others like it might be entertained. One might, for example, require that whenever some set is introduced, so too are all its subsets: possible worlds will then have to satisfy this closure condition, resembling the familiar V_αs, and the above question will be answered negatively in general. Another, more liberal option would be to answer in the affirmative in all cases. This corresponds to allowing set introduction in a maximally 'piecemeal' fashion: given any things, you can introduce just the set of them and nothing else.

However this question is answered by the potentialist, the actualist may introduce a corresponding formula, $World(x)$, defined so as to pick out those sets that the potentialist thinks of as *possible worlds*, or in this context simply as sets that the potentialist might allow to be all the sets. The actualist will then use that definition, along with quantification over such worlds, to interpret the potentialist's claims of possibility and necessity.

Formally, this means our translation will have to carry formulas φ in the modal language (which may be sentences) to formulas $\varphi^{\exists}(w)$ in the first order language with a free 'world' variable w. Key clauses of the translation are things like

$$(\forall x\varphi)^{\exists}(w) := \forall x \in w\varphi^{\exists}(w)$$
$$(\forall X\varphi)^{\exists}(w) := \forall x \subseteq w\varphi^{\exists}(w)$$
$$(\Box\varphi)^{\exists}(w) := \forall u[World(u) \wedge w \subseteq u \rightarrow \varphi^{\exists}(u)]$$

where $World(u)$ is the definition of world at issue.

For the sake of concreteness, let us take a maximally liberal approach to set formation where given any things one is allowed to introduce the set of those things and nothing else. A 'possible world' will then be any transitive set. The potentialist's axioms for SHWP are then readily seen to be true, under the translation, in SOA. For HP, this is because every set is an element of a transitive set. For WP, this is because SOA proves all sets are countable. So for any given partial order, we may move to a transitive set that witnesses its countability, and then proceed to introduce a generic for it if need be.

3.3 Strong HWP

The theory I have just cited combines height and width potentialism in a straightforward way. The result is a theory that is (up to something like mutual interpretability) essentially just second order arithmetic. This is a pretty weak theory, although as we know from the reverse mathematics literature it is plenty strong enough to develop much of the mathematics needed in applications. Can the height and width potentialist do better?

There is indeed a reasonable way to proceed here. Our potentialist is interested in expansions of the universe along two 'directions'. One has the ability to extend the universe 'upwards' to create sets of higher rank; and one has the ability to extend it 'outwards' by forcing. By separating out these two possible methods of expansion, and by asserting strong axioms regarding what can be attained along the vertical dimension of expansion alone, much stronger HWP systems can be derived, indeed ones with all the power of ZFC and more. This extension makes the general picture of height and width potentialism much more philosophically interesting, since such versions have at least the consistency strength to recover all standard mathematics. But they also bring new conceptual problems with them, problems that will be exploited in one of the inconsistency arguments we will consider.

Let us first discuss in a little more detail how to implement this idea. We expand the language $\mathcal{L}_0$ to $\mathcal{L}_1$ by adding in two new modal operators, ⦶ and ⦸. ⦶ is to reflect possibility by *only* vertical expansion: one can think of this as possibility by *only* iteratively introducing new sets for given pluralities. ⦸, on the other hand, is to reflect possibility by *only* horizontal expansion: one can think of this, indifferently, as just by adding new generic filters, or adding new enumerating functions. (◫ and ⧅ are the duals.) $\Diamond$ remains in the language, and represents 'absolute' possibility, that is, possibility by either domain expansion method.

Here, in outline, is a way to axiomatise a 'strong' system of this kind. (More details can be found in Scambler, 2021.)

We begin by modifying HP and WP.

HP $\Box\forall X$ ⦶ $\exists x\forall y[y \in x \leftrightarrow Xy]$

WP $\Box\forall\mathbb{P}\forall X[D(X,\mathbb{P}) \rightarrow$ ⦸$\exists g(F_{\cap}(g,\mathbb{P},X))]$

Next, we add rules to reflect the generality of $\Diamond$.

Gh ⦶$\varphi \rightarrow \Diamond\varphi$
Gw ⦸$\varphi \rightarrow \Diamond\varphi$

Finally, add the following restricted version of the powerset axiom to vertical possibility.

r-Pow $\Box\forall x$ ⦶ $\exists y$ ◫ $\forall z[z \in y \leftrightarrow z \subseteq x]$

r-Pow says that it is always possible by vertical expansion to get all the subsets of any given set, *so long as one disregards the subsets you can get by forcing*. This is a kind of local powerset axiom: in intuitive (procedural) terms, it says that you can always eventually get all the subsets of a given set you can get without forcing, if you go on introducing sets long enough.

It is useful to compare r-Pow with the corresponding 'unrestricted' version

Pow $\Box\forall x$ ⦶ $\exists y\Box\forall z[z \in y \leftrightarrow z \subseteq x]$

which is provably inconsistent with **WP** over the rest of the theory. Pow says that it is possible for there to be such things as *all possible* subsets of any given set; but given the existence of an infinite set (as we are guaranteed), and the universal possibility of forcing, enshrined in **WP**, this cannot be, since we can always force to add new subsets to any given set.[2]

Let the extension of the core theory by the above principles (but not, of course, Pow) be called HWP. The following facts come easily.

Theorem 12.4 *There is a translation* $\cdot^{⦶} : \mathcal{L}_{\in} \to \mathcal{L}_1$ *that preserves theoremhood from ZFC to HWP.*

Theorem 12.5 *There is a translation* $\cdot^{\Diamond} : \mathcal{L}_{\in} \to \mathcal{L}_1$ *that preserves theoremhood from SOA to HWP.*

In each case, the result follows much as it did in the previous case. In fact Theorem 12.5 really is just Theorem 12.2. The first uses the same translation but with ⦶ in place of $\Diamond$ and ⎅ in place of $\Box$ everywhere. What the theorem goes through is a known theorem of Linnebo in a slightly different key.

The resulting theory has some intriguing features. If one restricts one's attention to vertical possibility and necessity, one can construct objects that satisfy anything you can get in ZFC, for example the existence of uncountable cardinals, beth fixed points, and so on. But these large cardinals (to describe them a little provocatively) are always mirages; and the mirage can always be revealed by appeal to ⦸-possibility. Anything that might in some possible world satisfy the formula saying that it is (say) ω_1 will also not satisfy that description in some other possible world. There are no *absolute* uncountables, only relative pretenders.

The chief advantage HWP has over SHWP is interpretative power: it can interpret all of ZFC, notwithstanding its ultimate commitment to the countability of all things. This gives HWP the capacity, at least once motivational details are filled in, to potentially offer a countabilist foundation for all mathematics. This is one of the principal interests of the combination of height and width potentialism.

I have said that the chief advantage of HWP over SHWP is interpretative power. But how much of that does it have? Does it have too much? We now turn to addressing these issues of consistency in HWP.

[2] The philosopher interested in questions of motivation may wonder why, once Pow has been shown to be inconsistent (given **WP**), we should nevertheless be interested in r-Pow. The answer here is simple in the first case: mathematical strength, and the desire to accommodate as much mathematics after Cantor as possible, compatibly with **WP**. In the second instance, the connection to topological regularity properties (unfolded later in the chapter) provides further scope for conceptual motivation, although the full exploration of these connections is postponed for further work.

It turns out that the methods employed in the previous section generalise fairly naturally to cater to systems like HWP and even further extensions. The generalisation involves extending second order arithmetic to include 'topological regularity axioms', and then giving a more nuanced definition of possible world in terms of such an extension.

Let's start with the extension of second order arithmetic involved, to include so-called topological regularity axioms. What are those?

Well, there are certain 'nice' topological properties of sets of reals—things like being Lebesgue measurable, or having the perfect set property—that cannot hold everywhere, at least given the axiom of choice. The issue (repeatedly) is that the axiom of choice allows you to well-order the reals and then construct barbaric sets of various kinds, ones that don't have the nice features, by exploiting the well-order.

There's a general feeling in set theory that the 'nice' properties should all hold of easily definable sets of reals, and that the 'nasty' counterexample sets should all be pretty complicated to define (in terms of the usual analytical hierarchy). For example, if $V = L$, then there are easily definable nasty sets of reals, and this is generally taken to be a mark against the principle.

Since $V = L$ implies there are easily definable but nasty sets, it follows that principles asserting that easily definable sets are nice must in some cases go beyond SOA (which is of course consistent with $V = L$). Principles of this kind are what I'm calling *topological regularity axioms*. They assert that good-behaviour properties hold for certain easily definable classes of reals, even when this is not provable in SOA (or ZFC).

The weakest such axiom is something called the Π^1_1-Perfect Set Property (PSP), which says that every uncountable set of reals definable by a Π^1_1 formula is either countable or has a perfect subset.[3] It turns out that this minimal extension of SOA is enough to secure the interpretability of HWP.

Theorem 12.6 *There is a map* $\cdot^{\exists} : \mathcal{L}_1 \to \mathcal{L}_{\in}$ *that preserves theoremhood from HWP to SOA +* Π^1_1*-PSP.*

The proof uses similar ideas to those employed for SOA and SHWP in the previous section, but as I said, one has to take a more nuanced definition of 'possible world'. In particular, it turns out that here our definition of a world will need to be 'doubly parameterised', that is, we will need *two* free variables in $\varphi^{\exists}$.

---

[3] A perfect set is a closed set with no isolated points.

Why? Let us think about how our possible worlds will have to look to get an interpretation going. In order for r-Pow to be true, there will have to be worlds w at which the set x of all ◇-possible sets of numbers exist. These worlds are worlds where any plurality of numbers, in the sense of w, forms a set. And yet, since we can always force to add subsets to any infinite set, there must still be further possible sets of numbers. These sets of numbers have numbers as their members. But we said that all pluralities of numbers at w formed sets at w!

The only way out of this tangle for the HWP proponent is to accept indefinite extendability with respect to both the singular domain and the plural domain (holding the singular domain fixed). The introduction of new generics witnesses, on this view, the extension of the plural domain (over a fixed) singular domain. Indefinite extendability 'runs in two dimensions', according to the present picture. Hence, our possible worlds will be doubly parameterised, with one parameter representing the singular domain, and the other parameter representing the plural.[4] Expansions will then be possible, under the proposed interpretation, in both directions.

Let's now look in more detail at how this works. We will make use of the following fact, whose proof is originally due to Solovay.

Fact 12.7 Over SOA, the Π^1_1-PSP is equivalent to the claim that, for every real number r, there are only countably many reals in $L[r]$.

Thus, in effect, the Fact says that $L[r]$ always falls badly short of containing all reals if we have Π^1_1-PSP. It is thus a strong form of $V \neq L$.

It is fairly easy to see that the fact has the following corollary.

Corollary 12.8 *In SOA + Π^1_1-PSP, $L[r]$ is a (class) model of ZFC for every real r.*

The interested reader can find a short proof of this fact, by Dmytro Taranovsky, online.[5]

This now gives us the tools we need to implement our parametrised possible worlds strategy. We define, in SOA + Π^1_1-PSP, our 'possible worlds' to comprise a transitive set t (representing the first order domain) and a real number r

[4] Strictly speaking, the other parameter does not just represent the plural domain, but also encodes information about future plural domains as well. Two 'second parameters', r and s, in our proposed model theory, may in fact give the same plural domain over a given singular domain, but differ with respect to the plural domains they will eventually yield under height expansion.

[5] See https://web.mit.edu/dmytro/www/other/PerfectSubsetsAndZFC.htm.

(representing the second order domain). We require that $t \in L[r]$, and will say that plural quantification in a possible world is always restricted to subsets of $L[r]$. A possible world t, r may then be extended along two parameters. 'Vertical' expansion expands the t parameter, staying within $L[r]$. 'Horizontal' expansion expands the r parameter, and in so doing accommodates the needed growth in the plural domain even holding the first order domain fixed. t will in general have different subsets in $L[r]$ than $L[s]$, and *these* are the first order theorist's interpretation for talk of plurals in HWP.

To implement this formally, our mapping $\varphi \mapsto \varphi^{\exists}$ will therefore need to carry $\varphi \in \mathcal{L}_1$ to a formula in $\mathcal{L}_{\in}$ with not just one but two free variables: $\varphi \mapsto \varphi^{\exists}(t, r)$. It will contain clauses like

$$
\begin{aligned}
(\forall x\varphi)^{\exists}(t, r) &:= \forall x \in t\varphi^{\exists}(t, r) \\
(\forall X\varphi)^{\exists}(t, r) &:= \forall x \subseteq t[x \in L[r] \rightarrow \varphi^{\exists}(t, r)] \\
(\boxed{\,|\,}\varphi)^{\exists}(t, r) &:= \forall u \in L[r][Tran(u) \wedge t \subseteq u \wedge u \in L[r] \rightarrow \varphi^{\exists}(u, r)] \\
(\Box\varphi)^{\exists}(t, r) &:= \forall s \geq r \forall u \in L[s][Tran(u) \wedge t \subseteq u \rightarrow \varphi^{\exists}(u, s)]
\end{aligned}
$$

(In the final expression, $s \geq r$ means that r is constructible from s.)

With the full translation in hand it is a fairly straightforward matter to prove that every theorem of HWP comes good in SOA + Π^1_1-PSP under the translation: $\varphi \mapsto \forall t, rTran(t) \wedge t \in L[r][\varphi^{\exists}(t, r)]$. (The fact that $L[r]$ models ZFC in SOA + Π^1_1-PSP is needed to get the translations of the ⦶ axioms to come good.) In fact, one can show that HWP has Π^1_1-PSP$^{\Diamond}$ as a theorem, and prove the same kind of tight proof-theoretic equivalence obtains here between HWP and SOA + Π^1_1-PSP as did between SHWP and SOA simpliciter. (Proofs of these claims are provided in the appendix.)

That concludes the (outline of the) formal arguments in favour of the consistency of height and width potentialism. We have seen, in outline, that there are arguments for the consistency of height and width potentialism in both a strong and a weak form, relative to extensions of SOA by topological regularity axioms.

Let us now turn to the arguments alleging inconsistency between height and width potentialism, and see how they relate to the approaches sketched up to now.

4 Roberts

Right from the off it is perhaps worth mentioning that Roberts' argument applies to HWP only. This is not a big deal: he is assuming that the height potentialist will be committed at least to the recovery of ZFC in terms of pure height potentialism, so that any commitment to width potentialism will require a bimodal treatment of the sort described in the last section. But it is worth mentioning that a less grand form of height and width potentialism can still be viable, in the form of SHWP.

Here then is the argument.[6] Let HWP$^+$ be the result of adding all instances of comp,

$$\exists X \forall x [Xx \equiv \varphi]$$

where φ may include free variables, to HWP. Then:

Theorem 12.9 *HWP$^+$ is inconsistent.*

Proof HWP can be used to derive[7]

$$\Diamond \exists x [\Diamond(\exists y [y \subseteq x \wedge y = z]) \wedge \boxed{\text{I}}(\neg \exists y [y = z])] \tag{12.1}$$

by letting x be any infinite set, and y a possible enumeration of the set of $\textcircled{\text{I}}$-possible subsets of x.

Abbreviate (12.1) as $\Diamond \exists x \Psi(x, z)$. By existential instantiation and the comprehension instance

$$\exists X \forall w [Xw \equiv \Diamond w \in z] \tag{12.2}$$

in parameter z we get:

$$\Diamond(Ex \wedge \Psi(x, z) \wedge \exists X \forall w [Xw \equiv \Diamond w \in z]) \tag{12.3}$$

---

[6] This is not exactly the argument I saw in the most recent draft of Roberts' paper. Roberts gives a version of the argument that does not straightforwardly use the free variable version of comprehension that plays a starring role in this argument, but instead uses versions of the principles that are closed but in which all the modal operators are replaced by actualised counterparts— Roberts adds an actuality operator @ to the language, and uses comprehension in the form $@\Box \forall \vec{x} @ \Box \exists X \forall y [Xy \equiv \varphi(y, \vec{x})]$. However, comprehension in this form is not derivable from the straightforward combination of an actuality operator to the language without using instances of free variable comprehension. Consequently, even in that more complicated setting I think the issues turn on the question of which version of comprehension to accept, and I have therefore opted to consider this simpler version of the argument.

[7] This is a little misleading, since the last line of the derivation is an application of existential instantiation. So the below is not a theorem, and neither (of course) is its universal generalisation.

(Here E is an existence predicate, say $\exists y[y = x]$ or suitable alphabetic variant.) Then by HP,

$$\Diamond(Ex \wedge \Psi(x, z) \wedge \Phi\exists u\forall w[w \in u \equiv \Diamond w \in z]) \tag{12.4}$$

one can then instantiate on u and prove $u = z$, using SetExt.

But this is a contradiction.

□

Naturally, given this argument, we should expect the proof of consistency implicit in the previous section to break down with strong comprehension. And this is precisely what happens in the interpretation offered for HWP in SOA $+\Pi^1_1$-PSP. Consider, for example, the following simple case. Let w be the world that has ω for its first order domain and $P(\omega)^L$ for its plural domain. The interpretation for closed plural comprehension at this world amounts to the claim that for any y and Y *in* L, the subset of ω for which $\varphi(n, y, Y)$ holds is again in $P(\omega)^L$, which is obviously true. But if we remove the restriction to closed formulas, we are saying that for any φ, y and Y we choose (whether in L or not), the set of n with $\varphi(n, y, Y)$ is in L. But this is provably false in SOA + Π^1_1-PSP.

The question about the consistency of HWP that arises from this argument concerns, then, the status of the 'strong' open version of plural comprehension. Formally speaking, the path towards saving HWP is clear: accept only the closed instances of comprehension.

There will only be an incompatibility between height potentialism and width potentialism, then, if the open instances of comprehension are mandatory. Is this so?

It is true that real-world height potentialists do seem to subscribe to the open instances as a matter of fact. But then, circumstances to doubt the principle haven't really arisen, because generally speaking height potentialists have not considered the prospects for width potentialism explicitly in the literature, still less defended it. So if the argument is to establish the incompatibility of height and width potentialism in any meaningful sense, there must be deep reasons for a height potentialist to endorse the open scheme, and not simply rest content with the closed one.

One reason for this might be that, in general, any reasonable plural logic will contain the open comprehension instances. But I see no good reason for thinking this is so. Indeed, I would argue that the intuitive validity of the instances of comprehension that are used in the cited argument is legitimately (and independently) questionable, whether or not one is a potentialist. For example, those of a Stalnakerian actualist bent in modal metaphysics might find the open plural comprehension schema doubtful, by analogy with things they already think about property comprehension.

Let me explain. Stalnakerian actualism involves contingentism about (among other things) *properties*. For example, in the event that I should not exist, the property of not being identical to me would also fail to exist according to Stalnaker. Of course, given that I exist, it makes sense to look upon some other circumstances in which I don't exist, and truly affirm of various things that they are not identical to me. But this does not mean that the circumstances in question would be ones in which various things had the property of not being identical to me, since in those circumstances that property does not exist.

Whatever the merits of this view in general—I myself happen to be sympathetic to it—it is clearly coherent and a reasonable idea to consider in modal metaphysics. At its core, from a formal point of view, is precisely a denial of arbitrary instances of comprehension for properties. It is not true, on this picture, that arbitrary worlds will satisfy comprehension for properties on formulas like 'being identical to x', though they will have the closed analogue (for every x there is a property of being identical to x). Similar points go for 'higher order' instances of comprehension, like 'being identical to the property P'.

For a Stalnakerian, the problem with open instances of property comprehension is that they allow worlds illicit access to things in *other* worlds in the model. In our formal model, we have a bunch of worlds with separate domains, and we can name all the things in the domain of any model and ask questions about whether one thing that exists according to one world exists according to another. But these questions are only askable with our external, model-theoretic perspective. When z gets assigned something that doesn't exist at w, inhabitants of w do not have the resources to refer to z, and there in a certain sense simply are no comprehension instances involving z at w. Such formulas, under interpretation in the model, just don't correspond to real propositions that exist at the worlds at which truth is to be evaluated. There is no thought there to be thought there.

The view naturally goes along with an analogous one that denies the relevant instances of comprehension for plurals. Suppose, for example, that we think of a plurality as fundamentally tied to an individuating property, roughly along the lines that a plurality is just the extension of a property (the things instantiating it). We still hold—since pluralities are *extensions*—that plural membership is rigid, extensional, and all the other things involved in HWP. Only now, our basic comprehension principle for plurals is of the form: for any property P, there are some things such that x is one of them if and only if Px. The plurality of Ps is, on this view, 'rigidly' the things that happen to instantiate P.

Such an account of pluralities would support failures of comprehension on non-existent parameters of the kind needed to defend HWP, given the corresponding failures for property comprehension. For example, on this view, in circumstances where I did not exist there would be no guarantee there were such things as *my* molecules, unless some other property (not involving me) happened to have them for its extension; and, this is so even if each individual molecule that would comprise me nevertheless existed.[8] The existence of each thing, as it were, is not enough for the existence of the many: there must in addition be 'resources' to single them out. Similarly, in circumstances where a generic filter for a partial order (is possible but) does not exist, there would be no guarantee that there exist such things as 'its' elements, whether or not each element we will eventually recognise 'it' to have in fact exists: for there may be no property, like 'being an element of *it*', that singles 'its' elements out. (Note that denying precisely this instance of comprehension would suffice to block the argument above.)

This seems to me to be a clear enough rationale for holding only closed comprehension to be valid, compatibly with the other tenets of height (and width) potentialism. We still have, for example, the pluralities we'd like to secure the translations of separation in set theory; and we still have reasons for holding membership in a plurality to be strongly extensional. But we have reason to reject comprehension on parameters that don't exist in our formal theory, since in fact in a certain sense we believe there *are* no such instances, when it comes to the semantics proper, in line with Stalnakerian actualism.

Another possibility would be that there are reasons specific to height potentialism that somehow require the open instances. For example, one might say that the height potentialist is someone who thinks that possible co-existence suffices for the possible existence of a set—a position explicitly endorsed by Linnebo in Linnebo (2017, p151)—and then argue that failures of open instances of comprehension are incompatible with such a view, because in cases where some possible sub-plurality of a given plurality fails to exist, the

[8] In this case it is plausible that there would be properties not involving me that would fit the bill, since for each molecule m_i in me there would be the property of being m_i, and hence the disjunctive property of being either m_1 or m_2 or ... or m_k, where k is the number of molecules comprising me. But this sort of argument is not available in cases like the next one I mention and like those relevant to HWP, where infinitely many things are involved.

co-existence of the things in the possible sub-plurality does not suffice for the possible existence of a set.

In reply, one might question the precise details of this characterisation of height potentialism; in other places height potentialists, including Linnebo himself (see, e.g., Linnebo, 2010, p219), have explicated the view directly in terms of any possible plurality possibly forming a set, a characterisation that does not lead to this apparent problem. But even granted the characterisation of height potentialism, I would argue that the problem can be seen to be merely apparent. After all, the characterisation is one on which possible co-existence suffices for the *possible* existence of the set, and in the sense of the outer modality $\Diamond$ in HWP, it is at least arguably true that the co-existence of the things in question suffices for the possible existence of a set containing them; even if, in the construction of the set in question, forcing must first be done to introduce the things in question as a plurality, and then simple set construction applied to get the set.[9]

I conclude that Roberts' argument does not show HWP is inconsistent. Instead, it serves to emphasise the importance of the restriction to closed comprehension in the formulation of HWP, and highlights the need for a corresponding conception of plurality according to which only the closed instances are supported, such as the one given above. (There are in fact others besides, salient among them a predicativist conception of pluralities, but the one above is simpler to explain.)

I think that a useful way to understand Roberts' argument is as showing that HWP is inconsistent *given the 'nothing over and above' conception of plurals.* This is the popular conception according to which pluralities are the sorts of things you get 'for free' when you have each individual comprising them.[10] It is crystal clear that this take on plurals is incompatible with HWP: for example, the nothing over and above conception would clearly mandate acceptance of

BF $\Box\forall Y(\Diamond\exists X \sqsubseteq Y[X = Z] \supset \exists X \sqsubseteq Y[X = Z])$

where $X \sqsubseteq Y$ abbreviates $\forall x[Xx \supset Yx]$. This says that given pluralities cannot gain sub-pluralities, which is clearly true if the existence of each component of a plurality implies the existence of the plurality; moreover, it

[9]It is actually a little unclear whether a width potentialist should concede co-existence of the relevant things. The question here is whether *the members of the relevant plurality* all co-exist, and given that the plurality itself does not exist this question is a little subtle, given considerations of the preceding paragraphs.

[10]This must not be confused with the view, accepted on all sides, that a plurality is not a single thing separate from the many comprising it.

is violated in HWP, since there even the sub-pluralities of the natural numbers are indefinitely extendible.

Roberts and I agree here: this 'nothing over and above' conception and HWP cannot live together. Perhaps this was all Roberts intended by his argument that height and width potentialism were incompatible; if so, we completely agree, although I think the description in terms of that incompatibility is misleading. In my view, the message is that the proponent of HWP must reject the nothing over and above conception.

There is, I would argue, a general moral in the vicinity. HWP uses a cluster of ideas: modality, set membership, quantification, plurality, and so on. None of them is an absolute fixed point, and there are few conceptual fixed points within them. It is a matter of finding the right configuration of these things to do justice to the guiding picture, and then assessing the resulting tableau on its own merit. That is why, in my view, it is highly unlikely that any serious inconsistency argument for the combination of height and width potentialism will be forthcoming: pretty much any derived inconsistency in a formal system would as naturally be taken to show the formalisation was not up to the task.[11]

5 Brauer

Brauer's argument is quite closely related to Roberts', and in a sense the same key logical observations are at work in each. But Brauer's argument brings out some interesting issues distinctive to the foundations of set theory that are not really brought up by Roberts, and for that reason warrants separate discussion.

The core argument is simple.

P1 Height potentialism, as motivated by the iterative concept of set, sanctions only convergent extensions of the universe: if it could be that the set x existed, and it could be that some other set y existed, then it could be that x and y both exist together.

P2 Width potentialism, as motivated by the idea of forcing potentialism, requires allowing for divergent extensions of the universe: for there are (or at least could be) partial orders P and possible generic filters G_1 and G_2 such that G_1 and G_2 could not exist together.

C Height and width potentialism (so understood) are inconsistent.

[11] This generally seems to be the right response to these sorts of inconsistency proofs. Compare, e.g., Russell's paradox and the concept of set, or Berkeley's critique of the notion of the infinitesimal.

I'm not completely sold on either premise. In the case of P1, Brauer gives an argument which he glosses informally as follows:

> The basic idea of the argument is straightforward: if you have a universe V_0, you can get different extensions V_1 and V_2 when V_1 is formed by bringing together some collections X into sets and V_2 is formed by bringing some other collections Y into sets; but nothing in the process of forming sets out of the X collections would then prevent you from later forming sets out of the Y collections.

So, we are to conclude, that according to height potentialism under the guise of the iterative concept of set, only convergent extensions of the universe are possible.

My doubts about this form of argument concern whether it is OK to conclude that quite generally the iterative concept of set is incompatible with divergent extensions of the universe. After all, set theory is not everything, and although there may be nothing in the simple process of forming sets from given things that prevents re-convergence, nevertheless other things may do so.

For example, suppose I have a handle, two blades, some permanently binding superglue, and a universe of sets V. I can then form the knife K_1 and form its singleton getting a universe of sets V_1. Or I can form the knife K_2 and form *its* singleton getting V_2. But these can never be unified.

Or, for a more (and yet less) pointed example—one in the spirit of the 'second approach' to width potentialism discussed by Brauer on p7—suppose I have two machines. One takes in some things and gives back a set with them as its elements. The other takes in a partial order, some dense sets in it, and gives back a plurality that intersects all the given dense sets. Then, given suitable background assumptions (including but not by any means restricted to P2), we will be in a structurally analogous situation to the blade-handle case: we might find it is possible to produce a set whose elements are G_1, and also possible to find a set whose elements are G_2, without these admitting a unification. But this is no different from the previous case, which seems unproblematic.

P2 is equally if not more problematic in the present context: for, from the perspective of (S)HWP, the arguments standardly given in favour of the non-convergence of forcing look shaky. For example, under the interpretations for (S)HWP given here, it is easy to see that forcing *is* convergent, in the sense that for any possible generic g_1, and any other possible generic g_2, there is a possible world at which they both exist.[12] Thus the modal logic of forcing, in this setting, can be taken to converge, and indeed is naturally taken to.

Brauer's argument on the other hand is based on the familiar result from ZFC saying that, in any given countable transitive model for ZFC, one can find a partial order and a pair of generics for it that can't exist together in a model of ZFC. The argument uses the meta-mathematical fact that any countable model can be enumerated in the metatheory, and one can then encode said enumeration into a real number in an absolute way. That real number then cannot live in the relevant countable model, since if it did the model would be able to furnish a bijection between its ordinals and ω. But that of course cannot hold in any model of ZFC.

This is clearly correct, and is indeed a fact that the proponent of HWP can recognise: they can see, for example, that it is vertically possible that there are models for (fragments of) ZFC, and partial orders they contain, and such that no possible model for the same theory contains both generics. But from the point of view of HWP, this is just because you are ignoring some universes of sets, namely those wherein everything under discussion is in fact countable. More generally, according to them, it is always possible for generics to co-exist, even if sometimes they never cohabitate in a model of ZFC.

The point is that HWP is associated with a radically different conception of the mathematical universe than that associated with ZFC. Effectively, it locates all mathematics in second order arithmetic, taking all sets to be countable, and interprets ZFC only in restricted inner models. As a result, the whole meta-mathematical landscape looks somewhat different, and standard results about models of ZFC now have quite a different significance. In particular, the standard argument for non-convergence does not, from this point of view, show in any deep sense that the modal logic of forcing fails to converge, at least in the most general sense at issue. It only shows it holds for a certain restricted class of models.

I should say that, as with Roberts previously, there is a sense in which I do agree with Brauer. Insofar as we view height potentialism and the iterative concept of set as essentially associated with principles related to ZFC, then what he says is correct: width (or forcing) potentialism is incompatible with these things. For example, if our height potentialist requires (as Brauer suggests they should) that *all possible* sets over a given domain get introduced when any set at all gets introduced, then (as with the discussion of Pow on p7) we will be able to prove the negation of width potentialism. But I take it that this is

[12] For, in $L[g_1, g_2]$, both will occur as elements of a transitive set eventually.

clearly not a deep obstacle to a height and width potentialist theory, since this commitment of height potentialism is optional. One could, as Brauer notes, think of sets as the sort of thing that could be introduced 'one at a time', and such a view is compatible with both width potentialism and its negation.

Analogous things stand to be said about the iterative concept of set, as Brauer notes again. There are really two ideas, a weak version which says that a set is something obtainable from already given things, and a strong version which says that in addition the powerset is always obtainable from the set. HWP is compatible with the iterative concept in the first form, but not the second.

It is also true that the stronger form of the iterative concept of set, as associated with ZFC, is at once standardly accepted as basic in set theory and rejected in HWP. As Brauer emphasises, there therefore remain significant questions about the extent to which this revisionary position is defensible, and questions about what the underlying motivation for the view is (in analogy to the familiar iterative concept); these are questions on which Brauer himself has illuminating ideas, although I think ideas along the lines of the theories set out here (of a more radical kind, to be sure) are also possible. One such account, which I am currently working on with Neil Barton, is a foundational picture according to which the notion of real number is basic, and the concept of set (like other mathematical concepts) is held to be reducible to it, rather than the other way around. Formally, you can think of this as an account where the true axioms for set theory are really just extensions of SOA by topological regularity, rather than ZFC by large cardinals. There is no loss of mathematical power, though, because in effect you can have all the things you usually want in 'inner models'. But the details, and defence, are for another time.

So as before, I think there is value in Brauer's argument: not just in bringing the interesting considerations about the modal logic of forcing to the fore, but also in really emphasising the need to understand set theory quite differently from usual in order to support HWP. But I do not think that in either case there is anything that presents a severe impediment to the combination of height and width potentialism, or to the combination of the iterative concept of set and forcing potentialism, in anything other than unreasonably restricted readings of these terms.

With that said, I understand that to some extent this is a terminological point: others may feel differently about what is essential to 'the iterative concept of set', or forcing potentialism, or whatever. But the terminological points are not important: what is important, and I hope clear from this discussion, is that there is at least a coherent position that incorporates a lot of the traditional ideas in the iterative conception, understood in height potentialist terms, but that amalgamates width potentialism as well.

The foundations of set theory is at present rather like the foundations of quantum mechanics. The discipline itself has thrown up results that our ordinary mathematical tools are incapable—or seem incapable—of resolving. Foundational researchers have therefore been led to look for different ways to understand what's going on, different ways of organising the data presented in mathematical research into a conceptual scheme. Height and width potentialism combined seems to me to be a reasonable player here that has clear benefits when it comes to understanding the centrality of forcing practice to contemporary set theory, one of the big open problems in the area. As the considerations here attest, it is also logically and metaphysically subtle, and conceptually revisionary. But let us not discard it outright for that.

Appendix

In this appendix I will refer to the theory HWP as H, and SOA + Π^1_1-PSP as SOA^+. I will construe the latter theory as ZFC without powerset together with the axiom that all sets are hereditarily countable, and that any Π^1_1 subset of ω has a perfect subset.

Necessary Lemmas

Lemma 12.10 (Mirroring) *Let $\Gamma \cup \phi$ be a set of formulas in the first order language of set theory. Then $\Gamma \vdash \phi$ in first order logic if and only if $\Gamma^\Diamond \vdash_H \phi^\Diamond$. The same is true for* ⏀.

Proof See, for example, Linnebo (2018, Chapter 12). □

Lemma 12.11 (Bounded Modal Absoluteness) *Let ϕ be a formula in the first order language of set theory with only bounded quantifiers. Then $H \vdash \phi \leftrightarrow \phi^\Diamond$. The same is true for* ⏀.

Proof An induction on the complexity of ϕ. See Lemma 12.2 of Linnebo (2018). □

Lemma 12.12 (**Δ_1 Absoluteness**) *Let ϕ be a formula that is provably Δ_1 over ZFC without powerset. Then we have:*

$$H \vdash \phi^{\Diamond} \leftrightarrow \phi^{\obar}$$

Proof Using the previous two lemmas, we see our assumption implies that there are Δ_0 formulas ψ, θ for which

$$H \vdash \phi^{\Diamond} \leftrightarrow \Box \forall x \psi^{\Diamond} \leftrightarrow \Diamond \exists x \theta^{\Diamond}$$

and similarly

$$H \vdash \phi^{\obar} \leftrightarrow \boxbar \forall x \psi^{\obar} \leftrightarrow \obar \exists x \theta^{\obar}$$

by the previous lemma, these are each equivalent to

$$H \vdash \phi^{\Diamond} \leftrightarrow \Box \forall x \psi \leftrightarrow \Diamond \exists x \theta$$
$$H \vdash \phi^{\obar} \leftrightarrow \boxbar \forall x \psi \leftrightarrow \obar \exists x \theta$$

respectively. But weakening implies that $\obar \exists x \theta \rightarrow \Diamond \exists x \theta$ and similarly that $\Box \forall x \psi \rightarrow \boxbar \forall x \psi$. The result follows. □

Theorems

Theorem 12.13 *There is an interpretation $\exists : \mathcal{L}_1 \rightarrow \mathcal{L}_{\in}$ that preserves theoremhood from H to SOA^+.*

Definition 12.14 ($\varphi_{\exists}$) The translation $\varphi \mapsto \varphi_{\exists}$ is defined by the following clauses. In each case, d_1, d_2 are the least variables not occurring in φ, and $\leq$ is the relation of relative constructibility for reals.

- $x \in y_{\exists} := x \in y \wedge \bigwedge_i d_i = d_i$
- $Xx_{\exists} := Xx \wedge \bigwedge_i d_i = d_i$
- $(\neg \varphi)_{\exists} := \neg \varphi_{\exists}(d_1, d_2)$
- $(\varphi \vee \psi)_{\exists} := \varphi_{\exists}(d_1, d_2) \vee \psi_{\exists}(d_1, d_2)$
- $(\exists x \varphi)_{\exists} := \exists x \in d_1[\varphi_{\exists}(d_1, d_2)]$
- $(\exists X \varphi)_{\exists} := \exists x \subseteq d_1 \wedge x \in L[d_2][\varphi_{\exists}(d_1, d_2)]$
- $(\obar \varphi)_{\exists} := \exists e[e \supseteq d_1 \wedge Tran(e) \wedge e \in L[d_2] \wedge \varphi_{\exists}(e, d_2)]$
- $(\Diamond \varphi)_{\exists} := \exists e_1, e_2[d_2 \leq e_2 \wedge d_1 \subseteq e_1 \in L[e_2] \wedge Tran(e_1) \wedge \varphi_{\exists}(e_1, e_2)]$
- $(\oslash \varphi)_{\exists} := (\Diamond \varphi)_{\exists}$

In these, $\varphi_\exists(e_1, e_2)$ represents the result of substituting e_1, e_2 for d_1, d_2 in $\varphi_\exists$.

We then set $\exists(\varphi) := Tran(d_1) \wedge Real(d_2) \supset \varphi_\exists$.

Proof The propositional tautologies and quantifier logic principles are straightforward.

The laws of S4.2 for ⦶ are easy.

As for $\Diamond$, it is again easy to see the axioms of S4 come good. For 0.2, suppose $(\Diamond\Box\varphi)_\exists$. Then there is a real e_2 with d_2 constructible from e_2, and transitive e_1 a superset of d_1 and element of $L[e_2]$, such that $(\Box\varphi)_\exists(e_1, e_2)$. That in turn means that for every real e with e_2 constructible from e, and every transitive set d containing e_1 and an element of $L[e]$, we have $\varphi_\exists(d, e)$.

So suppose given arbitrary d, e with d_2 constructible from e, d a superset of d_1 and an element of $L[e]$ We must find a real r with e constructible from r, and transitive set t a superset of d and element of $L[r]$ such that $\varphi_\exists(t, r)$. The obvious candidates are $t := e_1 \cup d$, and $r := e_2 * e$. (Here $*$ can be any function taking a pair of reals to a real they are both constructible from.) It is not hard to see they have the required features.

All the plural axioms are straightforward. For comprehension, we use separation in $L[r]$. Note that only the closed version is valid, since the open version might involve us with parameters from outside $L[r]$.

On to the set-theoretic axioms. Extensionality is a straightforward consequence of extensionality for sets. The axiom of choice similarly is a consequence of choice in SOA^+. The existence of an infinite set is also given.

For HP: suppose given a transitive extension e of d_1 in $L[d_2]$, and a subset X of e that is also an element of $L[d_2]$. Since $L[d_2]$ satisfies ZFC, it has that every set is an element of a transitive set in $L[d_2]$. Thus we can extend e to a transitive set e_1 in $L[d_2]$ that contains X as an element, and the result follows. Inf is pretty much the same as before.

For r-Pow, suppose given $d_4 \geq d_2$ and $d_3 \supseteq d_1$, $d_3 \in L[d_4]$, and $x \in d_3$. The needed result then follows from the powerset axiom in $L[d_4]$.

As to WP, by proposition 1 it is enough to show that any set in any $L[r]$ is countable in some extension to $L[s]$ where s is a real from which r is constructible. This follows immediately by winding in a real enumerating the set in question to r.

The modal version of replacement is just the $\Diamond$ and ⦶ translations of replacement; these follow again from replacement in the metatheory. □

Theorem 12.15 *There is an interpretation* $\Diamond : \mathcal{L}_\in \rightarrow \mathcal{L}_1$ *that preserves theoremhood from* SOA^+ *to* H on first order formulas.

Proof The only non-trivial part is to show that Π^1_1-PSP$^\Diamond$ holds.

Using Fact 12.7 and the mirroring theorem, to establish the desired conclusion it is sufficient to show the $\Diamond$-translation of the right hand side of the biconditional in Fact 12.7. That is, assuming given an arbitrary real r, we must show that is possible to produce a function on the natural numbers such that, necessarily, if s is a real constructible from r, then s is in the range of f.

The strategy for doing so is simple: we first show that, given any real r, it is possible to produce the set of all possible reals constructible from r in H (in the sense of the modality $\Diamond$). Since H proves all sets are countable (with respect to the $\Diamond$ translation) this implies the desired result.

In more detail, we first observe that by the φ-translation of ZFC, we have the φ translation of the assertion that for any real r, the set of reals constructible from r exists, namely,

$$\varphi\, \exists x[\mathbb{R}^{L[r]}(x)^{\varphi}] \tag{12.5}$$

Where in (12.5) $\mathbb{R}^{L[r]}(x)^{\varphi}$ is the φ-translation of the first order formula asserting that x is the set of reals constructible from r. The problem now is to derive from this that

$$\Diamond\exists x[\mathbb{R}^{L[r]}(x)^{\Diamond}] \tag{12.6}$$

since it is only with respect to $\Diamond$, and not φ, that we have the countability of all sets in H.

To move from (12.5) to (12.6) we use the absoluteness Lemma 12.12. The formula asserting that x is $L[r]_\alpha$ is Δ_1 (in parameters r, α) over ZFC without power. Thus, for any x, α, r,

$$H \vdash (x = L[r]_\alpha)^{\Diamond} \leftrightarrow (x = L[r]_\alpha)^{\varphi}. \tag{12.7}$$

A simple induction shows that for any ordinal α, $\Diamond\exists x[x = \alpha]$ if and only if $\varphi\exists x[x = \alpha]$. Moreover, standard set-theoretic reasoning (together with mirroring) implies that $\mathbb{R}^{L[r]}$ exists is equivalent to $\omega_1^{L[r]}$ exists, relative to either modality. We show that ($\omega_1^{L[r]}$ exists)$^{\varphi}$ is equivalent to ($\omega_1^{L[r]}$ exists)$^{\Diamond}$, in contrast of course to the real ω_1.

For the non-trivial direction, suppose $\omega_1^{L[r]}$ exists$^{\varphi}$. This is equivalent to there being $L[r]$-*uncountable ordinals relative to* φ, in the sense that

$$\varphi\, \exists\alpha \in L[r]^{\varphi}\, \boxbar\, \forall f \in L[r]^{\varphi}[\neg(f : \alpha \twoheadrightarrow \mathbb{N})].$$

So assume the latter and instantiate such an α. By Lemma 12.12, $(\alpha \in L[r])^{\Diamond}$. Suppose α is not $L[r]$-uncountable relative to $\Diamond$, i.e. $\Diamond \exists f \in L[r]^{\Diamond}[f : \alpha \twoheadrightarrow \mathbb{N}]$ This implies $\Diamond \exists \beta, f \in L[r]^{\Diamond}_{\beta}[f : \alpha \twoheadrightarrow \mathbb{N}]$. But then using (12.7) and the fact that $\oplus\beta$ exists we may infer

$$\oplus \exists \beta, f \in L[r]^{\oplus}_{\beta}[f : \alpha \twoheadrightarrow \mathbb{N}]$$

contradicting our assumption that α *is* $L[r]$-*uncountable*$^{\oplus}$.

We may thus infer (12.6) from (12.5): (12.5) is equivalent to $(\omega_1^{L[r]} \text{ exists})^{\oplus}$, which is equivalent to $(\omega_1^{L[r]} \text{ exists})^{\Diamond}$, which latter is equivalent to (12.6) by standard set theory and mirroring.

But now since H proves all sets are countable in the sense of $\Diamond$, it follows that

$$(\exists x[\mathbb{R}^{L[r]}(x) \wedge \exists f : \mathbb{N} \twoheadrightarrow x])^{\Diamond} \tag{12.8}$$

which is just the $\Diamond$-translation of the claim that there are only countably many reals constructible from r. Hence, by an application of mirroring, we conclude $\Pi^1_1[r]$-$PSP^{\Diamond}$. The result follows.

□

References

Brauer, E.: Forth., What is forcing potentialism? Forthcoming in the Palgrave Companion to the Philosophy of Set Theory.

Builes, D., & Wilson, J. M. (2022). In defense of countabilism. *Philosophical Studies, 179*(7), 2199–2236.

Krapf, R. (2017). *Class forcing and second order arithmetic*, PhD Thesis, Universitat Bonn.

Linnebo, O. (2010). Pluralities and sets. *Journal of Philosophy, 107*(3), 144–164.

Linnebo, O. (2017). *Philosophy of mathematics*, Princeton University Press.

Linnebo, O. (2018). *Thin objects: An abstractionist account*. Oxford University Press.

Meadows, T. (2015). Naive infinitism: The case for an inconsistency approach to infinite collections. *Notre Dame Journal of Formal Logic, 56*(1), 191–212.

Pruss, A. R. (2020). Might all infinities be the same size?, *Australasian Journal of Philosophy, 98*(3), 604–617.

Roberts, S. (2024). Ultimate V. *unpublished manuscript*.

Scambler, C. J. (2021). Can all things be counted? *Journal of Philosophical Logic, 50*(5), 1079–1106.

13

Higher-Order Platonism and Multiversism

Claudio Ternullo

1 Hamkins' Multiverse and Platonism

The 'set-theoretic multiverse conception' holds that the subject matter of set theory is not a *single* universe (in particular, not the well-founded, cumulative hierarchy V), but *some* (or *all*) *models* of some set theory T (most commonly, of ZFC).[1]

The most lucid and radical expression of this conception can be found in Hamkins (2012). This is not to deny that alternative conceptions are possible (and, indeed, have been formulated); only, Hamkins' multiversism seems to

Department of Mathematics and Computer Science, University of Barcelona (Spain). claudio.ternullo@ub.edu. For the writing of this chapter, I have been supported by the AGAUR and the European Union through a Beatriu de Pinós (Marie Skłodowska-Curie COFUND) Fellowship (grant n. 00192 BP 2018). I am indebted to Edward Zalta for valuable comments and further clarifications of his theory, and to two anonymous reviewers for their extensive and detailed feedback.

[1] I assume that readers have some familiarity with 'models of set theory'. For a systematic treatment, which covers all models mentioned in this chapter, see Kunen (2011).

C. Ternullo (✉)
Department of Philosophy, University Babeş-Bolyai of Cluj-Napoca, Cluj-Napoca, Romania
e-mail: claudio.ternullo@ubb.cluj.ro

C. Antos et al. (eds.), *The Palgrave Companion to the Philosophy of Set Theory*, Palgrave Companions, https://doi.org/10.1007/978-3-031-62387-5_13

stand out as the most coherent, and self-conscious, available form of *set-theoretic pluralism*.[2]

Despite having the unquestionable merit of bringing about the emergence of a new approach to set-theoretic foundations, Hamkins (2012) plainly rests on philosophically controversial claims. The invocation of the existence of a *plethora* of universes of set theory looks especially controversial.

Moreover, as a preliminary characterization of the philosophy it aims to advocate throughout, the chapter makes clear, from the beginning, that the 'multiverse position' is:

> ..one of higher-order realism—Platonism about universes—and I defend it as a realist position asserting actual existence of the alternative set-theoretic universes into which our mathematical tools have allowed us to glimpse. The multiverse view, therefore, does not reduce via proof to a brand of formalism. In particular, we may prefer some of the universes in the multiverse to others, and there is no obligation to consider them all as somehow equal.[3]
> (Hamkins, 2012, p. 417)

In a fully consequential manner, a substantial amount of work in the chapter is, then, devoted to articulating the idea that *set-theoretic models*, in particular, models obtained through *forcing*, may be construed as 'objects' in the *platonic* sense.[4]

The goal of this chapter is to investigate whether and how Hamkins' multiverse conception may legitimately claim to be able to do that, and on what metaphysical grounds; a parallel question, which I will also investigate, is how well 'higher-order realism' serves Hamkins' own purpose of laying out a *pluralist* foundation of set theory.

In order to be clear about exactly what is the position I shall be dealing with, I shall first explicate what can be viewed as the two fundamental tenets of Hamkins' conception, which I shall henceforth indicate as Higher-Order Platonism (HOP). We may summarize them as follows:

---

[2]The literature on 'multiverse conceptions' is vast. A rough classification, and discussion of the main features of each conception, may be found in Antos et al. (2015). For an introduction to, and discussion of, the basics of 'set-theoretic pluralism', see Linnebo (2017), section 12.4.

[3]As pointed out by one reviewer, also formalists may want to pick out some particular universe as their *preferred* universe of sets. An instance of this attitude can be found, for instance, in Shelah (2003), p. 211f.

[4]For a technical exposition of *forcing*, cf. again the mentioned Kunen (2011), or Jech (2003), Ch. 14.

Platonism (PLAT) The models of set theory are *platonic* entities, that is, are self-standing, independently existing, physically acausal, mathematical objects.

Perspectivism (PERSP) The set-theoretic multiverse is always *relative to* a single model, that is, it is *describable* only from the point of view of one particular set-theoretic model.

(PLAT) already looks very peculiar, but I am inclined to see (PERSP) as being more crucial to the HOP-ist's purposes. Some further explication of (PERSP) may, therefore, already at this stage, be useful.

The view implies that set-theoretic structures are not definable ('observable') from the point of view of an absolute background structure.[5] For instance, one usually thinks of inner models like $L(\mathbb{R})$, HOD, etc., *just* as subclasses of V. But as Hamkins himself explains:

> I counter this attitude, however, by pointing out that much of our knowledge of these inner models has actually arisen by considering them inside various outer models. We understand the coquettish nature of HOD, for example, by observing it to embrace an entire forcing extension, where sets have been made definable, before relaxing again in a subsequent extension, where they are no longer definable (Hamkins, 2012, p. 418)

So, it is not simply through viewing them as *definable* subclasses of V that one comes to understand the very nature of inner models, but rather through studying their behavior within *other* models, such as, for instance, forcing extensions of another (provisional) background universe V. This is an instance of what I call 'perspectivism' with respect to models (with respect to set theory at large): a single model will display different, possibly more interesting, features if studied in the context of (as part of) other models.

In an effort to further clarify this view, more recently Hamkins has suggested the following characterization of (PERSP):

> The multiverse perspective ultimately provides what I view as an enlargement of the theory/metatheory distinction. There are not merely two sides of this distinction, the object theory and the meta-theory; rather, there is a vast hierarchy of metatheories. Every set-theoretic context, after all, provides in effect a metatheoretic background for the models and theories that exist in that

[5] The reader is warned that 'model' and 'structure' are used interchangeably throughout.

context—a model theory for the models and theories one finds there. (Hamkins, 2020, pp. 297-8)

An equally promising way of construing (PERSP), then, is through envisioning the existence of a *hierarchy* of set theories, each reflecting a (provisional) theory/metatheory distinction with respect to another one, such that: the model theory of a theory T (i.e., the 'multiverse' of T) is observed from the point of view of another theory (and provisional *metatheory*) T', the multiverse of T' from the point of view of the theory (and provisional *metametatheory*) T'' and so on.[6] On pain of contradicting the essence of (PERSP), there can be no *ultimate* level in this hierarchy, insofar as there is no ultimate background theory from whose point of view one can 'observe' the model theory of other theories.

Now, (PLAT) and (PERSP) may, to some extent, seem to contradict each other. After all, many 'classic' platonists take platonic entities to be *determinate* and *static*.[7] How is it, then, that they change their 'nature' depending on which model, as postulated by (PERSP), one observes them from? By (PERSP), for instance, a set-theoretic structure, say, $\mathfrak{A}$, could, if observed from the point of view of another structure $\mathfrak{A}'$, look different from the way it could look if observed from the perspective of one further structure $\mathfrak{A}''$.

One major issue relating to HOP is, as a consequence, its *coherence*, that is, whether it is able to yield a conceptual framework which makes sense of the co-existence between (PLAT) and (PERSP).

What I aim to do in this chapter is precisely to describe a metaphysical framework which does not only account for the coherence of HOP, but is also able to provide responses to several concerns about Hamkins' multiverse which have been left unanswered.

[6] Note that this hierarchy of theories should not be taken to correspond to the hierarchy of set theories linearly ordered by *consistency strength* as calibrated by large cardinal axioms.

[7] But then the issue of referential indeterminacy, as illustrated by Putnam (1980), arises for this kind of platonist. A quick response to Putnam's classic arguments would consist in taking platonic entities to be determinate in an idealized, pre-theoretic sense. Another one would be to thrive on the *categoricity* of higher-order axioms; for an exhaustive overview of, and troubles with, the latter strategy, see Button and Walsh (2018).

2 Higher-Order Platonism: Main Features

Before doing that, I need to discuss a few further features of HOP, in particular, I should make clear in what sense it could be seen as instantiating *mathematical (set-theoretic) platonism.*

Although accounts of platonism may differ, scholars usually agree on its basic tenets, which are taken to consist in the following claims:[8]

Existence (E) There *exist* mathematical objects.

Abstractness (A) Mathematical objects are *abstract* (as opposed to *physically instantiated, concrete*) objects.[9]

Independence (I) Mathematical objects are *independent* of minds, language, uses, and practices.

The E, A, I triad also seems to fit in with HOP: that there exist mathematical objects is implicit in Hamkins' 'platonism about universes', only the objects referred to by HOP are not the 'classic', first-order mathematical objects; models of set theory seem to be no less *abstract* than first-order objects, and, finally, Independence may carry over to HOP without much ado or, in any case, it does not seem to be less appropriate to HOP than it is to classic (first-order) platonism.

Some platonists also hold truth-value realism (TVR), the contention that all set-theoretic statements have a *determinate* truth-value. Now, if one takes TVR to be an essential component of classic platonism as much as the E, A, I triad, then HOP noticeably departs from classic platonism, since, as is clear, each set-theoretic model will fix the truth of some set-theoretic statements in a way which might differ from that of another model.[10] Hence, by HOP, it is not always the case that, for a given statement ϕ, ϕ is *true* or *false*: in many cases ϕ will *neither* be true *nor* false (it will be *indeterminate*).

---

[8] For an overview of platonism in mathematics, see Linnebo (2018), which I have used here to characterize the fundamentals of the position.

[9] Abstractness, construed as *non-spatiotemporality*, may not be a peculiar and exclusive feature of mathematical entities. In particular, also some physical particles, as, for instance, described in *quantum field* theories, may not be spatio-temporally located. I am indebted to an anonymous reviewer for pointing this to me.

[10] For noticeable examples of platonists also holding TVR, see Gödel (1947) and Martin (2001). Isaacson (2011), while subscribing to TVR up to the level of **ZFC** holds that at least *some* set-theoretic statements (e.g., the existence of large cardinals) are genuinely *indeterminate*.

However, one may attempt to rescue TVR by *relativizing* it to a specific structure. One could, that is, propose a revised version of TVR along the following lines:

RTVR (Relativized Truth-Value Realism) Each mathematical (set-theoretic) statement is *determinately* true or false *in* a given mathematical structure.

Moreover, one could also attempt to go beyond the Tarskian concept of 'truth in a model' and define a concept of 'truth *in all* models (of some kind)'. Such an account of truth is, for instance, described by Woodin (2011) in connection with the 'set-generic multiverse conception'. The multiverse concept of truth is as follows:

Multiverse Conception of Truth (Falsity) ϕ is true (false) iff it is true (false) in *all* members of the set-generic multiverse, that is, in the collection of all set-generic extensions of an initial model M, and of their grounds.[11]
However, this strategy does not seem to square well with HOP. This is clear if one considers this further feature of Hamkins' multiverse, as expressed in the following quote:

> The background idea of the multiverse, of course, is that there should be a large collection of universes, each a model of (some kind of) set theory. There seems to be no reason to restrict inclusion only to **ZFC** models, as we can include models of weaker theories **ZF**, $\mathbf{ZF}^-$, **KP**, and so on, perhaps even down to second-order number theory, as this is set-theoretic in a sense. (Hamkins, 2012, p. 436)

What can be gleaned from the quote above is that Hamkins' multiverse is not exhausted by any collection of set-theoretic models (of any kind). At this point, this would hardly strike anyone as surprising: already by (PERSP), every single universe must 'yield' its own multiverse, and it should not be possible to amalgamate all such multiverses into a unique, coherent model-theoretic super-structure. As a consequence, the Hamkinsian multiversist ought to abandon entirely the goal of defining 'multiverse truth' in the way sketched above, although, in principle, he might subscribe to RTVR.

But the quote also provides us with another fundamental piece of information about HOP: all models of *all theories* (of sets) are members of Hamkins' multiverse. This means that the Hamkinsian multiversist is not only a *semantic*

[11]This notion is best explicated in terms of 'satisfaction' in Woodin's own Ω-logic, '$\models_\Omega$', for which also see Woodin (2001) or Bagaria (2005).

pluralist, in the sense that he thinks that truth may vary across the multiverse, but also a *proof-theoretic pluralist*, insofar as he thinks that any collection of consistent set-theoretic axioms provides us with some version of 'set-theoretic truth' (that it meets, as Hamkins says himself, some 'concept of set').

I will discuss in depth, and actively exploit, the entangled character of the two positions later on; for the time being, it is worth summarizing this further fundamental feature of HOP as follows:

Proof-Theoretic Pluralism (PTP) All *consistent* theories of sets instantiate equally acceptable versions of set-theoretic truth ('concepts of set').

One fundamental complication relating to (PTP) is that, in violation of (PERSP), it could be taken to be an attempt to articulate the whole (or, for that matter, some) multiverse, but then the problem would arise of what the background theory is, from whose viewpoint (PTP) is formulated. This may have awkward consequences. For instance, from the point of view of the theory $\mathsf{ZF} + \neg Con(\mathsf{ZF})$, ZF has no (standard) models, so, from the point of view of that theory, very oddly, the multiverse would not contain any (standard) ZF-models.

One possible response to this issue is provided by Clarke-Doane (2022), which proposes to replace 'consistent' with 'arithmetically sound' in (PTP), a move which would prevent theories like $\mathsf{ZF} + \neg Con(\mathsf{ZF})$ from being viewed as acceptable.[12]

In Sect. 5, we shall see that the use of Object Theory can provide us with an alternative and, in my view, equally workable response to this problem. Meanwhile, I will maintain that, at this stage, (PTP) may be seen as enjoying a pre-formalized status which, *prima facie*, does not rule out inclusion of any theory (and models thereof) in the Hamkinsian multiverse.

Our description of HOP is now complete. We have seen that classic platonism's E, A, I triad is kept by HOP, whereas unrelativized semantic determinacy (TVR) is dropped. Moreover, HOP's *semantic pluralism* is also subtly related to *proof-theoretic pluralism*, in a word, the idea that there are as many set-theoretic concepts and versions of truth as, arguably, formal theories of sets.

[12] Cf. Clarke-Doane (2022), p. 34.

3 Object Theory

The metaphysical framework which seems most suitable to make sense of HOP is *plenitudinous platonism*. Several forms of plenitudinous platonism are available on the market.[13] All of these presuppose that, contrary to classic platonism, there exist *plenitudes* of platonic mathematical objects, possibly one plenitude for each item of formal languages: objects, theories, relationships, worlds, models, etc.

Moreover, the classic platonist is (mostly) construed as believing in a special form of *intuition*, which would provide her with knowledge of mathematical objects as certain (and determinate) as knowledge of physical objects.[14] Now, plenitudinous versions of platonism purport to be able to eliminate the somewhat mysterious character of such an intuition. This is because they hold that what one really has access to when one refers to *abstracta* are totalities of them, which are made epistemically accessible to one through some Comprehension Principle, or, more simply, by the Completeness Theorem for first-order theories.[15]

The specific form of plenitudinous platonism with which I will be concerned in this work is Object Theory (OT).[16] Its authors hold that knowledge of mathematical objects (as well as reference to them) is conveyed by *descriptions*. They say:

> Knowledge of particular abstract objects does not require any causal connection to them, but we know them on a one-to-one basis because de re knowledge of abstracta is by description. All one has to do to become so acquainted de re with an abstract object is to understand its descriptive, defining condition, for the properties that an abstract object encodes are precisely those expressed by their defining conditions. So our cognitive faculty for acquiring knowledge of abstracta is simply the one we use to understand the comprehension principle. (Linsky & Zalta, 1995, p. 547)

---

[13] Lewis (1986)'s modal realism, Balaguer (1995)'s full-blooded platonism, Shapiro (1997)'s *ante rem* structuralism, and Blechschmidt (2022)'s topos-theoretic multiversism may all be seen as instances of plenitudinous platonism.

[14] For classic platonism, see Bernays (1935) and Gödel (1947) (but also see Gödel's further arguments (and explanations) in Gödel, 1951). For a useful overview of Gödel's conception, see Parsons (1995).

[15] For the connections between consistency and plenitudinousness, cf., in particular, Balaguer (1995), pp. 304ff.

[16] The material which follows is mostly based on Zalta (1983), Linsky and Zalta (1995), and Nodelman and Zalta (2014). Further details on OT may be found in Zalta (2000) and Linsky and Zalta (2006). For an extensive discussion of OT, see Panza and Sereni (2013), section 5.2.

'Encoding', referred to in the quote above, lies at the heart of OT, so it should be explained straight away. This consists in a new form of *predication*, denoted 'xF', which means: 'x encodes F' (as opposed to the well-known 'Fx', which means: 'x exemplifies F'). The difference between exemplification and encoding can be explained as follows.

An object x exemplifies a property P if that property can be *predicated* of x: '2 is prime' is a case in point. An object x encodes a property P if x *is* that property. '2 encodes *primality*' is a case in point. Abstract objects will both exemplify and encode properties.[17]

Encoding is regulated by the following Comprehension Principle:[18]

$$(\exists x)(A!x \wedge \forall F(xF \leftrightarrow \phi)) \qquad \text{(Comp)}$$

where $A!x$ means: 'x is abstract', F is a property, ϕ is a set of conditions (expressing the property F). So, (Comp) asserts the existence of a plenitude of abstract objects, each of which encodes all and only those properties expressed by a set of conditions ϕ. In simpler terms, an abstract object is just an object which *encodes* a collection of properties.[19] (Comp) is OT's archetypal Plenitude Principle. It allows the formation of *any kind of* abstract objects, in particular, mathematical objects.

But OT is not exhausted by (Comp). The theory extends *abstractness*, and *plenitudinousness*, to all other 'objectifiable' parts of formal languages, among these, crucially, to *theories*.

Formally, a *theory* is just a collection of propositions. In turn, *propositions* are well-formed formulae, consisting of symbols for constants, predicates, relations, all of which are also distinct abstract objects. So, theories, qua abstract objects, are combinations of other abstract objects. More specifically, through the corresponding Comprehension Principles, one first forms a plenitude of *relations (predicates)*, then a plenitude of *propositions*, construed as 0-place relations. Finally, one forms a plenitude of theories by identifying one single theory T with its *theorems*, that is, with the closure set of the '$\vdash$' relationship,

[17] A way to characterize 'encoding', as opposed to 'exemplification', suggested to me by one reviewer, is through viewing encoding as expressing *essences*: so, 'x encodes P' may be paraphrased as expressing that 'x is some *essence* P'.

[18] OT's language is that of the second-order quantified S5 logic, enriched with the abstraction operator λ, the unary predicate $E!$, which, applied to the variable x, means: 'x is concrete' (predicates for 'abstract' and 'ordinary', resp. $A!$, and $O!$ are, in turn, obtained from $E!$) and the corresponding equality symbol $=_{E!}$. Cf. Nodelman and Zalta (2014), pp. 42-3.

[19] Alternatively, for any non-empty collection of properties, there is an abstract object which encodes them.

or, *semantically*, with the set of the *logical consequences* of T.[20] The resulting Comprehension Principle is as follows:

$$T \stackrel{\text{def}}{=} \iota x(A!x \wedge (\forall F)(xF \leftrightarrow \exists p(x \models p \wedge F = [\lambda y\ p]))) \qquad (\text{Comp}_{\text{T}})$$

that is, a theory T is the *only* abstract object which encodes all and only those properties asserted by its *true* propositions.[21]

(Comp_T) may also be seen as a Comprehension Principle for *structures* (i.e., for *models* of a theory T), taken to be the *referents* of 'true' theories. Thus, in practice, OT identifies theories with the *structures* which satisfy them. **PA** is, for instance, the collection of all propositions true of **PA**-*structures*, **ZFC** the collection of all propositions true of **ZFC**-structures, and so on. Also structures can be referred back to theories via (Comp_T). For instance, a model of **ZFC** which satisfies the Continuum Hypothesis (CH) will be interpreted as the abstract object corresponding to the theory **ZFC**+CH.

Overall, then, OT identifies models with *particular theories* (and vice versa), and, as a consequence, it also provides a careful explanation of why **ZFC** is, for instance, different from **ZFC**+CH: because the abstract object **ZFC**+CH *encodes* properties that are not encoded by the abstract object **ZFC**.

OT also envisages the existence of a plenitude of first-order mathematical objects. OT's treatment of such a notion follows in a fully consequential manner from the theory's presuppositions. Let κ be a mathematical object: the exact reference of κ is fixed by some theory T. Thus, there can be no object κ without an accompanying theory T (which fixes its reference). So, κ splits into a multitude of κ_T, one for each theory T. In formal terms:

$$\kappa_T \stackrel{\text{def}}{=} \iota y(A!y \wedge (\forall F)(yF \leftrightarrow T \models F(\kappa_T))) \qquad (\text{Comp}_{obj})$$

(Comp), (Comp_T), and (Comp_{obj}) are sufficient for my purposes. In fact, only (Comp_T) shall prove indispensable.

One last remark is in order. *Encoding* also differs from *exemplification*, insofar as abstract objects are 'encoding-incomplete' whilst being 'exemplification-complete'. This means that, say, a natural number κ_{PA} *encodes* all and only those properties that **PA** asserts κ to have. Therefore, it may be the case that

---

[20]The switch between syntax and semantics is fundamental for Linsky and Zalta's purposes. This can be accounted for, in OT, through using what the authors call Importation Rule, described in Nodelman and Zalta (2014), p. 48.

[21]The ι-operator means: 'the only', $[\lambda y\ p]$ denotes: '(being) the object y such that p', and $F = [\lambda y\ p]$ 'the property F constructed out of propositions p'.

κ_{PA} does not encode some property F, or its negation. On the other hand, for all F's, either κ_{PA} exemplifies it or its negation. Thus, the encoding-related features of abstract objects are also able to account for *set-theoretic referential indeterminacy*. Reference to objects is fixed by theories, and these are, by their nature, 'encoding-incomplete', so also first-order mathematical objects (*sets*) will be encoding-incomplete. Hence, theories (of sets) are about inherently *incomplete* objects, which are, however, liable to become 'more complete'. A classic example is the real continuum (the power-set of ω). The object $\mathcal{P}(\omega)_{\mathsf{ZFC}}$ is encoding-incomplete with respect to CH. But this doesn't mean that $\mathcal{P}(\omega)_T$ will, in general, be encoding-incomplete for any T. In fact, $\mathcal{P}(\omega)_{\mathsf{ZFC}+V=L}$ is encoding-complete with respect to CH.

4 Higher-Order Platonism Inside Object Theory

HOP can be adequately interpreted ('embedded') inside OT. This is the present work's main claim, which I will now proceed to discuss. But first I should clarify what I really mean by this claim.

My goal is not to suggest that Hamkins' multiverse is formally reducible (whatever that could mean) to a fragment of OT, nor do I posit that the Hamkinsian multiversist should think that his conception is just OT. What I actually think is that OT provides us with a detailed and coherent metaphysical account of HOP, which also enriches our understanding of Hamkins' multiverse.

In order to attain my goal, I will have to show that OT expresses the three main features of HOP, that is, (PLAT), (PERSP), and (PTP). This is a relatively trivial task as far as (PLAT) is concerned, but (PTP) and (PERSP) will need a more extended examination.

As far as (PLAT) is concerned: (Comp_T) envisages the existence of a plenitude of higher-order abstract objects, that is, *theories*. Moreover, as we have seen, (Comp_T) also entitles us to construing theories as structures satisfying them, and *vice versa*. It is, in my view, a noticeable strength of OT that it is also able to incorporate, and refer to, *models* of set theory. To this end, as we have seen, one just picks out the relevant abstract objects corresponding to theories.

(PTP) is also met by OT, since *any* theory of sets represents a self-standing abstract object. Again, it is a defining feature of OT that theories are taken to be platonic objects. All of them are acceptable, that is, *legitimate*, since OT's Comprehension Principles do not discriminate between 'acceptable' and 'unacceptable' *entities*. Moreover, that any abstract object corresponding to a theory of sets T reflects a 'different concept of set', as envisaged by Hamkins, is expressed, in OT, by the fact that theories *encode* different properties (in particular, they encode different theorems and truths about different structures). So, there is a clear and robust sense in which OT clearly asserts the existence of different 'concepts of set'.

Moreover, by using OT, one may have at hand a way to formulate (PTP) which avoids the objections in Sect. 2: (PTP) now just reduces to OT's assertion that there exists a plenitude of theories, in particular, by (Comp_T) all *conceivable* theories exist, so (PTP), in essence, just reduces to (Comp_T). In addition, one no longer needs to care about 'consistent theories'. In fact, some theories sanctioned by (Comp_T) will, in fact, be inconsistent, which just means, by OT, that they are not exemplified, that is, that they do not have *models*. I shall say more on this in Sect. 5.1.

Finally, I proceed to address (PERSP), which, as I have said in Sect. 1, is crucial to HOP's purposes. (PERSP) has been formulated as the view that the multiverse is always observed from the point of view of (is relative to) a specific model. In order to exemplify the position, I also mentioned Hamkins' own example of *inner models of* V's being best understood from the point of view of *forcing extensions*.

The other aspect of (PERSP) I have highlighted, and which is also to be reconsidered in light of OT, is that, within Hamkins' multiverse, one is supposed to constantly *jump* from one model to another, and this is also construed (more broadly) by Hamkins as 'an enlargement of the theory/metatheory distinction'.

In order to illustrate the first aspect, let us go back to the example of inner models inside forcing extensions of a provisional background universe V. One starts with constructing a forcing extension $V[G]$ of V (for our and Hamkins' purposes, this could just be any countable transitive model of **ZFC**). The corresponding Hamkins multiverse, by (PERSP), is, thus, the collection of all models of **ZFC** accessible from $V[G]$. Among these, one finds inner models such as L, HOD, $L[\mathbb{R}]$, etc. Any of these will now reveal, inside $V[G]$, features which, in turn, are contingent on the features of $V[G]$ itself.

Now, what kind of abstract objects and properties are, by OT, involved in observing, say, the constructible universe L from the point of view of $V[G]$? The first object that needs taking into account is $V[G]$ itself, a model of **ZFC**. Now, recall that, by (Comp_T), a model of **ZFC** is the *same abstract object* as **ZFC**. So, in particular, any property encoded by $V[G]$ will be *encoded* by the abstract object **ZFC**. But $V[G]$, presumably, encodes other properties that are not encoded by **ZFC**, for instance, $\neg$CH (presumably, that was also the reason why we picked out $V[G]$ in the first place). In that case, we should, more correctly, refer $V[G]$ back to the abstract object (and theory) **ZFC**+¬CH.

But now we need to express, in OT, that **ZFC**+¬CH does not only encode the features of $V[G]$, but also those of the L *inside* $V[G]$. So, our next object of scrutiny is the sentence:

(Φ). $V[G]$ is a model of **ZFC**+¬CH, and, moreover, $V[G] \models$ 'L has some set-theoretic property Ψ'.

where Ψ expresses some feature of L as seen from the point of view of $V[G]$.[22] Now, Φ is a metatheoretic statement about models of **ZFC**+¬CH, so we cannot expect that it is expressible in **ZFC**+¬CH. But, given the representability of **ZFC**+¬CH's metatheory in **ZFC**+¬CH itself, we may find, in the object theory, an equivalent statement, say, $\Phi_{\mathsf{ZFC}+\neg\mathsf{CH}}$ somehow corresponding to Φ. Now, $\Phi_{\mathsf{ZFC}+\neg\mathsf{CH}}$ is a *bona fide* abstract object. In particular, by (Comp$_T$), $\Phi_{\mathsf{ZFC}+\neg\mathsf{CH}}$ is a truth about some structure satisfying **ZFC**+¬CH itself and, thus, can, finally, be viewed as being encoded by **ZFC**+¬CH. In the end, in OT, one has that the abstract object **ZFC**+¬CH encodes a property which expresses a 'property of L *inside* a model $V[G]$ of **ZFC**+¬CH' (i.e., Ψ), and this is precisely what was required of us to establish.

In order to discuss the second aspect of (PERSP), that is, the idea of 'jumping' from one model to another, let us carry on with the example of L inside $V[G]$.

I seem to live in L (in fact, the L of $V[G]$), so my Hamkins multiverse consists of all models accessible from L itself. By OT's conception, this just means that now I am carrying out my reasoning from the point of view of a different theory, that is, **ZFC**+V=L, so, again, through (Comp$_T$), I am picking out a different abstract object, corresponding to the theory **ZFC**+V=L. In other terms, the jump from $V[G]$ to L can now be expressed as the shift from one abstract object to another, which is, in turn, tantamount to picking out (describing) *distinct* abstract objects. If we wished to jump from there to another model M (presumably of **ZFC**), then we would just have to pick out the abstract object corresponding to the theory whose theorems are *true* in M and, of course, this process can be iterated as many times as one pleases.[23]

[22] Hamkins (2012)'s example, on p. 418, is Cohen's construction of a $V[G]$ to get an inner model of ¬AC.

[23] I wish to stress that, on the one hand, 'true in M' should be taken to be equivalent, through (Comp$_T$), to 'true in (of) a theory T', wherein no specific model of T is presupposed, and, on the other, that the satisfaction relation should always be seen as relative to T itself (i.e., should be more accurately symbolized as follows: '$\models_T$') and, as a consequence, to the models of T.

One further technical remark is necessary. Of course we cannot expect **ZFC**+¬CH (or **ZFC**, for that matter) to literally prove that it has models (like $V[G]$) which satisfy ¬CH and further set-theoretic properties (of their L, for instance). The theory will, at most, prove that some arbitrarily finite fragment of the theory has a *countable* model with the mentioned properties, so the truth encoded by the abstract object corresponding to **ZFC**+¬CH will refer to such models. This is not ideal, but inevitable. On the other hand, if one used Hamkins' *toy model perspective*, whereby all models are countable from the beginning, then one would get that the multiverse, and all relevant metamathematical facts about it, would be encoded only by the abstract object **ZFC**.[24] But this approach would be awkward, for then, in particular, the jump from one model to another would just be a jump from one countable model of **ZFC** to another countable model of **ZFC**. Through upholding the principle of identifying a structure with the corresponding theory, we can, on the contrary, modulo the mentioned limitations, construe the jump from one model to another as picking out *distinct* theories (abstract objects), something which seems to be a lot more faithful to the motivation behind Hamkins' multiverse.

In sum, OT is able to interpret (PLAT), (PERSP), and (PTP), which, overall, means that OT is able to interpret HOP. Let me emphasize, again, that, in particular, OT is flexible enough to interpret the jumps from one model to another which are so characteristic of Hamkins' multiverse. The crux of OT's interpretation of (PERSP) consists in transforming metamathematical facts, on which the Hamkinsian multiversist thrives, into *properties* encoded by abstract objects corresponding to the relevant theories.

5 Three Problems: Articulation, Skepticism, Practice

Let's take stock. I have concluded on a positive note about HOP's interpretability inside OT. I have also made it clear that OT should be taken to be a 'companion' theory to, not a formalization of, HOP.

Now, HOP (and Hamkins' multiverse, for that matter) have sparked much controversy and raised concerns relating to the underlying logic and philosophy. In what follows, I will show that some of these concerns may be successfully addressed if one adopts the conception and tools afforded to us by OT.

[24] For a discussion of the toy model perspective, see Hamkins (2012), p. 436ff.

5.1 Articulation

Koellner (2013) raises a problem of 'articulation' for what he calls the 'broad multiverse conception', that is, Hamkins' conception.[25] The problem may be explained as follows.

The existence of a set-theoretic multiverse, that is, of all models of some theory of sets T, is secured through assuming the consistency of T. But, by Gödel's Second Incompleteness Theorem, one cannot hope to prove Con(T) in T, so, in order to get Con(T), one needs to use a stronger theory as a background theory. Now, suppose one has ascended to a stronger theory $T' = T + Con(T)$ and can, now, articulate the multiverse of T. This would clearly be insufficient for the Hamkinsian multiversist, as he still cannot refer to models of T', so one will have to ascend, again, to a stronger theory, $T'' = T' + Con(T + Con(T))$ to articulate the multiverse of T', and so on. Overall, this will result in an infinite regress through background theories of increasing consistency strength, none of which exhausts the 'whole' Hamkinsian multiverse.[26]

The response to Koellner's argument is, in the light of OT, relatively straightforward. The HOP-ist does not need to articulate the 'whole multiverse' in the way suggested by Koellner. In fact, by (PERSP), he doesn't even have a fixed concept of 'set-theoretic multiverse'. It is true that he needs to start from some theory T, in order to build an initial multiverse, but, by (Comp$_T$), members of such multiverse will come to the fore automatically, since they are, so to speak, inbuilt features of the abstract object corresponding to the theory itself.

So, what the HOP-ist needs to do to articulate *some* multiverse is just to pick out the abstract object corresponding to his initial theory T and the properties this encodes. Among such properties, in particular, he will find the relevant metamathematical facts about the models of T. Afterwards, he can either pick out another theory (and abstract object) T or, by (PERSP), he can progressively explore all other models accessible from the initial multiverse.

Now, the HOP-ist's (PTP) may be construed as contravening the strategy above and as, in fact, requiring the articulation of the *whole* multiverse. However, as has been made clear in the previous section, inside OT, (PTP) just reduces to (Comp$_T$). Moreover, it should be stressed, once again, that the

[25] In fact, Koellner (2013), pp. 4-5, distinguishes between the 'broad multiverse conception' and the 'relative broad multiverse conception' that is, the narrowing of the former to just one background theory T. For my purposes, it will just suffice to deal with the 'broad multiverse conception'.

[26] Koellner (2013), pp. 7-8. A similar argument pointing out, this time, a 'referential regress' in Hamkins' multiverse is in Barton (2016), p. 202.

HOP-ist does not need to take into account the issue of the consistency of T, insofar as OT does not discriminate between consistent and inconsistent abstract objects. In particular, some abstract objects will encode inconsistent theories. However, by OT, this does not prevent such theories from existing: only, there are *no* abstract objects which *exemplify* them, so in particular there are *no* structures which exemplify them.[27]

Koellner (2013) also lays emphasis on what he thinks are further paradoxical features of Hamkins' multiverse related to consistency.[28] Given any theory T (of sufficient consistency strength), Hamkins' multiverse will also contain models of $T + \neg Con(T)$. Now, in order to secure the existence of those models in the multiverse, one has to presuppose that T is, in fact, *consistent*, so there is a tension between the 'internal' and 'external' claims of consistency of T. But, again, my interpretation of HOP gets rid of the distinction between internal and external. Access to abstract objects corresponding to theories is carried out through *description*, so no *external* presupposition needs to be made concerning the properties encoded by those objects, in particular, as we have seen, concerning their consistency. The existence of a (non-standard) model of $T + \neg Con(T)$ entirely lies 'inside', and is encoded by, the abstract object $T + \neg Con(T)$, so the issue of the clash between the external and internal claims of consistency of such theory, practically, vanishes.

5.2 Skepticism

Does HOP instantiate a form of *skepticism* about set theory and mathematical knowledge at large?

In order to answer this question, one should, presumably, have at hand a transparent notion of what would count as 'skepticism', and the issue is too broad to even start examining it here. However, a few considerations on this issue may be inevitable, especially insofar as the charge has been leveled many times by different authors.[29]

---

[27]With reference to inconsistent objects, Linsky and Zalta (1995), p. 537, fn. 32, mentions the famous example of the 'round square': by OT, there exists a unique abstract object encoding these two *mutually inconsistent* properties, but there is *no* object which exemplifies them. Note that, in principle, it is conceivable that there are *no* abstract objects *exemplifying* consistent theories, although this is, presumably, contradicted by our experience of theories which we reasonably deem to be consistent, that is, such that these theories do, in fact, have *models*.

[28]Cf. Koellner (2013), pp. 8ff.

[29]Cf. Koellner (2013), pp. 9. Button and Walsh (2018), pp. 205ff., take Hamkins' multiversism to be a contemporary version of Skolem's 'model-theoretic skepticism', for which cf. Skolem (1967).

One straightforward way of arguing that HOP is a form of skepticism is through raising the issue of *referential indeterminacy* in connection with it. The argument, broadly, runs as follows: set-theorists seem to have *determinate* (*unique*) referents in mind for set-theoretic objects, whereas HOP subscribes to ($Comp_{obj}$), which implies, among other things, that referents of set-theoretic objects are *non-unique* in an essential way; hence HOP cannot represent an accurate interpretation of the set-theoretic discourse, let alone a foundation for it.

One possible response is that this argument may be thriving on a (subtle) confusion between 'relativism' and 'indeterminacy'. Granted, HOP is relativistic, since properties of set-theoretic objects are taken by it to be contingent on theories (and structures modeling them). Yet, it would not be correct to view HOP as supporting the referential indeterminacy of the set-theoretic discourse. HOP (and OT, for that matter) does not hold that the reference of, say, 1 is *indeterminate*; rather, they hold that it is *plural*. According to HOP-ists as well as OT-ists, 1 is an incomplete denotation of the object 1_T, that is, of 1 as defined within some theory T. Now, they may agree that, when mathematicians (and set-theorists) refer to 1, they refer, most of the times, to 1_{PA}, but will, as is clear, deny that this is inevitable. For instance, 1_{ZF} is a sensible alternative that comes with a lot of new theory.[30]

One could insist that the interpretation of arithmetical discourse is so plain and evident, that there is no reason to assume that the referent of a natural number is contingent on a specific theory, but I think that this only begs the question.

As a consequence, HOP-ists' appeal to the relativity of the set-theoretic discourse should be more correctly construed as expressing the belief in the *multiple realizability*, not in the *indeterminacy*, of the discourse itself. Hamkins has stressed this point many times:

> [...] let me remark that this multiverse vision, in contrast to the universe view with which we began this article, fosters an attitude that what set theory is about is the exploration of the extensive range of set-theoretic possibilities. (Hamkins, 2012, p. 440)

[30]However, the distinction between, say, κ_{T_1}, the object κ as seen from the point of view of a theory T_1, and κ_{T_2}, the object κ as seen from the point of view of another theory T_2, should, sometimes, be carefully delimited, if (PERSP) is to be implemented satisfactorily. For instance, one may want to say that κ_{T_1} is countable and κ_{T_2} uncountable. Then, for the purpose of comparing them, it would be desirable to have that ω_{T_1} is the same as ω_{T_2}. However, it is not necessary to further posit that '$\cong_{T_1}$' is the same as '$\cong_{T_2}$', as the use of the '$\cong$' relationship is clearly bounded by the domains of quantification of the respective theories.

As regards the issue of whether such a view instantiates any form of skepticism, he explains that:

> ..to my way of thinking, this label is misapplied, for the multiverse position is not especially or necessarily skeptical about set-theoretic realism. We do not describe a geometer who works freely sometimes in Euclidean and sometimes in non-Euclidean geometry as a geometry skeptic. (Hamkins, 2020, p. 295-6)

Now, it seems to me that my interpretation of HOP *within* OT adds further weight to this argument. Believers in the existence of all conceivable abstract objects do not deny the *existence* (or *knowability*) of (mathematical) reality. They think, instead, that mathematical reality consists of a very rich realm of realit*ies*, from which one can pick one's preferred mathematical concepts. Inside OT, this idea prominently features as the claim that mathematical objects are *non-spatiotemporal* and *sparse*.[31]

The accusation frequently leveled against *plenitudinous platonism* of being a brand of formalism, that is, of 'anti-realism', is not new in the debate, in any case, and discussing it in full is besides the scope of this chapter.[32] Presumably, whether one may have a coherent concept of 'multiple (mathematical) realities' also depends on what one takes reality to consist and on how the latter (causally or non-causally) interacts with our minds, but now, precisely on these grounds, some people will keep insisting that there is a *unique* mathematical reality and, therefore, that plenitudinous platonism (as well as HOP) has a skeptical attitude about set-theoretic reality.[33] But this is a claim we ought not to grant light-heartedly.

5.3 Practice

Our theorizing, so far, has been 'metaphysically thick'. By HOP, in order to understand what set-theorists really do, one does not only need to understand set theory mathematically, but also understand some accompanying explanatory theory of set-theoretic objects, theories, and structures like OT.

---

[31]On this, cf., Linsky and Zalta (1995), especially section VI.

[32]For instance, Field (2001), pp. 334ff. takes Balaguer's full-blooded platonism (FBP) to be of an 'anti-objectivistic' character, and Potter (2004), p. 11, even denies that FBP has anything to do with platonism. Linnebo (2018), p. 21, on the other hand, views such conceptions as FBP and OT as belonging 'somewhere in the territory between anti-nominalism and full-fledged platonism'.

[33]Cf., for instance, Martin (2001).

Thus, one further objection against HOP could be that one doesn't really need to do that. That is, one could say that, whatever the merits of the position, regardless of whether it's coherent or not, such an extreme form of set-theoretic realism really has no impact on set-theorists' work. Moreover, the multiversist seems to want to make things more complicated than they are, since taking onboard a complex, platonistically construed, multiverse ontology affords no potential gains in foundational terms.

Naturalists, such as Maddy, have expressed this point of view. In particular, Maddy has stressed that Universism *and* the theory **ZFC**+Large Cardinals are more than sufficient for all our mathematical and philosophical purposes, where the role of philosophy is taken, by Maddy, to consist in that of Second Philosophy, that is, of a philosophy which attends to the 'details of practice' and does not aim to reinterpret or influence in any way practitioners' work.[34]

For instance, Maddy (2017) denounces all metaphysical incursions into mathematical work by declaring that:

> [t]he metaphysics of abstracta or meanings or concepts are all really beside the point. The fundamental challenge these multiverse positions raise for the universe advocate is this: are there good reasons to pursue a single, preferred theory of sets that's as decisive as possible, or are there not? (Maddy, 2017, p. 316)

We know Maddy's own answer to the question: the reasons to pursue just one theory, **ZFC**+Large Cardinals and 'single-V', are more cogent than the reasons to pursue a multiverse theory, since **ZFC**+Large Cardinals and 'single-V' are taken to better express the foundational goals of set theory.

Now, although the universist orthodoxy, so to speak, is far from being overthrown, several works have appeared which put pressure on this conception. For instance, Ternullo (2019) suggests that the goals mentioned by Maddy might also be attained by a multiverse theory and that consideration of further fundamental set-theoretic goals, such as studying the relationships between models (theories) might, in fact, lead to prefer a multiverse over a single-universe theory.[35] Moreover, 'multiverse maths' has delivered many promising mathematical programs, and many more might appear in the next future.[36]

[34]For the articulation of Second Philosophy, cf. Maddy (2007).

[35]Cf. Ternullo (2019), pp. 57ff.

[36]An immediate example that comes to mind is set-theoretic 'geology', for which see Fuchs et al. (2015), now thought to be central to many set-theoretic undertakings.

More recently, Antos (2022) has pointed out that also a Second Philosopher could see models of set theory as being part of set theory's 'fundamental entities'. This is because:

> [i]ntroducing models as fundamental entities is part of set-theoretic methodology and not just a heuristic aid. Being about methodology, this claim then becomes eligible to inform further philosophical questions [...]. (Antos, 2022, p. 6)

Of course, at this point, one could object that one thing is to hold that the set-theoretic multiverse (or some conception thereof) might fit in with a purely naturalistic methodology, and another is to advocate a complex, practically dispensable, metaphysical theory such as OT.

In other terms, while the naturalist could, in principle, be open to embrace Mostowski (1967)'s formalistic pluralism about 'set theory', as expressed in this quote:

> Probably we shall have in the future essentially different intuitive notions of sets just as we have different notions of space, and will base our discussions of sets on axioms which correspond to the kind of sets which we want to study. (Mostowski, 1967, p. 94)

she could not possibly subscribe to the metaphysical theorizing inherent in HOP. At which point, the HOP-ist could reply that HOP is consistent with the kind of naturalism that includes both Mostowski's pluralism and Antos' suggestion that models are indispensable mathematical entities. As a consequence, the anti-platonist naturalist will now turn to launch a full-scale attack against HOP's platonistic doctrines, and the HOP-ist will respond, as Hamkins does, that platonism is 'natural', insofar as the *experience* of models of set theory is genuine, and cannot be explained away.[37] Then, the naturalist will respond that the HOP-ist's notion of 'experience' is not hers, and so on. This debate will, most likely, end in a stalemate.

It might, in any case, be too early to assess whether HOP's foundation of set theory is going to fully convince naturalistically minded philosophers and set-

---

[37] Hamkins (2012), especially p. 418. The point of view could be expressed more intelligibly as follows: no *alternative* explanation of the existence of a plurality of models, as good as the HOP-ist explanation, is currently available; in particular, universist explanations, which hold that models are just convenient fictions (inside a single V), won't do, as experience with models is so mathematically substantial that considering the existence of models a matter of fact becomes, even by the Maddian naturalist's lights, inevitable.

theorists (incidentally, this might also depend on the status of Universism, as well as of Universism-inspired mathematical programs, in the coming years).

What is certain is that, through embracing OT, HOP seems to have noticeable advantages over other conceptions: correct or not, it provides an *explanation* of set-theoretic *undecidability*, that is, of the fact that many set-theoretic statements cannot be decided by the current set-theoretic axioms, and of why undecidability will persist, by bringing out the *incomplete*, in particular, *encoding-incomplete*, character of set-theoretic concepts and theories qua OT's abstract objects, that is, the fact that these objects do not have enough mathematical 'content' to decide lots of set-theoretic statements. It might not be too much for a naturalist, but having at hand some robust explanation of set-theoretic undecidability will sound appealing to many.

Finally, Clarke-Doane (2022) has pointed out that mathematical pluralism (one would say, by now, HOP) is also able to account for the *reliability* of our mathematical knowledge. The 'reliability challenge' is the challenge to the view that we have *true* mathematical knowledge. Mathematical pluralism, in a sense, dissolves the problem: it simply isn't possible, in the light of mathematical pluralism, that 'we could have had *false* mathematical knowledge' since pluralists commit themselves to the idea that any consistent collection of set-theoretic propositions (theory) instantiates *some* truths.[38]

6 Concluding Remarks

I have shown that there is a way to make sense of 'higher-order realism' in the context of Hamkins' multiverse, by embedding what I have called HOP into a general theory of abstract objects, OT. Such an embedding does not require of one to see HOP as a 'fragment' of OT, only to acknowledge that HOP's platonistic conception can be adequately expressed by OT.

In particular, the Hamkinsian multiversist's typical statements that *models* are set-theorists' main objects of investigation, that we have a genuine experience of them, that set-theorists jump from one universe to another, become fully intelligible in the context of OT. Moreover, as we have seen, OT comes with an explanation of set-theoretic undecidability, which, in my view, makes the 'multiverse undertaking' even more persuasive. Of course, one can resist the force of the explanation, but, certainly, the latter stands Hamkins' multiversism in good stead.

[38] Cf. Clarke-Doane (2022), especially Ch. 3.

Finally, I have also shown that, through OT's reinterpretation of HOP, the Hamkinsian multiversist may also be able to respond to issues concerning the articulation of the multiverse and reject as, at least in part, inappropriate, the label of 'skepticism' for his conception.

OT presents itself as an elegant (and economic) solution to the interpretative problems posed by (PLAT), (PERSP), and (PTP). The basic objects are taken to be theories, and these are reduced to the structures which satisfy them. So the complex metatheory of the multiverse which is needed to express salient facts about the models of set theory can be incorporated, through the very flexible and apt notion of *encoding*, into the theories themselves.

As hinted at the beginning of Sect. 2 and in footnote 13, there are other accounts of higher-order platonism in the market, and it cannot be ruled out that further, equally suitable, platonistic accounts of models of set theory, of multiversism at large, will be found.[39] On the contrary, it would be entirely legitimate to wonder:

Question 13.1 In which ways could the study of the metaphysics of HOP be further developed, and with what foundational implications?

Once shown that it is concretely possible to bring the metaphysical features of the multiverse to bear on the foundations of set theory, where does all this leave us? Does such incursion into metaphysics imply some major revision of the accepted methodologies in set theory?

My answer has been, and still is, a 'no', especially insofar as HOP does not directly bear on the mathematical methodologies in any way. But I think that the potential usefulness of 'auxiliary conceptions' in the context of the foundations of mathematics should not be denied straight away. This is because, as Hamkins himself asserts:

> ..our foundational ideas in the mathematics and philosophy of set theory outstrip our formalism, mathematical issues become philosophical, and set theory increasingly finds itself in need of philosophical assistance. (Hamkins, 2012, p. 436)

although, admittedly, the dividing line between 'philosophical assistance' and 'philosophical reinterpretation of mathematics' could be very hard to draw.

---

[39] For instance, although not directly addressing set-theoretic pluralism, Horsten (2019)'s metaphysical account of 'arbitrary objects' seems very promising in this respect. For a comparison between Horsten's own approach and OT, see Sect. 8.3 of that work.

This last remark very aptly leads me to pose one further question:

Question 13.2 To what extent should extra-set-theoretic, in particular, metaphysical, resources be taken to bear on set-theoretic work?

At this stage, I could not say anything on this issue which would not look like a collection of superficial remarks. I will just limit myself to observing that the present work may contribute to the view that, once the extra-set-theoretic resources are properly formalized and adequately interpreted, they could not only play the role of a 'useful heuristic', in the Maddian sense, but could also become part of, and even help explain, the underlying logic and concepts of set theory in a more robust sense. But clearly it would still be a long way to go to make a fully convincing case for this claim.

References

Antos, C. (2022). Models as fundamental entities in set theory: A naturalistic and practice-based approach. *Erkenntnis, 89*, 1–28.

Antos, C., Friedman, S.-D., Honzik, R., & Ternullo, C. (2015). Multiverse Conceptions in set theory. *Synthese, 192*(8), 2463–2488.

Bagaria, J. (2005). Natural axioms of set theory and the continuum problem. In P. Hájek, L. Valdés-Villanueva, & D. Westerståhl (Eds.), *Logic, Methodology and Philosophy of Science: Proceedings of the Twelfth International Congress* (pp. 43–64). King's College Publications.

Balaguer, M. (1995). A platonist epistemology. *Synthese, 103*, 303–25.

Barton, N. (2016). Multiversism and concepts of set: How much relativism is acceptable? In F. Boccuni & A. Sereni (Eds.), *Objectivity, realism and proof* (pp. 189–209). Springer.

Bernays, P. (1935). Sur le Platonisme dans les Mathématiques. *L'Enseignement Mathématique, 34*, 52–69.

Blechschmidt, I. (2022). Exploring mathematical objects from custom-tailored mathematical universes. In G. Oliveri, C. Ternullo, & S. Boscolo (Eds.), *Objects, Structures and logics. filmat studies in the philosophy of mathematics*, Boston Studies in Philosophy of Science (pp. 63–95). Springer.

Button, T., & Walsh, S. (2018). *Philosophy and model theory*. Oxford University Press.

Clarke-Doane, J. (2022). *Mathematics and metaphilosophy*. Cambridge University Press.

Field, H. (2001). *Truth and absence of fact*. Oxford University Press.

Fuchs, G., Hamkins, J., & Reitz, J. (2015). Set-theoretic geology. *Annals of Pure and Applied Logic, 166*(4), 464–501.

Gödel, K. (1947). What is Cantor's continuum problem? *American Mathematical Monthly, 54,* 515–525.
Gödel, K. (1951). Some Basic Theorems on the Foundations of Mathematics and its Applications. *Gibbs Lecture.*
Hamkins, J. D. (2012). The set-theoretic multiverse. *The Review of Symbolic Logic, 5*(3), 416–449.
Hamkins, J. D. (2020). *Lectures on the philosophy of mathematics.* MIT Press.
Horsten, L. (2019). *The metaphysics and mathematics of arbitrary objects.* Cambridge University Press.
Isaacson, D. (2011). The reality of mathematics and the case of set theory. In A. Simonyi & Z. Novák (Ed.), *Truth, reference and realism* (pp. 1–75).
Jech, T. (2003). *Set theory.* Springer.
Koellner, P. (2013). *Hamkins on the Multiverse.* Pre-print. https://api.semanticscholar.org/CorpusID:3247247
Kunen, K. (2011). *Set theory. An introduction to independence proofs.* College Publications.
Lewis, D. (1986). *On the plurality of worlds.* Blackwell.
Linnebo, Ø. (2017). *Philosophy of mathematics.* Princeton University Press.
Linnebo, Ø. (2018). *Platonism in the Philosophy of Mathematics.*
Linsky, B., & Zalta, E. (1995). Naturalized platonism versus platonized naturalism. *Journal of Philosophy, XCI, 1*(10), 525–555.
Linsky, B., & Zalta, E. (2006). What is neologicism? *The Bulletin of Symbolic Logic, 12*(1), 60–99.
Maddy, P. (2007). *Second philosophy: A naturalistic method.* Oxford University Press.
Maddy, P. (2017). Set-theoretic foundations. In A. Caicedo, J. Cummings, P. Koellner, & P. B. Larson (Eds.), *Foundations of mathematics. Essays in honor of W. Hugh Woodin's 60th Birthday.* Contemporary Mathematics (Vol. 690, pp. 289–322). American Mathematical Society.
Martin, D. (2001). Multiple universes of sets and indeterminate truth values. *Topoi, 20,* 5–16.
Mostowski, A. (1967). Recent results in set theory. In I. Lakatos (Ed.), *Problems in the Philosophy of Mathematics. Proceedings of the International Colloquium in the Philosophy of Science, London, 1965* (Vol. 47, pp. 82–108). Elsevier.
Nodelman, U., & Zalta, E. (2014). Foundations for mathematical structuralism. *Mind, 123*(489), 39–78.
Panza, M., & Sereni, A. (2013). *Plato's problem: An introduction to mathematical platonism.* Palgrave-MacMillan.
Parsons, C. (1995). Platonism and mathematical intuition in kurt Gödel's thought. *Bulletin of Symbolic Logic, 1*(1), 44–74.
Potter, M. (2004). *Set theory and its philosophy.* Oxford University Press.
Putnam, H. (1980). Models and reality. *Journal of Symbolic Logic, 45*(3), 464–82.
Shapiro, S. (1997). *Philosophy of mathematics. Structure and ontology.* Oxford University Press.

Shelah, S. (2003). Logical dreams. *Bulletin of the American Mathematical Society, 40*(2), 203–228.
Skolem, T. (1967). Some remarks on axiomatized set theory. In J. van Heijenoort (Ed.), *From Frege to Gödel. A Source Book in Mathematical Logic, 1879-1931* (pp. 290–301). Harvard University Press.
Ternullo, C. (2019). Maddy on the multiverse. In S. Centrone, D. Kant, & D. Sarikaya (Eds.), *Reflections on the foundations of mathematics: Univalent foundations, set theory, and general thoughts* (pp. 43–78). Springer.
Woodin, W. H. (2001). The continuum hypothesis. *Notices of the American Mathematical Society*, Part 1: 48, 6, pp. 567–76; Part 2: 48, 7, pp. 681–90.
Woodin, W. H. (2011). The continuum hypothesis, the generic-multiverse of sets, and the Ω-Conjecture. In J. Kennedy & R. Kossak (Eds.), *Set theory, arithmetic, and foundations of mathematics: Theorems, philosophies* (pp. 13–42). Cambridge University Press.
Zalta, E. (1983). *Abstract objects: An introduction to axiomatic metaphysics*. Reidel.
Zalta, E. (2000). Neo-logicism? An ontological reduction of mathematics to metaphysics. *Erkenntnis, 53,* 219–65.

14

Language, Models, and Reality: Weak Existence and a Threefold Correspondence

Neil Barton and Giorgio Venturi

1 Introduction

This chapter addresses issues concerning our relationship between *language*, *models*, and *reality*. Much of the philosophy of language, logic, and metaphysics attempts to understand the relationship between the linguistic devices we employ and the underlying ontology. This is often done by providing a *semantics* that accounts for the *truth* of our claims about the world—we explain the truth of various claims by explaining how the language we use is interpreted.

Under one account, we just bluntly point to reality: What do the truth and falsity of our claims depend on? On the way the world is of course! A problem with this view is that often we make claims about reality that seem to capture some aspect of it, but are not *obviously true*. Some claims of theoretical physics, for example, do not clearly map on to *exactly* the way the world is, requiring idealisation. So, how should we account for this discrepancy?

An attractive response to this problem is to understand claims in the context of *models*. A claim is true in some class of models, and these models can

N. Barton
National University of Singapore, Singapore, Singapore

IFIKK, University of Oslo, Oslo, Norway
e-mail: n.a.barton@ifikk.uio.no

G. Venturi (✉)
Department of Civilizations and Forms of Knowledge, University of Pisa, Pisa, Italy

C. Antos et al. (eds.), *The Palgrave Companion to the Philosophy of Set Theory*, Palgrave Companions, https://doi.org/10.1007/978-3-031-62387-5_14

correspond better or worse to the way the world actually is. So the realist about the nature of physical reality can consider the class of models for some physical theory and can then try to come up with an account of "closeness to truth" or "verisimilitude" of particular models. Of course, saying what this "verisimilitude" amounts to is a difficult philosophical problem, but the shape of the account is at least plausible.

This problem and the proposed response are especially interesting in the case of set theory. There, the structure of the underlying ontology for set theory is especially controversial. This is particularly so in light of the independence phenomenon—there are many sentences of set theory that are neither provable nor refutable from the axioms, with no agreed-upon solution. When assessing a set-theoretic claim for its "truth", it is thus hard to know whether and how it conforms to set-theoretic reality.

However the modelling perspective is available here too. On this view (which we'll discuss in Sect. 2) we are able to view *models* of set theory as providing better or worse pictures of the way the sets might be. So perhaps there is a similar kind of view: Even if our set-theoretic discourse doesn't perfectly match up with set-theoretic reality, it is non-vacuous and informative for understanding the sets. In slogan form: *We can still learn about the sets by studying models of set theory.*

One problem with this position is that models of set theory are often taken just *to be* particular kinds of set. There then seems to be a kind of circularity in the above picture (see Sect. 3). For, we want to use the models to talk about the sets, but the models themselves *just are* kinds of set. Our intermediary—the models—seems to depend on the very thing we're trying to understand—the sets.

We think a response can be obtained by employing a notion of *weak existence* (see Sect. 5). Call the notion in which entities exist in fundamental reality (whatever it may be) *strong* existence. *Weak existence* is the sense in which something exists merely in virtue of the language we employ. This is a notion with a rich history and an increasing body of work studying it (see Sect. 4). We will argue that weak existence can be leveraged to obtain the existence of models for our set-theoretic claims. This, in turn, provides an account of enough models of set theory to do the modelling work and, we think, stands free of set theory itself. Putting all this together achieves our:

Main Aim Propose an account on which the relationship between set-theoretic claims and reality is understood as a *threefold* correspondence between (1) the language we use, (2) models for our theories of sets, and (3) the world itself.

We will consider some objections in Sect. 6. We end the chapter in Sect. 7 with a concluding summary and suggest a few applications of this approach to the philosophy of set theory for future development.

2 An Analogy: The Use of Models in Science and Set Theory

As we mentioned above, we will use models to explain the relationship between our language and the world. In order to do so, we therefore need to understand the role of the models in set-theoretic investigation. While the importance of a model-theoretic approach in set theory needs little or any justification (see Antos, 2022 for a thorough substantiation of this point), it will be useful for us to be clear on our theoretical and methodological background.

The starting point of our analysis is searching for cohesion between the natural and formal sciences. More concretely, we want to consider set theory via analogy with other scientific investigations. Of course any particular field has its peculiar tools, tricks, and rules of thumb, but if we step back enough to blur these specific features we can notice a substantial and fruitful unity within the different sciences. Important for us will be the notions of *model* and *modelling*. Let us now take a look at some general traits of models, outlining the aspects that will be of relevance.

Reality is complex and there is no better representation of it beyond reality itself. This is why science needs models: simplified versions of reality that can be more easily studied. We will argue that by assuming a similar theoretical background, we can more easily understand important ideas and debates in the philosophy of set theory.

Let us first introduce the main reason that pushed the set-theoretic community to focus on the study of models, namely, the widespread presence of independence.

Independence There are many statements in the language of set theory that are neither provable nor refutable from our "standard" axioms ZFC.

Examples of sentences independent from ZFC include variants of the Continuum Hypothesis (e.g. GCH, CH), axioms of definable determinacy (e.g. PD, $\mathsf{AD}^{L(\mathbb{R})}$), and large cardinal axioms (e.g. "There is an inaccessible cardinal", "There is a proper class of measurable cardinals").

The plethora of models that set theorists produced in showing the independence of the above sentences has motivated something of a model-theoretic turn in set theory over the last century. The study of models was already percolating beneath the surface in the 1920s and 1930s (e.g. in Zermelo's work on models of second-order set theory in Zermelo (1930)), but it was after the discovery of the independence of **CH** via Gödel (1940) and Cohen (1963) that the contemporary turn to models really took off. As Hamkins writes:

> Set theorists build models to order. As a result, the fundamental objects of study in set theory have become the models of set theory, and set theorists move with agility from one model to another. (Hamkins, 2012, p. 418)

The abundance of the models set theorists have obtained was made possible by the sharpening of model-theoretic techniques like forcing and inner model constructions.[1] As a matter of fact, different models of **ZFC** offered different mutually incompatible pictures of the way set-theoretic reality might be. One might ask: Which should we regard as (more) faithful?[2] How to reconcile the different pictures provided by the different models? What sentences are in fact true? What is the relationship between the models and set-theoretic reality? Do they really help us to describe it?

The point that we would like to stress here is that these are deep and important questions both for mathematicians and philosophers. For example, if one is a *descriptivist* about reference to the sets—that is, one believes that we essentially refer to the sets via description—then the existence of many plausible-looking models indicates that there may be many descriptions consistent with our set-theoretic thinking (at least insofar as we are concerned with *first-order* **ZFC**).[3] But one needn't be a descriptivist to find these questions pressing; as more and more independent sentences were discovered in a wide variety of different models, the prospects for discovering truths about the universe of set theory became less and less clear. The attempt to answer these questions and the consequent way of phrasing the problem of independence in terms of the study of the models of set theory is a theoretical move of some import. Here we find a resemblance with the question of the relationship

---

[1]Of course these are not the only techniques available, but the most well-known. Moreover, there is a rich body of work examining independence in theories weaker than **ZFC**. For example, the study of the independence of the axioms of **ZFC** from one another has produced interesting results, such as (i) within **ZFC** without Powerset Axiom there can be no uncountable cardinals, and (ii) neither Replacement nor Foundation are provable from the version of **ZFC** with these axioms removed. And this phenomenon is very general, it is instantiated everywhere from the highest reaches of large cardinal theory down to weak systems of arithmetic.

[2]See Barton (2020b) and Antos et al. (2021) for some discussion of this question.

[3]For an examination of descriptivism, see Zeynep Soysal's contribution to the present volume Soysal (2024).

between models and reality that we find in the natural sciences.[4] Far from being the only way to conceptualise this model-theoretic shift in set theory, this analogy suggests that we use the models of set theory to study the sets themselves.

On the basis of this intuition we would like to sketch a threefold relation between language, models, and reality and to suggest that this can help clarifying theoretical and methodological questions related to the study of set theory. The relation we want to describe can be depicted as follows.

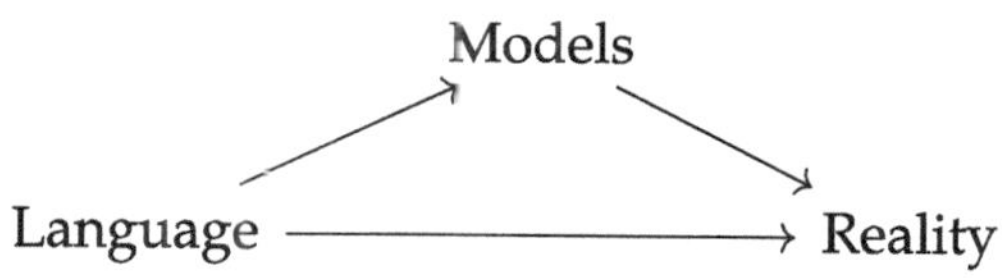

Language can be used to describe reality directly, in terms of sentences expressing some true or false propositions. This function of language is purely descriptive. However, if we utter or believe some false sentence(s), it is not the case that we simply *fail* to talk about reality (or can only be interpreted in some restricted interpretation). Rather we can view our theorising as *partially* corresponding to reality. By viewing models as intermediaries, we can lay the groundwork for such a proposal—we talk about models which can conform better or worse to the way the world is.

A further advantage of using models as an intermediary consists in allowing us to test our hypotheses about reality not in isolation, but in connection with other hypotheses and ideas. Many models (and model-theoretic constructions) provide us with "information-rich" contexts in which we can analyse particular principles (on the assumption that the principles themselves are consistent). For example, if one can construct an "L-like" model of $\mathsf{ZFC} + \phi$ for some suitable ϕ, then one knows that many statements of set theory are consistent with ϕ (e.g. as done with GCH and large cardinal axioms). In a similar vein, on the assumption that $\mathsf{ZFC} + \phi$ has a model, we can force over the model to obtain many other interrelations of set-theoretic principles (possibly not including ϕ, depending on the nature of the forcing). So a model does not just give us *one* picture of the way the world might be, but can allow us to generate multiple such pictures, each of which can be viewed as a possible candidate for representing reality more or less faithfully.[5]

[4]For a more general discussion on the relationship between the methodology of natural sciences and set theory see Barton et al. (2020).

[5]Some similar ideas are given in Arrigoni and Friedman (2013), but with the target of studying maximality principles rather than as a way of talking about the world.

In this sense models provide a medium that allows language to meaningfully correspond to reality beyond simple truth or falsity. The gain of this intermediate step is given by the structured way in which we can depict reality. As any model in natural sciences, also models of set theory capture some aspects of reality, while neglecting others. For example, any countable model clearly misses many reals and, because of its cardinality, offers a very pale image of the full richness of the sets. Although countable models do not provide fully faithful pictures, they nonetheless offer interesting insights on how the sets could or could not be.

Similarly, many entities of interest to the scientist—infinitely deep oceans, frictionless planes, miniature wind tunnel prototypes, and so forth—are clearly not perfect representations of reality. Others, perhaps, get closer (e.g. a suitable candidate mathematical model for the structure of the universe under general relativity). All, however, are useful given particular purposes—a model for general relativity isn't going to help much with trying to predict the behaviour of ocean waves for generating tidal power, the former is just too mathematically complex to provide useful predictions in that context. In the end, though, we would can view each as a description of an underlying subject matter and we would like some cohesion between the different models of reality.[6] And sometimes a slight dissimilarity from reality is the price we pay for the tractability of the relevant model. Again, these issues are of central importance to *scientific* descriptivists (those who think our reference to reality is mediated by the descriptions we provide), but one need not be a descriptivist to find questions concerning the faithfulness of models philosophically interesting.

Analogically, some models of set theory are more tractable than others. For example we can have models that are pointwise-definable (like the Shepherdson-Cohen minimal model—the smallest transitive model of **ZFC** under inclusion). Such models plausibly don't capture the full richness of set-theoretic reality, which we normally take to include non-definable objects (e.g. some real numbers) and many satisfy false statements (e.g. the Shepherdson-Cohen model satisfies $V = L$, which is usually taken to be false). Others might get closer, but might be more intractable, such as when we (schematically) assert that there is a V_δ elementary in the universe. One main goal of set theorists and philosophers is to produce a theory of how these models are together able to provide a coherent description of reality, despite their varying degrees of tractability and disagreements on many statements of set theory.

[6]Some scholars don't think full cohesion is desirable, for example, Nancy Cartwright's "patchwork" account of the philosophy of science (cf. Cartwright, 1999).

Of course, the nature of this modelling relation is responsible for some of the hardest problems in the philosophy of science. What does this *simplification* amount to? What is the *relation* between reality/phenomena and models thereof? How *faithful* are models? What conclusions can we draw from the study of *different* models? And *should* we really accept a descriptivism about reality, mediated by models, or should we just regard the above questions as important but not essential to the question of semantic reference? There is an enormous literature on the topic (see Frigg & Hartmann, 2020). On the one hand, these represent a cluster of challenges for the approach we outline here. On the other, it suggests that there are a wealth of resources that might be applied to the set-theoretic context.

Now is not the time to fully articulate the nature of the modelling relation between models and reality. We hope to have convinced the reader that there is a fruitful project to be examined here, and we leave the details to future work (we will consider some suggestions in Sect. 7). Whilst the approach is not yet fully developed, one can see how studying models in order to understand reality can be attractive. By the independence phenomenon, we know that **ZFC** fails to get much traction on set-theoretic reality. By studying models, we can hope to understand better the ways the world might be. As we'll now see, there's a powerful objection that threatens to scupper this project before it has had chance to begin its voyage. The rest of the chapter will be devoted to outlining this problem and providing a response, opening the door to further projects in this direction.

3 The Circularity Problem

From a philosophical perspective, we face an immediate challenge before even embarking on such a programme. This is because we often take the models of set theory to be *themselves* sets. In addressing Hamkins' claim that the fundamental objects of study in set theory are models, Koellner pushes back by insisting that sets and models should be put on a par.

> To this one is bound to protest that the fundamental objects are still what they always were, namely, *sets*. Some of those sets happen to be models of certain theories. (Koellner, 2013, p. 12)

Although Koellner may have slightly different aims by claiming that models are just particular kinds of set, we can extract an important problem for the threefold correspondence from his remarks. Suppose we want to use models of set theory in providing pictures of the way sets might be. Since (so the argument goes) models just are sets, we are using the very things we are meant to be clarifying in providing said clarity. One might immediately worry that this is somehow circular (more on this below). Call this the **Circularity Problem**.

Let's develop this in a little more detail. To this end, it's helpful to consider an analogy. One purpose of considering models in science (rather than reality itself) is to find a way of representing the world that can be agreed on by all parties. A realist about unobservable entities and a constructive empiricist can agree on the intrinsic facts about the models employed. What they disagree about is how those models and reality interrelate. But this shared point of contact allows both to consider the nature of the modelling relation.

If we have a simple identification of models with certain sets, the same does not apply. There *the very nature of the models* is dependent upon the nature of the sets. This has the unwelcome feature that we do not have the same kind of ontological neutrality as in the sciences. For a simple example, consider a realist and a nominalist about sets (and let us assume that neither has any complaints about set theory as a mathematical area of study, even if they disagree sharply on the ontology). It would be preferable if, ceteris paribus, the nominalist and realist could have some models as a point of contact. But they can't if we identify models with sets—the nominalist thinks that such things don't exist, so the question of how models and reality interrelate is moot. This problem is very general; any time there is significant disagreement on the nature of the underlying ontology (as there so often is in set theory) we run the risk of losing a shared grip on the nature of models. A slightly more complicated example involves contrasting the set-theoretic universist (who believes that there is an all-encompassing universe of sets) and a multiversist (who believes there is no such universe). How do theorists with such radically different views on the nature of sets come to agreement on the models? How much can be shared between the two views? (We shall provide a tentative proposal in Sect. 7.4.)

This metaphysical problem comes attached with an epistemological one. Even supposing that we're happy being set-theoretic realists, we may still worry about our epistemological access to the sets. They are, all things considered, pretty abstract and complicated objects. But now we have identified models with sets, it is unclear if there is any epistemological gain to be had. Contrast the case of models of relativistic theoretical physics. There we substitute one intractable object (the physical universe) with a much more tractable one (a particular kind of manifold). But given the identification of models with sets, we have substituted the study of one class of objects with the study objects of the very same kind, and so any epistemological problems of access with the former extend to the latter.

There's thus two dimensions to the **Circularity Problem**, and our response will address them both.

(i) We would like to ensure that there's enough things around to play the role of models, without depending on whatever theory we're engaged in modelling (in this case set theory, and in particular relatives of **ZFC**), and
(ii) We would like there to be models that are not simply the same kind of thing as the objects we are describing (in this case the sets).[7]

Our response will use the notion of *weak existence*—existence in virtue of language—to ground enough models to do the work we want that are in turn not simply sets. Models that exist in virtue of weak existence need not be ontologically dependent on set theory, and there are reasons to be hopeful that we have better epistemological access to them. We devote the next section to the task of showing that a close relation between language and existence is not new. On the contrary, weak existence is part of a rich philosophical tradition.

4 Existing Accounts of Weak Existence

In order to get a better handle on weak existence, we will start by considering some ideas as they appear in philosophy more broadly. Because of the many and heterogeneous ideas involved, it is hard to provide a comprehensive survey, and there may be a case to be made for including views that we have omitted.[8] However, we want to show that the possibility of a *direct* link between language and existence is an idea that has been advanced in many different contexts. Existence *merely in virtue of language* is what makes the resulting notion of existence "weak" rather than "strong", the latter is a notion concerning true or false propositions about existence in the *world as a whole*, weak existence is *merely linguistic*. This has substantial implications; language is a human tool, while existence, in a strong sense, is often understood as being independent from our cognitive, linguistic, or mental abilities.

[7]Thanks to an anonymous reviewer for suggesting this way of characterising the two dimensions of the **Circularity Problem**.

[8]For instance, anonymous reviewer suggests also considering Penelope Maddy's *arealism* and *thin realism* (see Maddy, 2011). Whilst we agree that there may be interesting points of contact here, Maddy seems to draw a close relationship between *linguistic* and *conventionalist* considerations and is careful to distinguish such views from her position (see, e.g. Maddy, 2011, pp. 5, 64–71). This complicates the interpretation, and so we've avoided in depth discussion of Maddy's views, despite their interest.

We will later see how this plays out in the specific case of set theory. For now, in order to delimit the field of investigation, we would like to review positions of mathematicians and philosophers that can be seen as implying such a notion of weak existence. We will discuss a cluster of accounts that, on closer inspection, are quite philosophically and chronologically disparate. If anything, this will show that this notion of weak existence is compatible with many different views concerning strong existence and indeed provides an undercurrent of much philosophical thought. This neutrality will be an important feature of our account of the relation between language and reality.

4.1 Consistency Implies Existence

An ideal starting point is Hilbert's view on existence. As is well known from the Frege-Hilbert controversy, Hilbert held the position that consistency implies existence.

> I found it very interesting to read this very sentence in your letter, for as long as I have been thinking, writing and lecturing on these things, I have been saying the exact reverse: if the arbitrarily given axioms do not contradict one another with all their consequences, then they are true and the things defined by the axioms exist. (Gabriel et al., 1980, p. 39)

Because of the syntactic character of the notion of consistency, we can safely consider his notion of existence as directly linked to language. Notice that, although Hilbert is often described as the champion of formalism, this direct connection between consistency and existence is compatible with very different views about mathematical objects. Indeed, consistency could be seen as constitutive of existence (as a formalist/nominalist reading would suggest) or as indicative of existence (as Cantor claimed, considering the perfect harmony between the immanent and the transcendent realities described in Cantor, 1883).[9]

Another important aspect of this Hilbertian notion of existence, besides its linguistic root, is its semantic realisation. Indeed, the independence results that Hilbert proved in the *Grundlagen der Geometrie* rest on the possibility of constructing different models for (non-Euclidean) geometries from one another. In this way the existence provided by consistency is realised in a

[9]A similar position has been proposed in modern times, with the name of full-blooded platonism, in Balaguer (1998), as arguably the most tenable form of mathematical platonism.

model-theoretic way and, therefore, this notion of weak existence is assimilated to that of existence in a model.[10]

4.2 Abstraction Principles

Another interesting case of a direct connection between language and weak existence is given by the use of abstraction principles, or abstraction more generally in the foundations of mathematics.

We can (of course) find very different views on abstraction that are compatible with different view on the ontology of mathematics. For example, Dedekind held that abstraction is able to create new mathematical objects. According to his view natural numbers are free creation of the human mind, while reals are abstracted from Dedekind's cuts. This creative use of abstraction found the fierce opposition of Frege, who could not accept any psychological interference in the platonic realm of thought. Nonetheless, Frege viewed the practice of definition in mathematics as still yielding something new, namely, the attribution of sense and reference to a new symbol through an act of declaration. The important difference with respect to other accounts of definition, however, lay in the possibility to constrain this form of creativity with clear criteria. In 1903, in the *Grundgesetze der Arithmetik*, Frege wrote:

> Can our procedure be called a creation? The discussion of this question can easily degenerate into a quarrel about words. In any case, our creation, if one wishes so to call it, is not unconstrained and arbitrary, but rather the way of proceeding, and its permissibility, is settled once and for all. Ebert and Rossberg (2013, §147)[11]

Whilst Frege did not have a theory/meta-theory distinction, the forms of neo-logicism that originated from the failure of Frege's foundational project have managed to show that abstraction principles weaker than Basic Law V have models witnessing their consistency (Heck, 2011).

In this tradition we find also positions like that of Øystein Linnebo on thin objects. Linnebo, inspired by Frege, proposes a realist view about mathematics

[10] Indeed, some people have argued that this is notion of existence in mathematics that should be attributed to Hilbert. See, for example, Doherty (2017). For a more general discussion of the creative role of implicit definitions, *à la* Hilbert see Schmidt and Venturi (2023).

[11] See Ebert and Rossberg (2019) for a discussion and an interpretation of these lines in the context of Frege's Platonism.

in which our linguistic practices contribute directly to the semantics of our mathematical sentences.

> In short, the assertibility conditions make a twofold contribution: to the determination of the semantic content of beliefs (and thus also to the truth of these beliefs), and to the formation of the beliefs. Linnebo (2018, p. 201)

The notion of mathematical object that emerges from this picture is that of thin objects, on which an object is the possible referent of a singular term. In this context, it is by reflecting on mathematical language and the various assertibility conditions that govern it that we get to know that mathematical objects exist. The epistemological twist of Linnebo's account makes it possible for him to hold a realist view of mathematical ontology. However, the semantic properties that emerge from our linguistic practices are of philosophical interest in themselves; they are used to back up existential claims.

Before we continue, we should note that not every neo-Fregean use of abstraction principles subscribes to such an idea. The position we have outlined is somewhat reminiscent of Field's interpretation of abstraction and implicit definition, an interpretation is critiqued by the likes of Hale and Wright (see Hale & Wright, 2009). Our claim is not that abstraction *necessarily* leads to notions of weak existence, but rather that it *can*.

4.3 Internal Questions and Easy Arguments

Another influential position that connected (weak) existential claims and linguistic practice(s) is that of Carnap. Specifically, the distinction between internal and external existential claims.

> If someone wishes to speak in his language about a new kind of entities, he has to introduce a system of new ways of speaking, subject to new rules; we should call this procedure the construction of a linguistic *framework* for the new entities in question. And now we must distinguish two kinds of questions of existence: first, questions of the existence of certain entities of the new kind *within the framework*; we call them *internal questions*; and second, questions concerning the existence or reality *of the system of entities as a whole*, called, *external questions*. (Carnap, 1950, pp. 21–22, original emphasis)

The standard reading of this distinction takes the internal question as motivating a notion of existence that is weaker than that investigated by metaphysicians (interested in external questions) and that can be simply answered on the base of the linguistic resources of the framework from which it emerges. Therefore, following Carnap's anti-metaphysical stance, the syntactic criteria provided by theoremhood offer clear criteria for a satisfiable answer to internal mathematical questions. For example, the question on whether there is a prime number between a n and $2n$ (for $n > 1$) is completely answered by the proof that this is indeed the case. On the contrary, whether natural numbers exist in general is an external question that arguably does not possess a clear (and relevant) answer. The weak existence provided by internal mathematical questions is, in Carnap's view, an anti-metaphysical perspective that collapses linguistic interpretation for singular terms and existence (in a weak sense).

More recently, a similar position has been developed by Amie Thomasson. In Thomasson (2014) she argues in favour of a deflationist notion of existence that, although compatible with realism, is rooted in our linguistic practices. The lightness of this new form of realism is attributed, by Thomasson, to the easy form that the corresponding ontological arguments assume. For example, if someone asserts that she ate two bagels, she can automatically commit to the existence of a natural number: the number two. There are affinities here with Carnap's ideas; to commit to some entity existing is just to commit to a linguistic framework in which such a commitment can be made. Space does not permit a full analysis of Thomasson's position, but one can see how there is an articulation of a close connection between language and its semantics. a connection able to account for a weak form of existence.

So, we see that weak existence is an idea with a rich history in philosophy. And as the reader may be able to guess, the idea that linguistic practice can underwrite existential claims may be useful in obtaining models for set theory, independent of set theory itself. To be clear, we do not think that weak existence need to take one of the forms adumbrated above, and we leave open the extent to which our approach below meshes with these accounts. But we do think that we can dissolve the **Circularity Problem** using the motivations of weak existence. Let's see how this plays out.

5 Underwriting Models with Weak Existence

Let's start by recalling the **Circularity Problem**. If we intend to study set-theoretic reality in terms of the models of set theory (partial descriptions of it), how are we to avoid the circularity brought about by the fact that models are (allegedly) just sets? In other terms, is there a way to account for models of set theory that does not view them as sets and that does not use features of mathematical objects that would be hard to justify outside set theory?

We will argue that one can do so by casting the existence of some models in a linguistic light. If our strategy is successful it will provide a linguistic approach to models that does not depend on set-theoretically loaded assumptions regarding the nature of set-theoretic reality. Notice that even if our strategy succeeds, we do not have to argue that models are *not* sets. What models "are" is, we think, ultimately unimportant. Our reasons for thinking so are roughly Benacerrafian; as far as models go it doesn't really matter what they're composed of, so long as one can think meaningfully of objects together satisfying certain theories, we are in the clear. And obviously if we *assume* set theory, we *can* encode models of set theory using sets. What we *do* want to do is identify a class of models that are sufficient for the needs of set theory and also obtainable via methods compatible with weak existence. In general, you can still view models as sets *if you like*, there is just no *obligation* to do so.

We will argue that models can be represented in linguistic terms and this allows us to separate enough models ontologically from the sets in order to answer the **Circularity Problem**.[12] The key observation is that some models of set theory can *themselves* be composed of syntax. This idea (that models can be thought of as composed of pieces of language) will be familiar to anyone who has studied proofs of the Completeness Theorem. We revisit some core moves made there to motivate this claim.

How does one prove the Completeness Theorem? As is well known, its standard formulation (i.e. every valid sentence has a proof) admits of an equivalent formulation that every consistent set of sentences is satisfiable. The textbook proof of this latter form proceeds by forming a *term model*. Given a consistent set of sentences Γ, we first add infinitely many constants $c_0, \ldots, c_n, \ldots$ to the language. Next, we carefully extend Γ to a maximal infinite consistent set Γ^* (in which quantified sentences are linked to constants via conditionals). We can then build a model for Γ by letting the domain of the

[12]One proposal that is parallel to our own, but we will only consider parenthetically, is Carolin Antos' recent idea that models should be regarded as fundamental entities in their own right (see Antos, 2022). Drawing on Penelope Maddy's naturalism/second philosophical approach (see Maddy, 1997 and Maddy, 2007) Antos argues that viewing models of set theory as fundamental (in addition to sets) provides an "effective means to particular desirable ends" (Maddy, 1997, p. 194) and on these grounds their fundamentality should be accepted. Antos' arguments are very helpful given our predicament, since they provide practice-based reasons to hold that models are sui generis entities. This is a possible response to the **Circularity Problem**, but our project is different. We want to show how (some) models can be postulated to exist in the light of weak existence. The advantage of this is that it is plausibly less ontologically demanding than the Antos picture and requires no recourse to Second Philosophy. Moreover, we claim that there is epistemic gain in viewing models as providing us with weak existence in set theory, since the models are explicitly constructed (rather their existence simply abductively inferred). But our proposal and Antos' need not be viewed as *competing* (indeed we view them as *complementary*).

model be the constant symbols and the predicates be interpreted via the atomic sentences in Γ^*.[13]

Notice that we can think of this proof as specifying a model via weak existence—the model we end up with is composed of *syntactic entities*. This yields a way of obtaining models of set theory. For, suppose we are considering some set Γ of sentences of **ZFC**. Then, so long as Γ is consistent, we can think of a model of Γ as composed of constants.

(An important remark, but one we'll make parenthetically for now: The eagle-eyed and fox-cunning reader may ask themselves at this point whether the fragment of the Completeness Theorem needed has reverse mathematical strength, and what amount of higher-order resources we're assuming in saying that the model is "composed" of constants. We will discuss this later (Sect. 6).)

The above strategy is sufficient to get us *a* model for any particular set theory we take to be consistent, but does it get us *enough* models for set theory to underwrite the *study of models of set theory* as it actually occurs in practice? There it is not just important that *there is* a model (for some theory **T**), but that the model bears the appropriate relationship to other models (e.g. that given a model M, we can move to the forcing extension $M[G]$, etc.). Can we get models bearing the appropriate relationships in this manner?

Our response will be two-pronged. First we'll note a cheap kind of response—by just increasing the strength of the theory we take to be consistent in the beginning, we can obtain weak existence models that are sufficient to do any particular construction we like. However, we'll then note a less cheap response, pointing out that one of the most important model-theoretic constructions in set theory (namely, forcing) can be interpreted in weak existence terms.

Starting with the "cheap" strategy, it bears mentioning that nothing in our construction of a weak existence model (e.g. as in the Completeness Theorem) depended upon the use of any particular theory or other. So long as we take a strong enough theory **T** that suffices to produce the models we need for some construction or other, we can turn Koellner's observation on its head. Since we *can* do the model theory of sets within set theory, a model of **T** obtained by weak existence will suffice for all model theory that can be done in **T**. We

[13]Things are a scooch more complicated if one has terms in the language. Since we're just working in the language of set theory (that has a single non-logical predicate symbol "$\in$") we don't have to worry about these subtleties, but since we think weak existence still suffices for other contexts too, it's worth at least providing a footnote on the matter. Where terms are involved, one has to instead think of the domain of the final term model as composed of sets of closed terms that are provably equal given Γ. As far as weak existence goes, the move from constants to these sets of terms strikes us as innocent, though we need not take a stand on the matter for the purposes of this chapter.

just live in this model of T and do all the required model theory purely set-theoretically there. To illustrate this, suppose I consider the model theory of ultrapowers using a measurable cardinal κ. Instead of getting a weak existence model for ZFC, I could simply argue (howsoever I choose to—now is not the time to adjudicate such matters) that I have good reason to believe that ZFC + "There is a measurable cardinal" is consistent. Let this be our theory T. By building a weak existence model for T (as above), I obtain a model M that satisfies ZFC and contains (what M takes to be) a measurable cardinal. I then just run the usual model-theoretic constructions in M. And this method is very general, applying to many model-theoretic constructions (e.g. inner models and forcing) that can be done within a particular consistent extension of ZFC.

Whilst this strategy is fine so far as it goes, we think that much stronger things can be said about weak existence in relation to many of the model building constructions present in set theory. In this way, the above response masks some important features of set theory in relation to weak existence that are more apparent if we approach the problem directly.

What we want to point out is that there's a kind of weak existence *relative* to a starting model that can be leveraged in favour of obtaining models for forcing extensions. Let's suppose we have some model or other M (which could, e.g. be obtained via weak existence). We want to force over M. How does this proceed? One way of doing so is to start with a particular kind of Boolean-algebra $\mathbb{B}$. We then (within M) define the model $V^{\mathbb{B}}$ as follows:

(i) $V_0^{\mathbb{B}} = \emptyset$
(ii) $V_{\alpha+1}^{\mathbb{B}} = \{f | "f \text{ is a function"} \wedge dom(f) \subseteq V_\alpha^{\mathbb{B}} \wedge ran(f) \subseteq B\}$ (for successor ordinal $\alpha + 1$).
(iii) $V_\lambda^{\mathbb{B}} = \bigcup_{\beta<\lambda} V_\beta^{\mathbb{B}}$ (for limit ordinal λ)

Another (equivalent) method is to start with a partial order $\mathbb{P}$ in M, and consider the class of $\mathbb{P}$-names—these are functions whose elements' first coordinate is another $\mathbb{P}$-name and whose second coordinate is some condition $p \in \mathbb{P}$ (the whole thing gets off the ground because $\emptyset$ is trivially a $\mathbb{P}$-name). Then, when presented with a generic filter G for $\mathbb{P}$ and M, we can recursively evaluate the $\mathbb{P}$-names to obtain the forcing extension $M[G]$. This is done for some $\mathbb{P}$-name σ by including the values of a name τ such that $\langle \tau, q \rangle \in \sigma$ if $q \in G$.

The important point that we wish to note is that the construction of the forcing language and assessment of what the forcing extension satisfies can be viewed as a particular kind of weak existence construction in M. We introduce pieces of *language* via the $\mathbb{P}$-names, and within $M^{\mathbb{B}}$ the $\mathbb{B}$-names are all viewed as sets existing side by side (though $M^{\mathbb{B}}$ might view some of them as identical). In this sense, we have weak existence *relative* to M.

One salient objection here is that the forcing language is *enormous*, since it has a constant for every name, it will be proper-class-sized. In the present context though this should not deter us. The background in which we are considering these extensions is just a countable model M obtained weakly via the same idea as the Completeness Theorem. So, though "proper-class-sized" relative to M, the forcing language we end up discussing is really just another countable language, if we can obtain every element of M as weakly existing, then so can we obtain every element of $M^{\mathbb{B}}$.

So, we can obtain weak existence models for a theory we take to be consistent, but we can also think of forcing as obtaining particular kinds of weak existence models given the specification of an initial background. We can thus obtain enough models to do the model theory of sets via weak existence (at least as far as forcing and inner models are concerned). We conjecture that many model-theoretic constructions (e.g. ultrapowers in general) can be conceived via weak existence methods, but we need not take a stand on this here.

Before we continue, we should pause to reflect on the exact strength of our claims. At face value, our claims are quite strong—we claim that models satisfying a wide class of set-theoretic constructions can be obtained via methods that are compatible with weak existence. This is the first important answer to the **Circularity Problem:** the circularity of studying sets with sets disappears once we realise that the models we use (although representable as sets) are more epistemically tractable than sets in general. Indeed, the medium offered by language allows us to have a direct access to models, which is not often the case with sets (if we do not accept any strong forms of intuition). However, a possible objection to this line of argument could consist in accepting that weak existence models allow us to disentangle set-theoretic reality from its representation, whilst still insisting that this is not enough for the study of set theory in its generality. We can easily respond to this further complaint noticing that we do *not* claim that *every* model of set theory can be found by weak existence. Instead, we only claim that one *can* find models in terms of methods that yield objects that weakly exist and that this is enough for the propose of doing set theory. In order to see this it is enough to consider that the Löwenheim-Skolem Theorem allows us to restrict our attention only to countable models. Indeed, assuming that our language is countable, given *any* model there will always be a countable model that is elementary equivalent to it.

6 Objections

Thus far, we've sketched an account of how our utterances regarding the sets might relate to reality. We speak of *models* (which in turn can be viewed as linguistically constructed), and then these models may conform better or worse to the way the world is. We now discuss a few objections that our account might face.

Why Not Just Use the Natural Numbers? First of all one might object that our appeal to weak existence is not necessary, since we could have just done everything using natural numbers. Indeed (so this objection runs) since we can always view the syntax of a countable recursive language as encoded within a suitable theory of arithmetic, then any talk about weak existence is (at the end of the day) just a way of talking about the natural numbers. And we know (by the Completeness Theorem) that we can build a model using natural numbers for any consistent countable theory.

Our response is epistemological in nature. Whilst (of course) we *could* view a consistent countable theory as modelled using natural numbers if we so desired, having a weak existence model is more epistemologically direct. In particular the process of forming a term-model is one of composing a model via syntax. Of course this is all *countable* and so can be *encoded* by natural numbers. But why go through this epistemic detour? By noting that languages and names for objects are employed as soon as one is discussing any subject matter, and that these can be mobilised in the service of generating models, we shorten the gap between our language and the representations we employ.

Moreover our view is able to account for the practice of several model-theoretic constructions—such as forcing constructions—in syntactic terms. We thus get a similarity between the construction *of* a model and the constructions done *over* a model. Whilst the natural number interpretation is fine insofar as it goes (and indeed, given our purposes here—the more models of different kinds the better!) it blurs epistemically relevant and interesting features.

Too Many (Linguistic) Objects? A different line of objection is to criticise our approach by accusing it of smuggling in infinitary resources beyond those licensed by the epistemologically safe environment provided by language. One might complain that in order to build the model, we had to add *infinitely* many constants to our language. On these grounds, one might complain that this stretches the use of "existence in virtue of the language employed" beyond sensible bounds.

We can provide a reasonably quick answer here that helps clarify our position. We do not claim that we obtain the required syntax by ordinary usage as it occurs in day-to-day life. In this sense, the linguistic resources to study set theory are, to some extent, idealised. However, the idealisation needed does not go beyond the limits of an epistemically tractable language and moreover within this language there *is* a kind of existence, appealed to across philosophy, that allows us to underwrite existence claims in mathematics (this was the point of surveying various accounts in §3). And it is this idealised notion, we argue, that can deliver the models required and has benefits beyond regarding models as only obtained within set theory.

Did We Smuggle in Infinitary Resources? There is a more subtle objection lurking here, however. If it could be further shown that there is an implicit use of some infinitary assumption in constructing our models, one might undermine our claim that we have divorced the existence of models of set theory from set theory itself.

We can go further by observing that we appear to be using higher-order resources (we thank the anonymous reviewers for pressing us on this issue). There is the clear point that whilst the models are "composed" of syntactic objects, this talk is higher order in nature. For all we've said, a model is not guaranteed to be *a* syntactic object, but rather a *plurality* of constants (or some other suitable second-order surrogate). In this respect we follow the tradition of using *higher-order* resources rather than *sets* (combined with our own linguistic twist) in providing a model theory for the higher infinite. We think we therefore avoid the **Circularity Objection**, our models are pluralities of syntactic objects (rather than sets) and so are of a different ontological kind.

The natural rejoinder is, of course, that in using these higher-order resources we commit ourselves to set theory anyway. Whilst it is tempting to view us as adding just one layer of sets on top of our syntactic objects, a view that would land us squarely back in the sights of the **Circularity Objection** (though there would be *some* gain over the position that views models as *very* large infinite objects), we aren't moved by this objection. In particular, we don't think that second-order logic is dependent on set theory, a fact witnessed by the large variety of interpretations available. But we also don't expect everyone to share this view. Thankfully our point can be strengthened: We do not *need* very much higher-order strength to do the work we want. Let's examine how much higher-order power is required.

First, what we do need: It is possible to prove the Compactness Theorem from the Completeness Theorem and the Soundness Theorem. And in turn, we can note that the Compactness Theorem is equivalent (modulo **ZF**) to various weak but non-trivial Choice-like principles, for example, Tychonoff's Theorem for Hausdorff spaces, the Boolean Prime Ideal Theorem, and the principle according to which every filter on a Boolean algebra can be extended to an ultrafilter.[14] Thus, we do need *some* higher-order resources. And one might try to press the point that we have used a non-negligible amount of the Axiom of Choice in supporting the existence of weak existence models, and thereby undermine our own argument.

We can respond to this by noting that for our purposes (getting models for **ZFC** with some consistent set of sentences of set theory added) we only need the Completeness Theorem for *countable* languages. This version of the Completeness Theorem is equivalent (modulo RCA_0) to Weak König's Lemma—the principles that every infinite subtree of the binary tree has an infinite branch. So we do not depend on strong set-theoretic background principles, or anything like the power of full plural/second-order logic over a suitable syntax theory, but rather only require a small fragment of second-order arithmetic. In fact we stay well within the bounds of what is often regarded as Hilbertian reductionism and equivalent (over RCA_0) to many natural statements of ordinary mathematics (e.g. the Jourdan Curve Theorem).[15] So really, we don't need very much, and whilst we acknowledge that some mathematical background is needed to run our arguments, it is too strong to say that the construction of weak existence models is "really" underwritten by set theory, rather than our ability to talk about the small amount of mathematics required for weak existence more generally.

7 Conclusions and Open Questions

We started this chapter with a remark about the nature of using models to talk about reality. In the context of set theory we face a challenge—how to account for the existence of models when they themselves are meant to be sets? Appealing to weak existence, we think, guarantees enough models of set theory to do the job without appeal to any further assumptions regarding set-

[14] For a survey, see Paseau and Leek (2022).

[15] See Simpson (2009) for a discussion of the reverse mathematics here, and in particular §IV.3 for a proof of the equivalence between the Completeness Theorem for countable languages and Weak-König's Lemma.

theoretic foundations. This then suggests that we can think of our set-theoretic talk as involving a threefold correspondence between our utterances, models of those utterances, and the world itself (in a manner consonant with many areas of science).

On the one hand, we think we've provided a satisfactory solution to the challenge we proposed. On the other hand, our proposal is merely *programmatic*, in that it establishes the possibility and attractiveness of pursuing a particular idea, rather than providing a full account of the details of how it is to be carried out. There are thus *many* questions left open by the approach. We wish to identify some of them here (and hope that the reader will allow the conclusion to be slightly longer than is usual).

7.1 Compatibility Between Different Accounts of Weak Existence

In Sect. 4 we surveyed some existing programmes that employ notions of weak existence. We then employed the idea of weak existence in justifying the existence of enough models to underpin the threefold correspondence. However, we have said very little about the relationships between different accounts. And it is unclear what the relationships are between these different accounts (including ours), and indeed whether or not they can coexist happily. For instance, an anonymous referee helpfully pointed out to us that abstraction principles are typically understood by Fregeans and neo-Fregeans to deliver a kind of objectivity about mathematics. However, we think that our view is compatible with both a large and small number of objective truths (more on this below). Similarly, the Carnapian programme of analysing existence claims as "internal questions" seems to be in tension with the way we have spoken of weak existence has having definitive theory-independent answers. We therefore ask:

Question What are the relationships between the different kinds of weak existence adumbrated, and which (if any) can support the use of our weak existence models in the threefold correspondence?

7.2 Verisimilitude of a Model

The question of "verisimilitude" helps to clarify exactly what we've achieved in this chapter. We have merely substantiated the claim that the threefold-correspondence view is attractive and plausible. However, more needs to be done to provide a full account of the view. In particular, a natural project is to try to ascertain which models of set theory are closer to truth than others. This approach is not without precedent in the philosophical literature. Saharon Shelah (in Shelah, 2002) has advanced a conception of set-theoretic truth which should be conceived of *measure-theoretically*, with different sentences of set theory assigned different measures. He writes:

> I do not agree with the pure Platonic view that the interesting problems in set theory can be decided, we just have to discover the additional axiom. My mental picture is that we have many possible set theories, all conforming to **ZFC**. I do not feel "a universe of **ZFC**" is like "the sun", it is rather like "a human being" or "a human being of some fixed nationality".... So my meaning in saying "why the hell should [$V = L$] be true", is not that it is provably false, just as "the national lottery in the last ten years was won successively in turn by the nephews of the manager, so we know that there was cheating" is mathematically not proved. Clearly L is very special, to some extent unique, thus, the statement $V = L$ should get probability zero (thought not being impossible). So L is certainly a citizen with full rights but a very atypical one. Also a typical citizen will not satisfy $(\forall\alpha)(2^{\aleph_\alpha} = \aleph_{\alpha+\alpha+7})$ but probably will satisfy $(\exists\alpha)(2^{\aleph_\alpha} = \aleph_{\alpha+\alpha+7})$. However, some statements do not seem to me clearly classified as typical or atypical. You may think "does **CH**, i.e., $2^{\aleph_0} = \aleph_1$ hold?" being like "can a typical American be Catholic". More reasonably **CH** has a small measure, still much much more than $V = L$. (Shelah, 2002, p. 12)

Shelah's suggestion, whilst tantalising, has not been worked out in full mathematical and philosophical detail. We therefore ask:

Question How should we think of *verisimilitude* for the advocate of the threefold-correspondence approach in set theory?

7.3 Which Theories are "Legitimate"?

Throughout this chapter, we have remained neutral on what theories practitioners "should" be interested in.

One might argue (and we thank an anonymous referee for pressing us on this) that we face a problem of discriminating the "foundationally interesting" theories (perhaps, e.g. **ZFC** plus large cardinals) from "uninteresting" ones (such as **ZFC** + $\neg Con$(**ZFC**)). If we accept that there's a distinction to be made between "interesting" and "uninteresting" theories, and if all we are doing is studying these weak existence models (which may be countable), then what separates the good from the bad?

This is closely related to the question versimilitude and models, concerned with to what *extent* different weak existence models relate to reality, rather than *how*. A multiversist in the style of Hamkins (2012) may regard both **ZFC** plus suitable large cardinals and **ZFC** + $\neg Con$(**ZFC**) as in *some* sense "on a par", but a universe theorist will likely not. Others may regard well-founded models as legitimate contenders, but want to stop there. We want to remain neutral on this issue. The approach of taking a threefold correspondence is not unique to any one position, and different weak existence models will come out as legitimate contenders on different pictures. Our chapter has just been directed at justifying the use of models as tools in this threefold correspondence, not which are better than others. But nonetheless we can ask:

Question What criteria (if any) distinguish the "legitimate" theories and associated weak existence models?

7.4 Characterising Debates and Communication

Indeed, using the notion of weak existence proposed here, we think we can make sense of the universe/multiverse dichotomy. Authors on both sides of the divide, we contend, are disagreeing about *strong* existence rather than *weak*. The latter is agreed upon by all parties. It is the former that is not. Multiversists in the style of Hamkins (2012) think that weak existence corresponds *exactly* to strong existence, the other thinks they radically come apart (with other intermediate positions on offer). Weak existence thus provides another perspective on this debate.

We conjecture that this way of looking at what is different between the two views helps to explain the (surprising!) levels of *agreement* between the groups. Each party to the debate is able to agree on much, for example, the correctness of proofs. Often such reasoning is presented in very semantic terms (e.g. by reasoning about the structure of "the universe"); set theorists are by and large not churning out first-order **ZFC**-derivations.[16] This presents a puzzle (along similar lines to the **Circularity Problem**): How is so much agreement possible when the reasoning seems to be semantic and their respective pictures of set-theoretic reality are so radically different? We can give an attractive answer:

[16]There is, of course, a deep question about the nature of this reasoning and its relationship to derivations. See Rav (1999), Azzouni (2004), Larvor (2016), and Tanswell (2024) for discussion and further references.

The parties agree on the nature of weak existence, and this provides a shared semantic content. What is disagreed upon is the philosophical question of how this weak existence maps on to strong existence.

Really substantiating this we leave to other work. There is much to this idea that, while promising, remains mere speculation. For example, the idea that there is significant semantic (as opposed to proof-theoretic) content to what set theorists are doing is highly controversial. Thus the nature of the puzzle itself, as well as the possible response, needs to be made out in more detail.

7.5 The Access Problem

In the philosophy of mathematics, there is the general problem of access. This is a problem with a long history, but is perhaps most famously pressed by Paul Benacerraf in Benacerraf (1973). How do we gain knowledge of mathematical objects, if they are non-causal and non-spatiotemporal, etc.?

A natural line of thought here is that we can think of various kinds of mathematical property as *universals* and we gain knowledge of the universals by studying their *instances* which *can* be concrete. For example, this idea is pressed by Giaquinto (2017) and Barton (2020a) regarding natural number cognition, drawing on work in the cognitive sciences (e.g. Dehaene, 1997; Carey, 2009; Nieder & Dehaene, 2009).

This strategy perhaps works with number theory; we have cognitive access to instances of small numbers and the general rules and processes that would generate larger ones. But it is very unclear how such a strategy could work in the case of set theory—the operations there seem too infinitary, too ontologically prodigious, to be so tractable.

Our work suggests a response to this issue, at least insofar as first-order set theory is concerned. We can have access to large infinitary objects via our access to particular linguistic representations of those objects in some weak existence model of ZFC. In other words, as concrete instances of number-theoretic phenomena can help accessing the abstract rules of number theory, in the same way concrete linguistic models of first-order theory can help accessing more abstract areas of mathematics like set theory. Therefore, the properties of abstract mathematical objects whose strong existence make them hard for us to reach can be better studied, understood, and known by means of the concrete character of the model-theoretic representation we can offer them through language. Again though, this suggestion, whilst temping, would require serious further work to be made out in detail.

7.6 Modelism

Next we consider *moderate modelism*. This is a view that again has a long history, but has been neatly isolated by Tim Button and Sean Walsh in Button and Walsh (2018). It comprises the following two theses:

(1) Talk of particular mathematical structures should be understood as claims about *isomorphism types* (this is the "modelism").
(2) No appeal to mathematical intuition is allowed in specifying our knowledge of these isomorphism types (this is the "moderate" part).

In particular Button and Walsh consider the idea that we know the isomorphism types via description. *Very* roughly put, they argue that the moderate modelist's position is self-undermining; in order to fix a logic with sufficient expressive resources to do the descriptive work the moderate modelist wants, they must appeal to resources they themselves have prohibited. Moderate modelism *requires* one be *immoderate*.

Whilst we agree with Button and Walsh's argument, we think that weak existence can inform this debate. Whilst we concur that some degree of immoderation is necessary, perhaps not a huge amount is needed. All we need to do in order to be able to talk about certain first-order theories is have confidence that they are consistent. Weak existence will then give us a structure in which that theory is realised and give us access to that structure. Of course this is a long way off isomorphism types, but for first-order structure the modelist can appeal to weak existence.

It's helpful here to consider a remark from Hilary Putnam who writes:

> Models are not lost noumenal waifs looking for someone to name them; they are constructions within our theory itself, and they have names from birth. (Putnam, 1980, p. 482)

Our arguments suggest that for a significant class of models (in particular the ones compatible with a weak form of existence) Putnam was on the right track. Indeed, models are constructions from within a theory. But they *can* be given names (indeed every element of a model can be viewed as a name) and this can provide an account of how we can be assured of their existence.

7.7 Weak Existence as a Motivation for Further Axioms?

Another possible gain of focusing on models of set theory as part of that threefold relation between language and reality is a clearer direction in the justification of new principles extending ZFC.

As a matter of fact, if models are only partial descriptions of a (possibly) independent reality, the way they represent that reality might influence the theory we want to build for it. In other words, if we assume that our theory of sets is meant to describe the universe, and that the construction of models of set theory is a constitutive part of our theorising, then the way we use models can influence directly the theory we choose. Accepting this might mean that model-theoretic considerations become relevant for extending ZFC with new axioms—we want axioms that give us the best pictures of the way the world might be.

To give a concrete example of how model-theoretic considerations might influence the choice and the justification of new axioms, consider the case of principles that ensure forms of generic absoluteness (i.e. invariance across the different models obtainable by set forcing—Bounded Forcing Axioms are one source of examples here, as are inner model hypotheses).[17] The justification of these axioms, and of the axiomatic approach they suggest to the problem of independence, can be justified from a model-theoretic perspective by asking for the different (partial) perspectives (i.e. the different models of set theory) to be cohesive/coherent and exhibit agreement. If we take different models to provide different pictures of a unique underlying reality, then we would like these different pictures to cohere. Axioms that postulate exactly this phenomenon might be supported on these grounds.[18]

This is just an example, but suggests that a clear separation between models and reality might help in a process of theory choice, also in the case of set theory.

7.8 Summing Up

To sum up: We think that the weak existence is a powerful tool for philosophy and mathematics. Viewing these kinds of models as imperfect but helpful

---

[17] See Bagaria (2005), Bagaria (2008), and Friedman (2016).

[18] A similar line of argument has been followed in Venturi and Viale (2023) to justify Forcing Axioms and Large Cardinals in terms of the nice model theoretic properties, like model-completeness, they provide for initial segments of the cumulative hierarchy.

"depictions" of strong existence helps to understand the central place of model theory in our reasoning about sets. But this strategy, whilst promising, raises a host of open questions that need to be tackled. In particular: What kinds of applications can we find for weak existence? How does the cornucopia of models we can obtain help us in our philosophical and mathematical theorising?

Acknowledgments The authors wish to thank Carolin Antos for much useful discussion and two anonymous reviewers for helpful comments.

Funder The author Neil Barton is grateful to the VolkswagenStiftung via the project *Forcing: Conceptual Change in the Foundations of Mathematics* and Norges Forskningsrådet (the Research Council of Norway) for their support via the project *Infinity and Intensionality: Towards A New Synthesis* (no. 314435) for their support.

References

Antos, C. (2022). Models as fundamental entities in set theory: A naturalistic and practice-based approach. *Erkenntnis, 89*, 1683–1710. https://philpapers.org/rec/ANTMAF

Antos, C., Barton, N., & Friedman, S.-D. (2021). Universism and extensions of V. *The Review of Symbolic Logic, 14*(1), 112–154.

Arrigoni, T., & Friedman, S.-D. (2013). The hyperuniverse program. *Bulletin of Symbolic Logic, 19*, 77–96.

Azzouni, J. (2004). The derivation-indicator view of mathematical practice †. *Philosophia Mathematica, 12*(2), 81–106.

Bagaria, J. (2005). Natural axioms of set theory and the continuum problem. In *Proceedings of the 12th International Congress of Logic, Methodology, and Philosophy of Science* (pp. 43–64). King's College London Publications.

Bagaria, J. (2008). Set theory. In *The princeton companion to mathematics* (pp. 302–321). Princeton University Press.

Balaguer, M. (1998). *Platonism and anti-platonism in mathematics.* Oxford University Press.

Barton, N. (2020a). Absence perception and the philosophy of zero. *Synthese, 197*(9), 3823–3850.

Barton, N. (2020b). Forcing and the universe of sets: Must we lose insight? *Journal of Philosophical Logic, 49*(4), 575–612.

Barton, N., Ternullo, C., & Venturi, G. (2020). On forms of justification in set theory. *The Australasian Journal of Logic, 17*(4), 158–200.

Benacerraf, P. (1973). Mathematical truth. *The Journal of Philosophy, 70*(19), 661–679.

Button, T., & Walsh, S. (2018). *Philosophy and model theory.* Oxford University Press.

Cantor, G. (1883). Foundations of a general theory of manifolds; A mathematico-philosophical investigation into the theory of the infinite. In W. Ewald (Ed.), *From Kant to Hilbert; A source book in the foundations of mathematics* (Vol. 2). Oxford University Press.
Carey, S. (2009). *The origin of concepts*. Oxford University Press.
Carnap, R. (1950). Empiricism, semantics, and ontology. *Revue Internationale de Philosophie, 4*(11), 20–40.
Cartwright, N. (1999). *The dappled world: A study of the boundaries of science*. Cambridge University Press.
Cohen, P. (1963). The independence of the continuum hypothesis. *Proceedings of the National Academy of Sciences of the United States of America, 50,* 1143–1148.
Dehaene, S. (1997). *The number sense: How the mind creates mathematics*. Oxford University Press.
Doherty, F. T. (2017). Hilbert on consistency as a guide to mathematical reality. *Logique Et Analyse, 237,* 107–128.
Ebert, P. A., & Rossberg, M. (2013). *Basic laws of arithmetic*. Oxford University Press.
Ebert, P. A., & Rossberg, M. (2019). Mathematical creation in frege's grundgesetze. In P. A. Ebert & M. Rossberg (Eds.), *Essays on Frege's basic laws of arithmetic* (pp. 325–342). Oxford University Press.
Friedman, S.-D. (2016). Evidence for set-theoretic truth and the hyperuniverse programme. *IFCoLog Journal of Logic and its Applications, 4*("Proof, Truth, Computation" 3), 517–555.
Frigg, R., & Hartmann, S. (2020). Models in science. In E. N. Zalta, (Ed.), *The Stanford encyclopedia of philosophy*. Metaphysics Research Lab, Stanford University, Spring 2020 edition.
Gabriel, G., Hermes, H., Kambartel, F., Thiel, C., Veraart, A., McGuinness, B., & Kaal, H. (1980). *Gottlob frege: Philosophical and mathematical correspondence*. Blackwell.
Giaquinto, M. (2017). Cognitive access to numbers: The philosophical significance of empirical findings about basic number abilities. *Philosophical Transactions of the Royal Society B: Biological Sciences, 373*(1740), 20160520.
Gödel, K. (1940). *The consistency of the continuum hypothesis*. Princeton University Press.
Hale, B., & Wright, C. (2009). The metaontology of abstraction. In D. Chalmers, D. Manley, & R. Wasserman (Eds.), *Metametaphysics: New essays on the foundations of ontology* (pp. 178–212). Oxford University Press.
Hamkins, J. D. (2012). The set-theoretic multiverse. *The Review of Symbolic Logic, 5*(3), 416–449.
Heck, R. G. (2011). *Frege's theorem*. Clarendon Press.
Koellner, P. (2013). Hamkins on the multiverse. In P. Koellner (Ed.), *Exploring the frontiers of incompleteness*.

Larvor, B. (2016). Why the naïve derivation recipe model cannot explain how mathematicians' proofs secure mathematical knowledge. *Philosophia Mathematica, 24*(3), 401–404.

Linnebo, O. (2018). *Thin objects: An abstractionist account*. Oxford University Press.

Maddy, P. (1997). *Naturalism in mathematics*. Oxford University Press.

Maddy, P. (2007). *Second philosophy*. Oxford University Press.

Maddy, P. (2011). *Defending the axioms*. Oxford University Press.

Nieder, A., & Dehaene, S. (2009). Representation of number in the brain. *Annual Review of Neuroscience, 32*(1), 185–208 (PMID: 19400715).

Paseau, A. C., & Leek, R. (2022). The compactness theorem. In *Internet encyclopedia of philosophy*. Available from https://iep.utm.edu/compactness-theorem/

Putnam, H. (1980). Models and reality. *The Journal of Symbolic Logic, 45*(3), 464–482.

Rav, Y. (1999). Why do we prove theorems? *Philosophia Mathematica, 7*(1), 5–41.

Schmidt, J.-V., & Venturi, G. (2023). Axioms and postulates as speech acts. *Erkenntnis*. Online first: https://doi.org/10.1007/s10670-023-00673-8

Shelah, S. (2002). Logical dreams. *arXiv:math/0211398v1 [math.LO]*.

Simpson, S. G. (2009). *Subsystems of second order arithmetic*. Perspectives in Logic (2nd ed.). Cambridge University Press.

Soysal, Z. (2024). The problem of existence for descriptivism about the reference of set-theoretic expressions. In C. Antos, N. Barton, G. Venturi (Eds.), *The Palgrave companion to the philosophy of set theory*. Palgrave Macmillan.

Tanswell, F. (2024). Go forth and multiply: On actions, instructions and imperatives in mathematical proofs. Forthcoming in J. Brown & O. Bueno (Eds.), *Essays on the philosophy of Jody Azzouni*. Springer.

Thomasson, A. L. (2014). *Ontology made easy*. Oup USA.

Venturi, G., & Viale, M. (2023). What model companionship can say about the continuum problem. *Review of Symbolic Logic, 17*. Online first: https://doi.org/10.1017/S1755020323000102

Zermelo, E. (1930). On boundary numbers and domains of sets. In Ewald, W. B. (Ed.) (1996). *From Kant to Hilbert. A source book in the foundations of mathematics* (Vol. I, Vol. 2, pp. 1208–1233). Oxford University Press.

Index

[1] Note: Page numbers followed by 'n' refer to notes.

C. Antos et al. (eds.), *The Palgrave Companion to the Philosophy of Set Theory*, Palgrave Companions, https://doi.org/10.1007/978-3-031-62387-5

G

H

I

J

K

L

M

Z